Integer Points in Polyhedra—Geometry, Number Theory, Algebra, Optimization

CONTEMPORARY MATHEMATICS

374

Integer Points in Polyhedra—Geometry, Number Theory, Algebra, Optimization

Proceedings of an AMS-IMS-SIAM
Joint Summer Research Conference
on Integer Points in Polyhedra
July 13–17, 2003
Snowbird, Utah

Alexander Barvinok
Matthias Beck
Christian Haase
Bruce Reznick
Volkmar Welker
Editors

American Mathematical Society
Providence, Rhode Island

The 2003 AMS-IMS-SIAM Joint Summer Research Conference on "Integer Points in Polyhedra" was held in Snowbird, Utah, July 13–17, 2003, with support from the National Science Foundation, grant DMS 9973450.

2000 *Mathematics Subject Classification.* Primary 52B20; Secondary 05A15, 11H06, 11P21, 13P10, 14M25, 52C07, 62H17, 68W30, 90C10.

Any opinions, findings, and conclusions or recommendations expressed in this material are those of the authors and do not necessarily reflect the views of the National Science Foundation.

Library of Congress Cataloging-in-Publication Data

AMS-IMS-SIAM Joint Summer Research Conference on Integer Points in Polyhedra (2003 : Snowbird, Utah)

Integer points in polyhedra : geometry, number theory, algebra, optimization : proceedings of an AMS-IMS-SIAM Joint Summer Research Conference on Integer Points in Polyhedra, July 13–17, 2003, Snowbird, Utah / Alexander Barvinok ... [et al.], editors.

p. cm. (Contemporary mathematics, ISSN 0271-4132 ; 374)

Includes bibliographical references.

ISBN 0-8218-3459-2 (acid-free paper)

1. Lattice theory. 2. Polyhedral functions. 3. Convex sets. 4. Number theory. 5. Polynomials. I. Barvinok, Alexander, 1963– II. Title. III. Contemporary mathematics (American Mathematical Society); v. 374.

QA171.5.A47 2003
516.3′5–dc22 2005040968

∞ The paper used in this book is acid-free and falls within the guidelines
established to ensure permanence and durability.
Visit the AMS home page at http://www.ams.org/

10 9 8 7 6 5 4 3 2 1 10 09 08 07 06 05

Contents

Preface

This volume contains original research and survey papers stemming from the AMS-IMS-SIAM Joint Summer Research Conference "Integer Points in Polyhedra. Geometry, Number Theory, Algebra, Optimization", which took place from July 13 to July 17, 2003, in Snowbird, Utah. The conference was organized by Alexander Barvinok, Matthias Beck, Christian Haase, Bruce Reznick, Michèle Vergne, and Volkmar Welker. The conference web site `http://math.sfsu.edu/beck/src.html` contains the program and pictures of the conference.

The volume is a cross section of recent advances connected to lattice-point questions. Similar to the talks given at the conference, topics range from Commutative Algebra to Optimization, from Discrete Geometry to Statistics, from Representation Theory to Symplectic Geometry.

We thank AMS, IMS, and SIAM for their generous support, which allowed us, among other things, to sponsor undergraduate and graduate students for the conference. We would like to express our deep gratitude to Wayne Drady, the local AMS organizer in Snowbird, who made sure–in a most pleasant way–that our conference ran smoothly in every possible aspect.

For their help and encouragement in the preparation stage, we are indebted to Jim Maxwell, Frank Sottile, and Sarah Witherspoon.

We are grateful to all contributors and the anonymous referees for invaluable help, and for keeping deadlines. Many thanks to Luann Cole, Christine Thivierge and the AMS staff for their assistance in preparing these proceedings.

Finally, thanks to all participants of the conference for making it such a pleasant and productive event.

Alexander Barvinok
Matthias Beck
Christian Haase
Bruce Reznick
Volkmar Welker

List of Participants

José Agapito
University of California at Santa Cruz
Santa Cruz, CA
jarpepe@math.ucsc.edu

Miklos Ajtai
IBM Almaden Research Center
San José, CA
ajtai@almaden.ibm.com

Alexander Barvinok
University of Michigan
Ann Arbor, MI
barvinok@math.lsa.umich.edu

Matthias Beck
San Francisco State University
San Francisco, CA
beck@math.sfsu.edu

Louis Billera
Cornell University
Ithaca, NY
billera@math.cornell.edu

Sylvain Cappell
Courant Institute
New York, NY
cappell@cims.nyu.edu

Beifang Chen
Hong Kong University
Hong Kong
mabfchen@ust.hk

Charles Cochet
Université Paris 7 - Denis Diderot
Paris, France
cochet@math.jussieu.fr

Jesús DeLoera
University of California at Davis
Davis, CA
deloera@math.ucdavis.edu

Maria Diansuy
University of the East
Philippines
ediansuy@lycos.com

Ian Dinwoodie
Duke University
Durham, NC
ihd@stat.duke.edu

David Einstein
Structured Decisions Corp.
West Newton, MA
Deinst@world.std.com

Lenny Fukshansky
Texas A&M University
College Station, TX
lenny@math.tamu.edu

Christian Haase
Duke University
Durham, NC
haase@math.duke.edu

David Haws
University of California at Davis
Davis, CA
dchaws@ucdavis.edu

Patricia Hersh
Indiana University
Bloomington, IN
phersh@indiana.edu

Serkan Hoşten
San Francisco State University
San Francisco, CA
serkan@math.sfsu.edu

Peter Huggins
University of California at Davis
Davis, CA
pmhuggs@earthlink.net

Jean-Michel Kantor
Université Paris 7 - Denis Diderot
Paris, France
kantor@math.jussieu.fr

Allen Knutson
University of California at Berkeley
Berkeley, CA
allenk@math.berkeley.edu

Yueh-er Kuo
University of Tennessee
Knoxville, TN
ykuo@math.utk.edu

Jeffrey Lagarias
University of Michigan
Ann Arbor, MI
lagarias@umich.edu

Jean Bernard Lasserre
LAAS-CNRS
Toulouse, France
lasserre@laas.fr

Diane Maclagan
Rutgers University
Picataway, NJ
maclagan@math.rutgers.edu

Tyrrell McAllister
University of California at Davis
Davis, CA
tmcal@math.ucdavis.edu

Isabella Novik
University of Washington
Seattle, WA
novik@math.washington.edu

Bruce Reznick
University of Illinois
Urbana, IL
reznick@math.uiuc.edu

Sinai Robins
Temple University
Philadelphia, PA
srobins@math.temple.edu

Francisco Santos
Universidad de Cantabria
Santander, Spain
santosf@unican.es

Achill Schürmann
University of Magdeburg
Magdeburg, Germany
achill@math.uni-magdeburg.de

Seth Sullivant
University of California at Berkeley
Berkeley, CA
seths@math.berkeley.edu

András Szenes
Budapest University
Budapest, Hungary
szenes@math.bme.hu

Wayne Tarrant
Western Kentucky University
Bowling Green, KY
wayne.tarrant@wku.edu

Rekha Thomas
University of Washington
Seattle, WA
thomas@math.washington.edu

Michèle Vergne
Ecole Polytechnique
Palaiseau, France
vergne@math.polytechnique.fr

Greg Warrington
Wake Forest University
Winston-Salem, NC
warrings@wfu.edu

Jonathan Weitsman
University of California at Santa Cruz
Santa Cruz, CA
weitsman@cats.ucsc.edu

Kevin Woods
University of California at Berkeley
Berkeley, CA
kwoods@math.berkeley.edu

Ruriko Yoshida
Duke University
Durham, NC
ruriko@math.duke.edu

Andrei Zelevinsky
Northeastern University
Boston, MA
andrei@neu.edu

Contemporary Mathematics
Volume **374**, 2005

A weighted version of quantization commutes with reduction for a toric manifold

José Agapito

ABSTRACT. We compute explicitly the equivariant Hirzebruch χ_y-characteristic of an equivariant complex line bundle over a toric manifold and state a weighted version of the quantization commutes with reduction principle in symplectic geometry. Then, we give a weighted decomposition formula for any simple polytope in $\mathbb{R}^n$. This formula generalizes a polytope decomposition due to Lawrence [**10**] and Varchenko [**14**] and extends a previous weighted version obtained by Karshon, Sternberg and Weitsman [**9**].

1. Introduction

A convex polytope is a convex bounded polyhedron in $\mathbb{R}^n$. We say that a polytope is simple if there are exactly n edges emanating from each of its vertices. In addition, if the lines along the edges of a simple polytope are generated by a $\mathbb{Z}$-basis of the lattice $\mathbb{Z}^n$, the polytope is called regular.

Let M be a toric manifold and let T be the torus acting on M. The image of M under the moment map of the torus action is a regular polytope Δ which lies on $\mathfrak{t}^*$ (the dual of the Lie algebra $\mathfrak{t}$ of T). Each integral lattice point in $\mathfrak{t}^*$ corresponds to a character of a representation of T. Counting lattice points in Δ is closely related to the quantization commutes with reduction principle. For the Dolbeault quantization, this principle states that each character α of T occurs in the quantization of M with multiplicity 1 or 0 depending on whether the lattice point representing α is in Δ or not. (c.f. [**5**], [**6**].)

We use the Atiyah-Bott Berline-Vergne localization formula in equivariant cohomology and a notion of polarization to give a weighted version of the quantization commutes with reduction principle for a toric manifold. This is expressed as a way of counting lattice points in Δ with *weights*. From a combinatorial point of view, we can count lattice points in Δ using a polytope decomposition due to Lawrence [**10**] and Varchenko [**14**]. (See Figure 1 in Subsection §§2.1.) We give a weighted decomposition formula for any simple polytope that generalizes the Lawrence and Varchenko polytope decomposition. This formula also extends a previous weighted version obtained by Karshon, Sternberg and Weitsman. (See [**8**] and [**9**].)

2000 *Mathematics Subject Classification.* 65B15 (Primary); 52B20, 53D20 (Secondary).

This work was partially supported by a UC Regents Fellowship the author received during Spring 2003.

Acknowledgements. I want to thank Jonathan Weitsman for his illuminating explanations and comments throughout this article. I would like to thank Shlomo Sternberg for his helpful comments on this paper. Finally, I would like to thank the referee for several useful suggestions.

2. Definitions and Facts

For the reader's convenience, we collect in this section some definitions, facts and remarks to be used later on in this paper.

2.1. Ordinary polar decomposition. A convex **polytope** Δ in $\mathbb{R}^n$ is a **convex bounded polyhedron** given by an intersection of half-spaces $\Delta = \cap H_i$, where

$$H_i = \{x \in \mathbb{R}^n \mid \langle x, u_i \rangle \geq \lambda_i \, , \lambda_i \in \mathbb{R}\} \qquad \text{for } i = 1, \cdots, N > n,$$

with $u_i \in (\mathbb{R}^n)^* \simeq \mathbb{R}^n$ and where N denotes the number of $(n-1)$-dimensional faces (the facets) of Δ. The vector u_i can be thought of as the inward normal to the i-th facet of Δ contained in the hyperplane $\langle x, u_i \rangle = \lambda_i$. We assume that Δ is obtained with the smallest possible N. We also assume that our polytopes are always compact and convex.

A polytope is **simple** if there are exactly n edges emanating from each vertex and is **regular**[1] if, in addition, the edges emanating from each vertex lie along lines which are generated by an integral basis of $\mathbb{Z}^n$. If the vertices of a polytope are in the lattice $\mathbb{Z}^n$, we call it **integral**.

Let Δ be a simple polytope in $\mathbb{R}^n$. Then, we have n edges emanating from each of its vertices. Let v be a vertex. We choose $\alpha_{1,v}, \ldots, \alpha_{n,v}$ in the direction of the edges pointing from v to other vertices of Δ. Up to a positive scalar and permutation, these *edge vectors* are uniquely determined. When Δ is regular and integral, we can choose $\alpha_{1,v}, \ldots, \alpha_{n,v}$ to have integer entries. In this case, we shall impose the additional normalization condition that the edge vectors be primitive elements of the lattice $\mathbb{Z}^n$; namely, none of them can be expressed as a multiple of a lattice element by an integer greater than one. This fixes our choice of edge vectors for such polytopes.

The tangent cone $\mathbf{C}_v$ at v is

$$\mathbf{C}_v = \{v + r(x - v) \mid r \geq 0, x \in \Delta\} = v + \sum_{j=1}^{n} \mathbb{R}_{\geq 0} \alpha_{j,v}.$$

Notice that we can always find a vector ξ such that $\langle \alpha_{j,v}, \xi \rangle \neq 0$ for all v and j. To see this, we can think of an edge vector as the normal vector of a hyperplane in $\mathbb{R}^n$ passing through the origin. Since there are a finite number of edge vectors, we have a finite number of such hyperplanes. They split $\mathbb{R}^n$ into several connected components. We pick ξ such that it is not perpendicular to the boundary of these components.

Definition 1. A **polarizing vector** ξ for a simple polytope $\Delta \subset \mathbb{R}^n$ is a vector in $\mathbb{R}^n$ such that $\langle \alpha_{j,v}, \xi \rangle \neq 0$ for all vertices v and all edge vectors $\alpha_{1,v}, \ldots, \alpha_{n,v}$.

---

[1]our regular polytopes are also known as smooth or Delzant polytopes [**4**], [**6**].

At each vertex v, we define the polarized vectors $\alpha^{\sharp}_{j,v}$ by

$$\alpha^{\sharp}_{j,v} = \begin{cases} \alpha_{j,v} & \text{if } \langle \alpha_{j,v}, \xi \rangle < 0 \quad \text{(unflipped)} \\ -\alpha_{j,v} & \text{if } \langle \alpha_{j,v}, \xi \rangle > 0 \quad \text{(flipped)} \end{cases} . \tag{1}$$

This process is called **polarization**. Then, the polarized tangent cone $\mathbf{C}^{\sharp}_{v}$ at v is

$$\mathbf{C}^{\sharp}_{v} = v + \sum_{j=1}^{n} \mathbb{R}_{\geq 0} \alpha^{\sharp}_{j,v}.$$

We have the following (adapted) decomposition of a simple polytope due to Lawrence [**10**] and Varchenko [**14**].

PROPOSITION 1. *Let Δ be a simple polytope in $\mathbb{R}^n$. For any choice of polarizing vector, we have*

$$\mathbf{1}_{\Delta} = \sum_{v} (-1)^{\#v} \mathbf{1}_{\mathbf{C}^{\sharp}_{v}}, \tag{2}$$

where the sum is over all vertices v of the polytope and the symbols $\mathbf{1}_{\Delta}$ and $\mathbf{1}_{\mathbf{C}^{\sharp}_{v}}$ denote the ordinary characteristic functions over Δ and over the polarized tangent cones $\mathbf{C}^{\sharp}_{v}$ (possibly with some facets removed). In this formula, $\#v$ denotes the number of edge vectors at v flipped by the polarization process (1).

Lawrence [**10**] and Varchenko [**14**] formulas actually hold for any simple convex polyhedron of full dimension in $\mathbb{R}^n$. In the case of lattice polytopes, $\Delta \cap \mathbb{Z}^n$, we avoid removing certain facets from the cones $\mathbf{C}^{\sharp}_{v}$, by *shifting* some of the cones. We illustrate this decomposition in Figure 1.

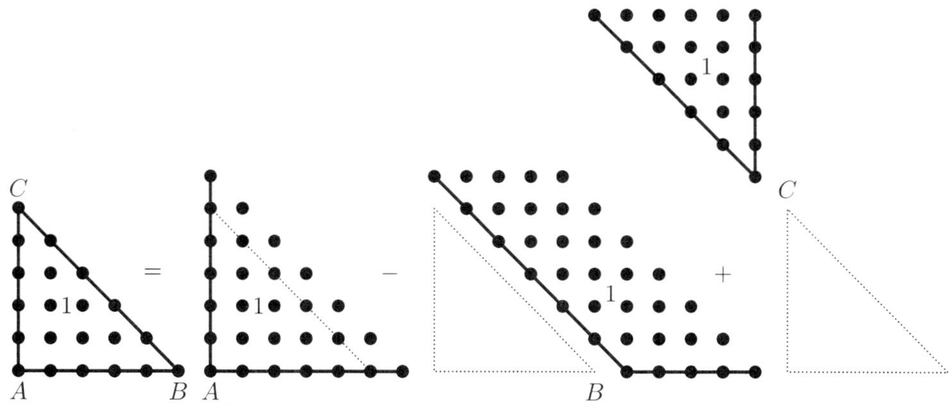

FIGURE 1. Ordinary polar decomposition of a lattice triangle

2.2. Some classical formal power series. We will pay attention to three formal power series.

The Todd function $\mathbf{Td}(x)$ is defined as

$$\mathbf{Td}(x) = \frac{x}{1 - e^{-x}} = 1 + \frac{1}{2}x + \frac{1}{12}x^2 - \frac{1}{720}x^4 + \cdots .$$

Since

$$\mathbf{Td}(-x) = \frac{x}{e^{x} - 1} = e^{-x}\mathbf{Td}(x),$$

we can formally average $\mathbf{Td}(x)$ and $\mathbf{Td}(-x)$ to get

$$\begin{aligned}\frac{1}{2}(\mathbf{Td}(x)+\mathbf{Td}(-x)) &= \frac{1}{2}\mathbf{Td}(x)(1+\mathrm{e}^{-x}) = (x/2)\frac{1+\mathrm{e}^{-x}}{1-\mathrm{e}^{-x}} \\ &= (x/2)\frac{\mathrm{e}^{x/2}+\mathrm{e}^{-x/2}}{\mathrm{e}^{x/2}-\mathrm{e}^{-x/2}} = \frac{x/2}{\tanh(x/2)},\end{aligned}$$

which is known as the $\hat{\mathbf{L}}(x)$ function. Both $\mathbf{Td}(x)$ and $\hat{\mathbf{L}}(x)$ are particular instances of the more general formal power series

$$Q(y,x) = \frac{x(1+y)}{1-\mathrm{e}^{-x(1+y)}} - yx = \frac{x(1+y\mathrm{e}^{-x(1+y)})}{1-\mathrm{e}^{-x(1+y)}}.$$

This is the Hirzebruch function. When $y=0$ we get $\mathbf{Td}(x)$ and if $y=1$ and x is replaced by $x/2$, we get $\hat{\mathbf{L}}(x)$. The Hirzebruch function can also be written as

$$Q(y,x) = \frac{1}{1+y}\mathbf{Td}(x(1+y)) + \frac{y}{1+y}\mathbf{Td}(-x(1+y)).$$

We replace x by $x/(1+y)$ above and denote the resulting formula by $\mathbf{Q}_y(x)$. We have a weighted average of $\mathbf{Td}(x)$ and $\mathbf{Td}(-x)$. For any $y \neq -1$ it follows that

$$\mathbf{Q}_y(x) = \frac{1}{1+y}\mathbf{Td}(x)(1+y\mathrm{e}^{-x}).$$

These classical formal power series define well known topological invariants; namely, characteristic classes associated to vector bundles over a manifold. Let M be a compact manifold and let $E \to M$ be a complex vector bundle over M. By the Splitting principle in topology (c.f. [**2**]), we can always assume that E splits into a direct sum of complex line bundles,

$$E = L_1 \oplus \cdots \oplus L_r.$$

Then the Todd class $\mathbf{Td}(E)$ of E is given by

$$\mathbf{Td}(E) = \prod_{k=1}^{r} \frac{c_1(L_k)}{1-\mathrm{e}^{-c_1(L_k)}}, \tag{3}$$

where $c_1(L_1), \ldots, c_1(L_r)$ are the first Chern classes of the line bundles $L_1, \ldots, L_r$. Similarly, the $\hat{\mathbf{L}}$-class of E and more generally the Hirzebruch $\mathbf{Q}_y$-class of E are respectively defined by

$$\begin{aligned}\hat{\mathbf{L}}(E) &= \mathbf{Td}(E)\prod_{k=1}^{r}\frac{1}{2}(1+\mathrm{e}^{-c_1(L_k)}) \quad \text{and} \\ \mathbf{Q}_y(E) &= \mathbf{Td}(E)\prod_{k=1}^{r}\frac{1}{1+y}(1+y\mathrm{e}^{-c_1(L_k)}).\end{aligned} \tag{4}$$

We say that a bundle $E \to M$ stably splits if $E \oplus \mathbb{C}^r \simeq \oplus L_j$, where $\mathbb{C}^r$ denotes a trivial bundle over M and $\oplus L_j$ is a direct sum of complex line bundles $L_j \to M$. Since $E \simeq \oplus L_k$ (by the Splitting principle) and $\mathbf{Td}(\mathbb{C}^r) = 1$, it follows from (3) that $\mathbf{Td}(E \oplus \mathbb{C}^r) = \mathbf{Td}(E)$. The same is true for $\mathbf{L}(E)$ and $\mathbf{Q}_y(E)$. Now, let T be a torus and let E be a T-equivariant vector bundle over M. We have the corresponding T-equivariant versions of (3) and (4),

$$\mathbf{Td}^{\mathrm{T}}(E) = \prod_{k=1}^{r} \frac{c_1^{\mathrm{T}}(L_k)}{1 - e^{-c_1^{\mathrm{T}}(L_k)}} \quad \text{and}$$

(5)

$$\mathbf{Q}_y^{\mathrm{T}}(E) = \mathbf{Td}^{\mathrm{T}}(E) \prod_{k=1}^{r} \frac{1}{1+y}(1 + y e^{-c_1^{\mathrm{T}}(L_k)}).$$

2.3. Toric manifolds and quantization commutes with reduction. Let $\mathrm{T} = (S^1)^n$ be a torus whose Lie algebra is denoted by $\mathfrak{t}$. A **toric manifold** is a space (M, ω, μ), where (M, ω) is a compact and connected symplectic $2n$-dimensional manifold on which T acts effectively and in a Hamiltonian fashion with moment map μ. This means that the action of T on M has trivial kernel, that the torus action leaves invariant the symplectic form ω and that M is provided with a T-equivariant map $\mu : M \to \mathfrak{t}^*$ such that $\imath_{X_\eta}\omega = d\langle \mu, \eta \rangle$ for any $\eta \in \mathfrak{t}$. The function $\langle \mu, \eta \rangle$ is the component of μ in the direction of η and X_η is the vector field on M generated by η. Examples of toric manifolds are the complex projective spaces $\mathbb{C}\mathrm{P}^n$ and blow ups, $\mathbb{C}\mathrm{P}^n \# \overline{\mathbb{C}\mathrm{P}^n} \# \ldots \# \overline{\mathbb{C}\mathrm{P}^n}$.

Under the moment map, toric manifolds (smooth toric varieties) correspond to regular polytopes and toric orbifolds (singular toric varieties) correspond to simple polytopes, modulo equivariant symplectomorphisms and translations respectively (c.f. [**4**], [**5**], [**12**].) The image of a toric manifold under the moment map is called moment polytope and its vertices correspond to the fixed point set of the torus action over the manifold. In general, we say that a symplectic manifold (M, ω) is **prequantizable** if there is a Hermitian complex line bundle $\mathbb{L} \to M$ such that $c_1(\mathbb{L}) = [\omega]$. Such a line bundle is a **prequantization** of (M, ω). In the case of a toric manifold, M is prequantizable if and only if the vertices of its moment polytope (possibly shifted) are integral lattice points. (c.f. [**5**], [**6**].)

Let (M, ω, μ) be a toric manifold and let $\mathbb{L} \to M$ be a prequantization of M. The theory of geometric quantization associates a virtual representation $\mathcal{Q}(M)$ of T to M together with $\mathbb{L}$. The virtual vector space $\mathcal{Q}(M)$ is called the **quantization** of the action of T on $(M, \mathbb{L})$. For instance, for the Dolbeault quantization we have $\mathcal{Q}(M) = \operatorname{kernel} D_{\mathbb{L}} \oplus -\operatorname{cokernel} D_{\mathbb{L}}$, where $D_{\mathbb{L}}$ is the Dolbeault operator defined on the complex $\Omega^*(M)$ of differential forms on M twisted by $\mathbb{L}$. (c.f. [**5**], [**6**].) In this case, the dimension of $\mathcal{Q}(M)$ can be computed by the Atiyah-Singer index theorem,

$$\dim \mathcal{Q}(M) := \operatorname{ind}(D_{\mathbb{L}}) = \int_M e^{c_1(\mathbb{L})} \mathbf{Td}(M), \tag{6}$$

where $\mathbf{Td}(M)$ is the Todd class of the tangent bundle of M and $c_1(\mathbb{L})$ is the first Chern class of the prequantization line bundle $\mathbb{L} \to M$.

The integer $\dim \mathcal{Q}(M)$ in (6) is equal to the number of integral lattice points in $\Delta = \mu(M)$ [**3**]. This is a manifestation of the **quantization commutes with reduction** principle in symplectic geometry. To be more precise, if α is a weight of the torus representation on $\mathcal{Q}(M)$, we define $\mathcal{Q}(M)^\alpha$ as the subspace of $\mathcal{Q}(M)$ on which T acts via the character given by α. The *quantization commutes with reduction* principle is stated as

$$\dim \mathcal{Q}(M)^\alpha = \dim \mathcal{Q}(M_\alpha), \tag{7}$$

where M_α is the reduced space of M at α. The number $\dim \mathcal{Q}(M)^\alpha$ is the **multiplicity** of the character given by α in $\mathcal{Q}(M)$ [**6**]. For the Dolbeault quantization, this multiplicity is 1 or 0 depending on whether α is in Δ or not. Since M is a toric manifold, the reduced space M_α is a point when α is in Δ and is empty otherwise. Therefore, we have

$$\dim \mathcal{Q}(M) = \#\text{ of integer lattice points in } \Delta = \sum_{\alpha\in\Delta\cap\mathbb{Z}^n} \dim \mathcal{Q}(M_\alpha).$$

Counting multiplicities boils down to counting integral lattice points in Δ. We can use Proposition 1 to do this count in terms of an alternating sum over shifted polarized cones.

Now, assume that $\mathbb{L}$ is a T-equivariant prequantization line bundle for (M,ω,μ). This means that the T-action on M lifts to a T-action on $\mathbb{L}$ by bundle automorphisms and also that $\mathbb{L}$ is provided with a connection ∇ such that its equivariant curvature is $\omega - i\mu$. This implies that $c_1^{\mathrm{T}}(\mathbb{L}) = [\omega - i\mu]$.[2] On the other hand, by a T-equivariant version of (6) (see Bismut [**1**]), the character χ around $0 \in \mathfrak{t}$ of this representation of T on $\mathcal{Q}(M)$ is

$$\chi \circ exp = \int_M \mathrm{e}^{c_1^{\mathrm{T}}(\mathbb{L})}\mathbf{Td}^{\mathrm{T}}(M), \tag{8}$$

where exp denotes the exponential map from $\mathfrak{t}$ to T. Using the Atiyah-Bott Berline-Vergne localization formula in equivariant cohomology (c.f. [**5**]), the right hand side of (8) can be expressed as a finite sum over the fixed points F of the torus action on M,

$$\int_M \mathrm{e}^{c_1^{\mathrm{T}}(\mathbb{L})}\mathbf{Td}^{\mathrm{T}}(M) = \sum_F \mathrm{e}^{-i\langle\mu(F),u\rangle}\prod_{j=1}^{n}\frac{1}{1-\mathrm{e}^{-i\langle\alpha_{j,F},u\rangle}}. \tag{9}$$

Combining (9) with the notion of polarization gives us a way of counting the lattice points of the moment polytope of a toric manifold in terms of the lattice points of cones based on the vertices of the polytope. This insight provided by the localization formula in the context of quantization commutes with reduction for toric manifolds inspires us to give a weighted polar decomposition for any simple polytope, not necessarily integral. The proof of this result is completely independent from the use of the localization formula and is shown in Section §4.

3. Weighted quantization commutes with reduction

DEFINITION 2. [**7**] The equivariant Hirzebruch χ_y-characteristic of an equivariant complex line bundle $\mathbb{L}$ over a toric manifold M is

$$\chi_y(M,\mathbb{L}) = \int_M \mathrm{e}^{c_1^{\mathrm{T}}(\mathbb{L})}\mathbf{Q}_y^{\mathrm{T}}(M). \tag{10}$$

Motivated by (9), we replace $\mathbf{Td}^{\mathrm{T}}(M)$ by $\mathbf{Q}_y^{\mathrm{T}}(M)$ in (8) and use the Atiyah-Bott Berline-Vergne localization formula again to compute (10).

Let (M,ω,μ) be a dimension $2n$ toric manifold with moment polytope $\mu(M)= \Delta$. Assume that the vertices of Δ are in $\mathbb{Z}^n$, then M is prequantizable. Let $\mathbb{L}\to M$ be a prequantization complex line bundle over M and assume furthermore that $\mathbb{L}$ is T-equivariant. The equivariant Chern class $c_1^{\mathrm{T}}(\mathbb{L})$ is equal to $[\omega - i\mu]$. The complexified tangent bundle of M stably splits into a direct sum of complex line

[2] the first T-equivariant Chern class $c_1^{\mathrm{T}}(\mathbb{L})$ is of the form $[\omega+\mu]$ up to a constant. We use the convention $c_1^{\mathrm{T}}(\mathbb{L}) = [\omega - i\mu]$ for technical reasons.

bundles $\oplus L_j$ which we assume are also T-equivariant. Then the T-equivariant Hirzebruch class $\mathbf{Q}_y^{\mathrm{T}}(M)$ is

$$\mathbf{Q}_y^{\mathrm{T}}(M) = \mathbf{Td}^{\mathrm{T}}(M) \prod_{j=1}^{n} \frac{1}{1+y}(1 + y\mathrm{e}^{-c_1^{\mathrm{T}}(L_j)}),$$

where y is any real number not equal to -1. By the Atiyah-Bott Berline-Vergne localization formula, we get

$$\begin{aligned} &\int_M \mathrm{e}^{c_1^{\mathrm{T}}(\mathbb{L})} \mathbf{Q}_y^{\mathrm{T}}(M) \\ &\qquad = \sum_{F \in M^T} \frac{\mathrm{e}^{c_1^{\mathrm{T}}(\mathbb{L})}|_F (\mathbf{Td}^{\mathrm{T}}(M) \prod_{j=1}^{n} \frac{1}{1+y}(1 + y\mathrm{e}^{-c_1^{\mathrm{T}}(L_j)}))|_F}{e^{\mathrm{T}}(M)|_F} \\ &\qquad = \sum_{F \in M^T} \mathrm{e}^{c_1^{\mathrm{T}}(\mathbb{L})}|_F \prod_{j=1}^{n} \left(\frac{1}{1+y} \left(\frac{1 + y\mathrm{e}^{-c_1^{\mathrm{T}}(L_j)}}{1 - \mathrm{e}^{-c_1^{\mathrm{T}}(L_j)}} \right) \right) \Bigg|_F . \end{aligned} \tag{11}$$

Let u be a generic element of the Lie algebra $\mathfrak{t} \cong i\mathbb{R}^n$ and $v = \mu(F)$ be the vertex of the moment polytope Δ corresponding to the fixed point F. For $1 \leq j \leq n$, let $\alpha_{j,v}$ be edge vectors at v normalized as in Subsection §§2.1. This implies that our choice of such vectors is unique. These edge vectors can be thought of as the weights of the isotropic representation of T on the normal bundle $N_F = T_F M$ at F [**6**]. The restriction of the equivariant Chern class $c_1^{\mathrm{T}}(L_j)$ to F is $c_1^{\mathrm{T}}(L_j)|_F = i\langle \alpha_{j,v}, u \rangle$ for $1 \leq j \leq n$. Similarly, the restriction of $c_1^{\mathrm{T}}(\mathbb{L})$ to F is $c_1(\mathbb{L})|_F = -i\langle v, u \rangle$. Then, (11) can be written as

$$\int_M \mathrm{e}^{c_1^{\mathrm{T}}(\mathbb{L})} \mathbf{Q}_y^{\mathrm{T}}(M) = \sum_v \mathrm{e}^{-i\langle v,u \rangle} \prod_{j=1}^{n} \left(\frac{1}{1+y} \left(\frac{1 + y\mathrm{e}^{-i\langle \alpha_{j,v}, u \rangle}}{1 - \mathrm{e}^{-i\langle \alpha_{j,v}, u \rangle}} \right) \right), \tag{12}$$

where the sum is over all vertices of Δ. We pick $\xi \in \mathbb{R}^n \setminus \{0\}$ such that $\langle \alpha_{j,v}, \xi \rangle \neq 0$ for $1 \leq j \leq n$ and all vertices v in Δ. As seen on Subsection §§2.1, we can always find such a ξ. Letting $u = i\xi$, we get

$$\begin{aligned} &\sum_v \mathrm{e}^{-i\langle v, i\xi \rangle} \prod_{j=1}^{n} \left(\frac{1}{1+y} \left(\frac{1 + y\mathrm{e}^{-i\langle \alpha_{j,v}, i\xi \rangle}}{1 - \mathrm{e}^{-i\langle \alpha_{j,v}, i\xi \rangle}} \right) \right) \\ &\qquad = \sum_v \mathrm{e}^{\langle v, \xi \rangle} \prod_{j=1}^{n} \left(\left[\frac{1}{1+y} + \frac{y\mathrm{e}^{\langle \alpha_{j,v}, \xi \rangle}}{1+y} \right] \frac{1}{1 - \mathrm{e}^{\langle \alpha_{j,v}, \xi \rangle}} \right). \end{aligned} \tag{13}$$

Denote $\mathrm{e}^{\langle \alpha_{j,v}, \xi \rangle}$ by $z_{j,v}$. If $\langle \alpha_{j,v}, \xi \rangle < 0$, we set $\alpha_{j,v}^{\sharp} = \alpha_{j,v}$ and have $z_{j,v} = \mathrm{e}^{\langle \alpha_{j,v}^{\sharp}, \xi \rangle}$, so that $|z_{j,v}| < 1$ and

$$\frac{1}{1 - z_{j,v}} = 1 + z_{j,v} + z_{j,v}^2 + \cdots = \sum_{m=0}^{\infty} z_{j,v}^m.$$

If $\langle \alpha_{j,v}, \xi \rangle > 0$, we set $\alpha_{j,v}^{\sharp} = -\alpha_{j,v}$ and have $z_{j,v}^{-1} = \mathrm{e}^{\langle \alpha_{j,v}^{\sharp}, \xi \rangle}$, so that $|z_{jv}^{-1}| < 1$ and

$$\frac{1}{1 - z_{j,v}} = -z_{j,v}^{-1} \frac{1}{1 - z_{j,v}^{-1}} = -z_{j,v}^{-1}(1 + z_{j,v}^{-1} + z_{j,v}^{-2} + \cdots) = -\sum_{m=1}^{\infty} z_{j,v}^{-m}.$$

Thus, when $|z_{j,v}| < 1$,

$$\left(\frac{1}{1+y} + \frac{yz_{j,v}}{1+y}\right)\frac{1}{1-z_{j,v}} = \frac{1}{1+y} + \sum_{m=1}^{\infty} z_{j,v}^{m}, \tag{14}$$

whereas, if $|z_{j,v}^{-1}| < 1$,

$$\left(\frac{1}{1+y} + \frac{yz_{j,v}}{1+y}\right)\frac{1}{1-z_{j,v}} = -\left(\frac{y}{1+y} + \sum_{m=1}^{\infty} z_{j,v}^{-m}\right). \tag{15}$$

Now, denote $e^{\langle \alpha_{j,v}^{\sharp},\xi\rangle}$ by $z_{j,v}$, regardless $\alpha_{j,v}^{\sharp} = \alpha_{j,v}$ or $\alpha_{j,v}^{\sharp} = -\alpha_{j,v}$. Plugging (14) and (15) in (13), we obtain

$$\sum_{v}(-1)^{\#v}e^{\langle v,\xi\rangle}\prod_{k=1}^{n-\#v}\left(\frac{1}{1+y}+\sum_{m_k=1}^{\infty} z_{j_k,v}^{m_k}\right)\prod_{l=1}^{\#v}\left(\frac{y}{1+y}+\sum_{m_l=1}^{\infty} z_{j_l,v}^{m_l}\right), \tag{16}$$

where $(-1)^{\#v}$ denotes the number of $\alpha_{j,v}$ flipped by the polarizing vector ξ. Setting $v = (v_1, \cdots, v_n)$, $\xi = (\xi_1, \cdots, \xi_n)$ and $\alpha_{j,v}^{\sharp} = (\alpha_{j,v}^{\sharp 1}, \cdots, \alpha_{j,v}^{\sharp n})$, we have

$$e^{\langle v,\xi\rangle} = e^{v_1\xi_1+\cdots+v_n\xi_n} \quad \text{and}$$
$$z_{j_k,v}^{m_k} = e^{m_k\langle \alpha_{j_k,v}^{\sharp},\xi\rangle} = e^{m_k(\alpha_{j,v}^{\sharp 1}\xi_1+\cdots+\alpha_{j,v}^{\sharp n}\xi_n)}.$$

Let $z_j = e^{\xi_j}$. Formula (16) is then equal to

$$\begin{aligned}
&\sum_{v}(-1)^{\#v}z_1^{v_1}\cdots z_n^{v_n}\prod_{k=1}^{n-\#v}\left(\frac{1}{1+y}+\sum_{m_k=1}^{\infty} z_1^{m_k\alpha_{j_k,v}^{\sharp 1}}\cdots z_n^{m_k\alpha_{j_k,v}^{\sharp n}}\right)\\
&\qquad\prod_{l=1}^{\#v}\left(\frac{y}{1+y}+\sum_{m_l=1}^{\infty} z_1^{m_l\alpha_{j_l,v}^{\sharp 1}}\cdots z_n^{m_l\alpha_{j_l,v}^{\sharp n}}\right)\\
&= \sum_{v}(-1)^{\#v}z_1^{v_1}\cdots z_n^{v_n}\sum_{m} w_v(m) z_1^{m_1\alpha_{1,v}^{\sharp 1}+\cdots+m_n\alpha_{n,v}^{\sharp 1}}\\
&\qquad z_2^{m_1\alpha_{1,v}^{\sharp 2}+\cdots+m_n\alpha_{n,v}^{\sharp 2}}\\
&\qquad \vdots\\
&\qquad z_n^{m_1\alpha_{1,v}^{\sharp n}+\cdots+m_n\alpha_{n,v}^{\sharp n}},
\end{aligned} \tag{17}$$

where $m = (m_1, \cdots, m_n) \in \mathbb{Z}_{\geq 0}^n$ and $w_v(m)$ is equal to

$$w_v(m) = \left(\frac{1}{1+y}\right)^{r_{1,v}(m)}\left(\frac{y}{1+y}\right)^{r_{2,v}(m)}. \tag{18}$$

The exponent $r_{1,v}(m)$ above is the number of m_j in m which correspond to $\alpha_{j,v}^{\sharp} = \alpha_{j,v}$ and $r_{2,v}(m)$ is the number of m_j in m that correspond to $\alpha_{j,v}^{\sharp} = -\alpha_{j,v}$. We write $z_1^{v_1}\cdots z_n^{v_n}$ as z^v and the long factor in the right hand side of Equation (17) as

$$z_1^{m_1\alpha_{1,v}^{\sharp 1}+\cdots+m_n\alpha_{n,v}^{\sharp 1}} z_2^{m_1\alpha_{1,v}^{\sharp 2}+\cdots+m_n\alpha_{n,v}^{\sharp 2}}\cdots z_n^{m_1\alpha_{1,v}^{\sharp n}+\cdots+m_n\alpha_{n,v}^{\sharp n}} = z^{\sum_j m_j\alpha_{j,v}^{\sharp}}.$$

Note that each point $p = (p_1, \ldots, p_n)$ in $\mathbf{C}_v^\sharp \cap \mathbb{Z}^n$ can be written as $p = v + \sum_j m_j \alpha_{j,v}^\sharp$ with $m_j \in \mathbb{Z}_{\geq 0}$ for $1 \leq j \leq n$, since $\{\alpha_{j,v}^\sharp\}_{j=1}^n$ form a $\mathbb{Z}$-basis of $\mathbb{Z}^n$. We write z^p for $z_1^{p_1} \ldots z_n^{p_n}$ and set $r_{1,v}(p) = r_{1,v}(m)$ and $r_{2,v}(p) = r_{2,v}(m)$ so that $w_v(p) = w_v(m)$. If $p \notin \mathbf{C}_v^\sharp$ we set $w_v(p) = 0$. Note that $r_{1,v}(p) + r_{2,v}(p)$ is equal to the codimension of the smallest dimensional face in $\mathbf{C}_v^\sharp$ containing p. With this notation, we can write (17) as

$$\sum_v (-1)^{\# v} \sum_{p \in \mathbf{C}_v^\sharp \cap \mathbb{Z}^n} w_v(p) z^p. \tag{19}$$

Let $p \in \Delta \cap \mathbb{Z}^n$ and let ξ_1 be a polarizing vector for Δ. The point p may belong to more than one polarized tangent cone. Since p is fixed, we have

$$\sum_v (-1)^{\# v} w_v(p) z^p = \left(\sum_v (-1)^{\# v} w_v(p) \right) z^p. \tag{20}$$

To determine the second sum in (20), we define $c(p)$ to be the codimension of the smallest dimensional face in Δ containing p. If p is in the interior of Δ we have $c(p) = 0$. Polarizing with respect to ξ_1 (in fact, with respect to any polarizing vector), we have that the polarized tangent cones of Δ are all pointing away from p except one which contains p in its interior. Suppose that such cone is $\mathbf{C}_v^\sharp$ for some vertex v of Δ. Then $r_{1,v}(p) = 0$ and $r_{2,v}(p) = 0$ since $r_{1,v}(p) + r_{2,v}(p) = c(p) = 0$. If $c(p) > 0$, the point p may belong to more than one polarized tangent cone. When $c(p) = 1$, there are two cases: the sum $\sum_v (-1)^{\# v} w_v(p)$ is equal to $\frac{1}{1+y} + zeros$ or equal to $1 - \frac{y}{1+y} + zeros$. In either case, we have $\sum_v (-1)^{\# v} w_v(p) = \frac{1}{1+y}$. In general, we can see that (after simplifications) there are basically three cases: the sum $\sum_v (-1)^{\# v} w_v(p)$ is equal to $(\frac{1}{1+y})^{c(p)} + zeros$ or equal to $(1 - \frac{y}{1+y})^{c(p)} + zeros$ or equal to $(\frac{1}{1+y})^{c(p)-1} - (\frac{1}{1+y})^{c(p)-1} (\frac{y}{1+y}) + zeros$. In any case, we have $\sum_v (-1)^{\# v} w_v(p) = (\frac{1}{1+y})^{c(p)}$. A different polarizing vector ξ_2 have the effect of just permuting the type of polarized tangent cones (that is, cones with no edges flipped, with one edge flipped, with two edges flipped and so on) over the vertices of Δ. In conclusion, regardless the polarizing vector ξ, we always obtain

$$\sum_v (-1)^{\# v} w_v(p) = \left(\frac{1}{1+y} \right)^{c(p)}. \tag{21}$$

From (17) on, our arguments apply to any regular integral polytope in $\mathbb{R}^n$. Therefore, by summing over all the points of $\Delta \cap \mathbb{Z}^n$, we have proved the following statement.

THEOREM 1. *Let Δ be a regular integral polytope in $\mathbb{R}^n$. For any polarizing vector $\xi \in \mathbb{R}^n$, we have*

$$\sum_v (-1)^{\# v} \sum_{p \in \mathbf{C}_v^\sharp \cap \mathbb{Z}^n} w_v(p) z^p = \sum_{p \in \Delta \cap \mathbb{Z}^n} \left(\frac{1}{1+y} \right)^{c(p)} z^p. \tag{22}$$

By considering equations from (10) to (22) together with the work in between, we get

$$\chi_y(M, \mathbb{L}) = \int_M e^{c_1^{\mathrm{T}}(\mathbb{L})} \mathbf{Q}_y^{\mathrm{T}}(M) = \sum_{p \in \Delta \cap \mathbb{Z}^n} \left(\frac{1}{1+y} \right)^{c(p)} z^p. \tag{23}$$

The equivariant Hirzebruch χ_y-characteristic of an equivariant complex line bundle $\mathbb{L}$ over a toric manifold M is in general a polynomial in $\mathbb{C}[z_1^{\pm 1}, \ldots, z_n^{\pm 1}]$, as we can see from (23). This equivariant characteristic number is not in general the virtual index of an elliptic operator on M. However, when $y = 0$, we have $\mathbf{Q}_y^{\mathrm{T}}(M) = \mathbf{Td}^{\mathrm{T}}(M)$ and (10) is the equivariant index of the twisted Dolbeault operator $D_{\mathbb{L}}$ mentioned in Subsection §2.3. When $y = 1$, we have $\mathbf{Q}_y^{\mathrm{T}}(M) = \mathbf{L}^{\mathrm{T}}(M)$ and (10) is in this case the equivariant index of the twisted signature operator on M [**11**]. Because of this, we can interpret (23) as a weighted version of the quantization commutes with reduction principle for a toric manifold.

THEOREM 2. *Let M be any toric manifold whose moment polytope is denoted by Δ and let $\alpha \in \Delta \cap \mathbb{Z}^n$ be any weight lattice of the corresponding torus action on M. Then*

$$\chi_y(M, \mathbb{L})^\alpha = \begin{cases} \left(\frac{1}{1+y}\right)^{c(\alpha)} & \text{if } \alpha \in \Delta \\ 0 & \text{if } \alpha \notin \Delta \end{cases}. \tag{24}$$

PROOF. It follows from (23) and the preceding discussion. □

REMARK 1. In [**13**], Y. Tian and W. Zhang perform an analytic approach to the quantization commutes with reduction principle and get a related weighted multiplicity formula for Spin^c-complexes twisted by certain exterior power bundles.

4. Weighted polar decomposition for a simple polytope

Formula (22) is valid for any regular integral polytope in $\mathbb{R}^n$. In analogy with (2), we can write (22) using characteristic functions and extend the result to any simple polytope, not necessarily integral.

Let Δ be a simple polytope in $\mathbb{R}^n$ and let x be any point in this polytope. Let $c(x)$ be the codimension of the smallest dimensional face in Δ containing x. We define the following weighted characteristic function over Δ:

$$\mathbf{1}_\Delta^w(x) = \begin{cases} \left(\frac{1}{1+y}\right)^{c(x)} & \text{if } x \in \Delta \\ 0 & \text{if } x \notin \Delta \end{cases}. \tag{25}$$

Given a polarizing vector ξ in $\mathbb{R}^n$, let $\mathbf{C}_v^\sharp$ be the polarized cone corresponding to the vertex v. If x is in $\mathbf{C}_v^\sharp$, let $c_v(x)$ be the codimension of the smallest dimensional face F in $\mathbf{C}_v^\sharp$ containing x. We define $r_{1,v}(x)$ and $r_{2,v}(x)$ as before; namely, $r_{1,v}(x)$ is the number of unflipped edges emanating from v which do not belong to F, and $r_{2,v}(x)$ is the number of flipped edges emanating from v which are not in F. Therefore, $r_{1,v}(x) + r_{2,v}(x) = c_v(x)$. For any $y \neq -1$, we set

$$w_v(x) = \left(\frac{1}{1+y}\right)^{r_1(x)} \left(\frac{y}{1+y}\right)^{r_2(x)}. \tag{26}$$

We define the following weighted characteristic function over $\mathbf{C}_v^\sharp$:

$$\mathbf{1}_{\mathbf{C}_v^\sharp}^{w_v}(x) = \begin{cases} w_v(x) & \text{if } x \in \mathbf{C}_v^\sharp \\ 0 & \text{if } x \notin \mathbf{C}_v^\sharp \end{cases}. \tag{27}$$

PROPOSITION 2. *For any simple polytope Δ and for any choice of polarizing vector, we have*

$$\mathbf{1}^w_\Delta = \sum_v (-1)^{\#v} \mathbf{1}^{w_v}_{\mathbf{C}^\sharp_v}, \tag{28}$$

where $\mathbf{1}^w_\Delta$ and $\mathbf{1}^{w_v}_{\mathbf{C}^\sharp_v}$ are the weighted characteristic functions over Δ and $\mathbf{C}^\sharp_v$ defined in (25) and (27), where the sum is over the vertices of Δ, and where $\#v$ denotes the number of edge vectors at v flipped by the polarizing process (1).

The proof of this result collects the main ideas from the proof of Theorem 1. Notice that when $y = 0$, we obtain the ordinary polar decomposition stated in Proposition 1. When $y = 1$, the weighted characteristic functions of the polarized tangent cones are the same at all vertices. In this case, we can write (28) as in [**8**] and [**9**],

$$\mathbf{1}^w_\Delta = \sum_v (-1)^{\#v} \mathbf{1}^{w}_{\mathbf{C}^\sharp_v}.$$

We illustrate Proposition 2 for a triangle in Figure 2.

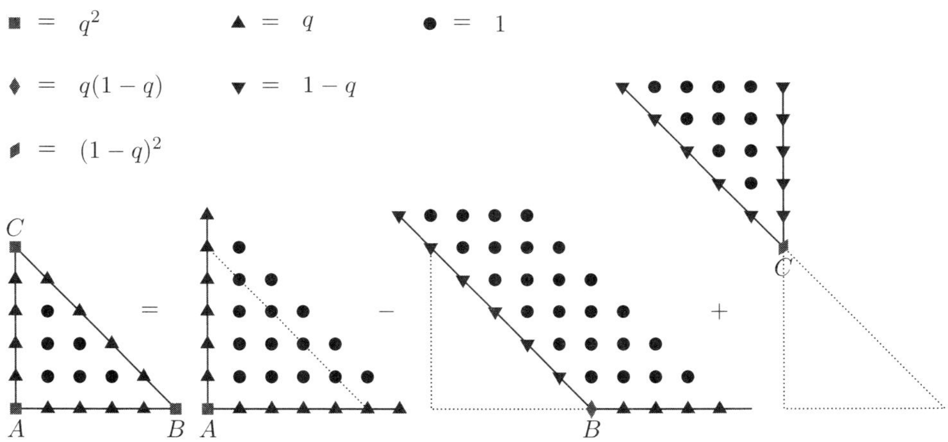

FIGURE 2. Weighted polar decomposition of a lattice triangle

Proof. The proof has two steps. First, for each $x \in \mathbb{R}^n$ we choose a polarizing vector ξ_x such that (28) holds. For this, we show that x appears in at most one polarized tangent cone in the corresponding polar decomposition of Δ given by ξ_x. Second, we show that the right hand side of (28) is independent of the choice of polarization.

Notice that for any polarization there is always a unique cone with no edges flipped at all. More explicitly, if x is in the exterior of Δ, we can take the closest vertex of Δ to x (if there are more than one such a vertex, pick any one of them). Say we take v_1. Then, we set $\mathbf{C}^\sharp_{v_1} = \mathbf{C}_{v_1}$. Clearly x is not in $\mathbf{C}^\sharp_{v_1}$. We polarize according to this cone. Namely, let ξ_x be a polarizing vector in $\mathbb{R}^n$ that makes $\mathbf{C}^\sharp_{v_1} = \mathbf{C}_{v_1}$ and let E_{ξ_x} be a codimension one plane with normal vector parallel to ξ_x. Moving this plane in a parallel way through the polytope, we flip the edges of the tangent cones whose vertices lie below E_{ξ_x} (according to ξ_x) until we get to the farthest vertex v_2 away from v_1 in the direction of ξ_x. There is only one such a vertex since ξ_x is a polarizing vector. The corresponding polarized tangent cone $\mathbf{C}^\sharp_{v_2}$ has all their edges flipped with respect to $\mathbf{C}_{v_2}$ and it does not contain x. In general, a polarizing vector flips an edge if this edge is shared with a previous polarized tangent cone and leaves that edge unflipped otherwise. Then, for any vertex other than v_1 and v_2, the polarized tangent cones have at least one edge flipped and they all appear pointing away from x. We conclude that x appears in no polarized tangent cone.

If x is in Δ, we can pick any vertex v in the face F whose relative interior contains x. Then, we set the polarized tangent cone $\mathbf{C}^\sharp_v$ equal to $\mathbf{C}_v$ and choose a polarizing vector ξ_x that makes $\mathbf{C}^\sharp_v = \mathbf{C}_v$. We now repeat the process explained above. Again, all other polarized tangent cones will point away from x, with the cone at the farthest vertex away from v having all their edges flipped.

From here on, we follow [**8**] closely with some modifications. Let

$$E_1, \cdots, E_N$$

denote all the different codimension one subspaces of $\mathbb{R}^{n*}$ that are equal to

$$\alpha_{j,v}^\perp = \{\eta \in \mathbb{R}^{n*} \mid \langle \eta, \alpha_{j,v} \rangle = 0\}$$

for some j and v. If no two edges of Δ are parallel, then the number N of such hyperplanes is equal to the number of edges of Δ. A vector ξ is a *polarizing vector* if and only if it does not belong to any E_j. The *polarized tangent cones* $\mathbf{C}^\sharp_v$ only depend on the connected component of the complement

$$\mathbb{R}^{n*} \smallsetminus (E_1 \cup \cdots \cup E_N)$$

in which ξ lies. Any two polarizing vectors can be connected by a path ξ_t in $\mathbb{R}^{n*}$ which crosses the *walls* E_j one at a time.

The second part of the proof consists of showing that the right hand side of formula (28) does not change when the polarizing vector ξ_t crosses a single wall, E_k.

As ξ_t crosses the wall E_k, the sign of the pairing $\langle \xi_t, \alpha_{j,v} \rangle$ changes exactly if $E_k = \alpha_{j,v}^\perp$. For each vertex v, denote by $S_v(x)$ and $S'_v(x)$ its contributions to the right hand side of formula (28) before and after ξ_t crosses the wall. The vertices for which these contributions differ are exactly those vertices that lie on edges e of Δ which are perpendicular to E_k. They come in pairs because each edge has two endpoints.

Let us concentrate on one such an edge, e, with endpoints, say, u and v. Let α_e denote an edge vector at v that points from v to u along e. Suppose that the pairing $\langle \xi_t, \alpha_e \rangle$ flips its sign from negative to positive as ξ_t crosses the wall; (otherwise we switch the roles of v and u). The *polarized tangent cones* to Δ at v, before and after ξ_t crosses the wall, are

$$\mathbf{C}^\sharp_v = v + \sum_{j \in I_e} \mathbb{R}_{\geq 0} \alpha^\sharp_{j,v} + \mathbb{R}_{\geq 0} \alpha_e \quad \text{and} \quad (\mathbf{C}^\sharp_v)' = v + \sum_{j \in I_e} \mathbb{R}_{\geq 0} \alpha^\sharp_{j,v} - \mathbb{R}_{\geq 0} \alpha_e,$$

where $I_e \subset \{1, \cdots, N\}$ encodes the facets that contain e. (The $\alpha^\sharp_{j,v}$ are the same for the different ξ_t's because the pairings $\langle \xi_t, \alpha^\sharp_{j,v} \rangle$ do not flip sign when ξ_t crosses the wall for $j \in I_e$.) Notice that the cones $\mathbf{C}^\sharp_v$ and $(\mathbf{C}^\sharp_v)'$ have a common facet and their union only depends on the edge e and not on the endpoint v. (This uses the assumption that the polytope Δ is simple and follows from the fact that $\alpha_{j,u} \in \mathbb{R}_{\geq 0} \alpha_{j,v} + \mathbb{R} \alpha_e$.)

To determine the contributions of v to the right hand side of (28) before and after ξ_t crosses the wall, we have two cases to analyze. In the first case, the smallest dimensional face F containing x also contains the edge e. Then, the contributions are

$$S_v(x) = \varepsilon\left(\frac{1}{1+y}\right)^{r_1(x)}\left(\frac{y}{1+y}\right)^{r_2(x)} \quad \text{and} \quad S_v'(x) = -\varepsilon \cdot 0,$$

where $\varepsilon \in \{-1, 1\}$. Their difference is

$$S_v(x) - S_v'(x) = \varepsilon\left(\frac{1}{1+y}\right)^{r_1(x)}\left(\frac{y}{1+y}\right)^{r_2(x)}.$$

The contributions of the other endpoint, u, are

$$S_u(x) = -\varepsilon \cdot 0 \quad \text{and} \quad S_u'(x) = \varepsilon\left(\frac{1}{1+y}\right)^{r_1(x)}\left(\frac{y}{1+y}\right)^{r_2(x)},$$

and their difference is

$$S_u(x) - S_u'(x) = -\varepsilon\left(\frac{1}{1+y}\right)^{r_1(x)}\left(\frac{y}{1+y}\right)^{r_2(x)}.$$

Notice that the numbers $r_1(x)$ and $r_2(x)$ remain the same for both endpoints because F contains the edge e. Hence, the differences $S_v(x)-S_v'(x)$ and $S_u(x)-S_u'(x)$, for the two endpoints u and v of e, sum to zero.

In the second case, the edge e is not contained in F. Then, the contributions are

$$S_v(x) = \varepsilon\left(\frac{1}{1+y}\right)^{r_1(x)}\left(\frac{y}{1+y}\right)^{r_2(x)} \quad \text{and}$$
$$S_v'(x) = -\varepsilon\left(\frac{1}{1+y}\right)^{r_1(x)-1}\left(\frac{y}{1+y}\right)^{r_2(x)+1},$$

and their difference is

$$S_v(x) - S_v'(x) = \varepsilon\left(\frac{1}{1+y}\right)^{r_1(x)-1}\left(\frac{y}{1+y}\right)^{r_2(x)}.$$

On the other hand, after crossing the wall, the smallest dimensional face F' in $\mathbf{C}_u^\sharp$ containing x has codimension one less than the codimension of F in $\mathbf{C}_v^\sharp$, before crossing the wall. Then, a straightforward calculation shows that the contributions of the other endpoint, u, are

$$S_u(x) = -\varepsilon \cdot 0 \quad \text{and} \quad S_u'(x) = \varepsilon\left(\frac{1}{1+y}\right)^{r_1(x)-1}\left(\frac{y}{1+y}\right)^{r_2(x)},$$

and their difference is

$$S_u(x) - S_u'(x) = -\varepsilon\left(\frac{1}{1+y}\right)^{r_1(x)-1}\left(\frac{y}{1+y}\right)^{r_2(x)}.$$

Hence, the differences $S_v(x) - S_v'(x)$ and $S_u(x) - S_u'(x)$, for the two endpoints u and v of e, sum to zero again. $\square$

References

[1] J. M. Bismut, *Localization formulas, superconnections, and the index theorem for families*, Comm. Math. Phys. **103** (1986), 127–176.

[2] Raoul Bott and Loring Tu, *Differential Forms in Algebraic Topology*, third ed., Graduate Texts in Mathematics, vol. 82, Springer-Verlag, 1982.

[3] V.I. Danilov, *The geometry of toric varieties*, Russian Math. Surveys **33** (1978), no. 2, 97–154.

[4] T. Delzant, *Hamiltoniens périodiques et image convexe de l'application moment*, Bulletin de la Société Mathématique de France **116** (1988), 315–339.

[5] Viktor L. Ginzburg, Victor Guillemin, and Yael Karshon, *Moment maps, cobordism and Hamiltonian group actions*, American Mathematical Society, 2002.

[6] Victor Guillemin, *Moment maps and Combinatorial Invariants of Hamiltonian T^n-spaces*, Progress in Mathematics, vol. 122, Birkhäuser, 1994.

[7] Friedrich Hirzebruch, *Topological Methods in Algebraic Geometry*, third ed., Die grundlehren der mathematischen wissenschaften in eizenldarstellungen, vol. 131, Springer-Verlag, 1966.

[8] Yael Karshon, Shlomo Sternberg, and Jonathan Weitsman, *The Euler-Maclaurin formula with remainder for a simple integral polytope*, (2003), arXiv:math.CO/0307125 v1.

[9] ———, *The Euler-Maclaurin formula for simple integral polytopes*, PNAS **100** (2003), no. 2, 426–433.

[10] J. Lawrence, *Polytope volume computation*, Math. Comp. **57** (1991), no. 195, 259–271.

[11] H. Lawson and Michelsohn M-L, *Spin Geometry*, Princeton Mathematical Series, vol. 38, Princeton University Press, 1989.

[12] T. Oda, *Convex Bodies and Algebraic Geometry*, Ergebnisse der Mathematik, Springer, 1988.

[13] Youliang Tian and Weiping Zhang, *Symplectic reduction and a weighted multiplicity formula for twisted Spin^c-Dirac operators*, Asian Journal of Mathematics **2** (1998), no. 3, 591–608, arXiv:math.GT/9901092v1.

[14] A.N. Varchenko, *Combinatorics and topology of the arrangement of affine hyperplanes in the real space*, (English translation) Functional Anal. Appl. **21** (1987), no. 1, 9–19.

Department of Mathematics, UC Santa Cruz, Santa Cruz, CA 95064, USA.
E-mail address: `jarpepe@math.ucsc.edu`

Contemporary Mathematics
Volume **374**, 2005

Coefficients and Roots of Ehrhart Polynomials

M. Beck, J. A. De Loera, M. Develin, J. Pfeifle, and R. P. Stanley

ABSTRACT. The Ehrhart polynomial of a convex lattice polytope counts integer points in integral dilates of the polytope. We present new linear inequalities satisfied by the coefficients of Ehrhart polynomials and relate them to known inequalities. We also investigate the roots of Ehrhart polynomials. We prove that for fixed d, there exists a bounded region of $\mathbb{C}$ containing all roots of Ehrhart polynomials of d-polytopes, and that all real roots of these polynomials lie in $[-d, \lfloor d/2 \rfloor)$. In contrast, we prove that when the dimension d is not fixed the positive real roots can be arbitrarily large. We finish with an experimental investigation of the Ehrhart polynomials of cyclic polytopes and 0/1-polytopes.

1. Introduction

In this article, a *lattice polytope* $P \subset \mathbb{R}^d$ is a convex polytope whose vertices have integral coordinates. (For all notions regarding convex polytopes we refer to [**25**].) In 1967 Eugène Ehrhart proved that the function which counts the lattice points in the n-fold dilated copy of P,

$$i_P : \mathbb{N} \to \mathbb{N}, \qquad i_P(n) \;=\; \# \left(nP \cap \mathbb{Z}^d \right) ,$$

is a polynomial in n (see [**6, 7**] and the description in [**8**]). In particular, i_P can be naturally extended to all complex numbers n. In this paper we investigate linear inequalities satisfied by the coefficients of Ehrhart polynomials and the distribution of the roots of Ehrhart polynomials in the complex plane.

The coefficients of Ehrhart polynomials are very special. For example, it is well known that the leading term of $i_P(n)$ equals the volume of P, normalized with respect to the sublattice $\mathbb{Z}^d \cap \operatorname{aff}(P)$. The second term of $i_P(t)$ equals half the surface area of P normalized with respect to the sublattice on each facet of P, and the constant term equals 1. Moreover, the function $i_P^\circ(n)$ counting the number of interior lattice points in nP satisfies the *reciprocity law* $i_P(-n) = (-1)^{\dim P} i_P^\circ(n)$ [**8, 14, 18**].

2000 *Mathematics Subject Classification.* Primary 52C07; Secondary 05A15, 52B20.

Our first contribution is to establish new linear relations satisfied by the coefficients of all Ehrhart polynomials. This is a continuation of the pioneering work of Stanley, Betke & McMullen, and Hibi [**19, 21, 1, 11**], who established several families of linear inequalities for the coefficients (see Theorems 3.1 and 3.4). If we think of an Ehrhart polynomial $i_P(n) = c_d x^d + c_{d-1}x^{d-1} + \cdots + c_1 x + 1$ as a point in d-space, given by the coefficient vector $(c_d, c_{d-1}, \ldots, c_1)$, their results imply that the Ehrhart polynomials of all d-polytopes lie in a certain polyhedral complex. Betke and McMullen raised the issue [**1**, page 262] of whether other linear inequalities are possible. We were indeed able to find such new inequalities in the form of bounds for the *k-th difference* of the Ehrhart polynomial $i_P(n)$. These are defined recursively via

$$\Delta i_P(n) \;=\; i_P(n+1) - i_P(n)$$

and

$$\Delta^k i_P(n) \;=\; \Delta\left(\Delta^{k-1} i_P(n)\right) \quad \text{for } k \geq 1 \quad \text{and} \quad \Delta^0 i_P(n) = i_P(n).$$

Our first result (proved in Section 3) is as follows.

THEOREM 1.1. *If the lattice d-polytope $P \subset \mathbb{R}^d$ has Ehrhart polynomial $i_P(n) = c_d\, n^d + \cdots + c_0$, then*

$$\binom{d}{\ell}\Delta^k i_P(0) \;\leq\; \binom{d}{k}\Delta^\ell i_P(0) \qquad \text{for } 0 \leq k < \ell \leq d.$$

In particular (put $k = 0$ resp. $\ell = d$),

$$\binom{d}{k} \;\leq\; \Delta^k i_P(0) \;\leq\; \binom{d}{k} d!\, c_d \qquad \text{for } 0 \leq k \leq d.$$

In Section 3 we give a proof of Theorem 1.1 using the language of rational generating functions as established in [**1, 22**], and make a summary of known linear constraints and their strength.

The relation between the coefficients and the roots of polynomials, via elementary symmetric functions, suggests that once we understand the size of the coefficients of Ehrhart polynomials we could predict the distribution of their roots in the complex plane. The second contribution of this paper is a general study of the roots of Ehrhart polynomials.

There is clearly something special about the roots of Ehrhart polynomials. Take for instance the integer roots: Since a lattice polytope always contains some integer points (namely, its vertices), all integer roots of its Ehrhart polynomial are negative. More precisely, by the reciprocity law, the integer roots of an Ehrhart polynomial are those $-n$ for which the open polytope nP° contains no lattice point. For instance, the Ehrhart polynomial $\binom{n+d}{d}$ of the *standard simplex* in $\mathbb{R}^d$ (with vertices at the origin and the unit vectors on the coordinate axes) has integer roots at $n = -d, -d+1, \ldots, -1$.

The roots of the Ehrhart polynomial of the cross polytope

$$\mathcal{O}^d \;=\; \left\{(x_1, \ldots, x_d) \in \mathbb{R}^d : \; |x_1| + \cdots + |x_d| \leq 1\right\},$$

also exhibit special behavior: Bump et al. [**3**] and Rodriguez [**16**] proved that the zeros of $i_{\mathcal{O}^d}$ all have real parts equal to $-1/2$.

Using classical results from complex analysis and the linear inequalities of Theorem 3.5, we derive in Section 4 the following theorems:

THEOREM 1.2. (a) *The roots of Ehrhart polynomials of lattice d-polytopes are bounded in norm by $1 + (d+1)!$.*
(b) *All real roots of Ehrhart polynomials of d-dimensional lattice polytopes lie in the half-open interval $[-d, \lfloor d/2 \rfloor)$.*

The upper bound we present in Theorem 1.2 (b) is not tight. For example, in Proposition 4.8, we give a very short self-contained proof (e.g. not relying on Theorem 1.2) of the fact that Ehrhart polynomials for polytopes of dimension $d \leq 4$ have real roots in the interval $[-4, 1)$. In contrast with the above theorem we can also prove the following result:

THEOREM 1.3. *For any positive real number t there exist an Ehrhart polynomial of sufficiently large degree with a real root strictly larger than t. In fact, for every d there is a d-dimensional $0/1$-polytope whose Ehrhart polynomial has a real zero α_d such that* $\lim_{d \to \infty} \alpha_d / d = 1/(2\pi e) = 0.0585 \cdots$.

Our third contribution is an experimental study of the roots and coefficients of Ehrhart polynomials of concrete families of lattice polytopes. Our investigations and conjectures are supported by computer experimentation using `LattE` [**4, 5**] and `polymake` [**12**]. For the complex roots, we offer the following conjecture, based on experimental data.

CONJECTURE 1.4. *All roots α of Ehrhart polynomials of lattice d-polytopes satisfy* $-d \leq \operatorname{Re} \alpha \leq d - 1$.

We also computed the Ehrhart polynomials of all $0/1$-polytopes of dimension less than or equal to 4 and for many cyclic polytopes:

CONJECTURE 1.5. *For the cyclic polytope $C(n, d)$ realized with integral vertices on the moment curve* $\nu_d(t) := (t, t^2, \ldots, t^d)$,

$$i_{C(n,d)}(m) = \operatorname{vol}(C(n,d))\, m^d + i_{C(n,d-1)}(m).$$

Equivalently,

$$i_{C(n,d)}(m) = \sum_{k=0}^{d} \operatorname{vol}_k(C(n,k))\, m^k.$$

We have experimentally verified this conjecture in many cases.

2. An appetizer: dimension two

Since Ehrhart polynomials of lattice 1-polytopes (segments) are of the form $\ell n + 1$, where ℓ is the length of the segment, we know everything about their coefficients and roots: the set of roots of these polynomials is $\{-1/\ell : \ell \geq 1\} \subset [-1, 0)$.

The first interesting case is dimension $d = 2$. Pick's Theorem tells us that the Ehrhart polynomial of a lattice 2-polytope P is

$$i_P(n) = c_2\, n^2 + c_1\, n + 1 \ ,$$

where c_2 is the area of P and c_1 equals $1/2$ times the number of boundary integer points of P. In 1976, Scott established the following linear relations. Two polytopes are *unimodularly equivalent* if there is a function which maps one to the other and which preserves the integer lattice.

THEOREM 2.1. [**17**] *Let $i_P(n) = c_2\, n^2 + c_1\, n + 1$ be the Ehrhart polynomial of the lattice 2-polytope P. If P contains an interior integer point, and P is not unimodularly equivalent to* $\operatorname{conv}\{(0,0),(3,0),(0,3)\}$*, then*

$$c_1 \ \le \ \frac{1}{2}c_2 + 2 \ .$$

By Pick's Theorem, for 2-polytopes with no interior lattice points, we have $c_1 = c_2 + 1$. For $P = \operatorname{conv}\{(0,0),(3,0),(0,3)\}$, we obtain $i_P(n) = 9/2\, n^2 + 9/2\, n + 1$.

It is interesting to ask which degree-2 polynomials can possibly be Ehrhart polynomials. Since the constant term has to be 1, we can think of such a polynomial as a point (c_2, c_1) in the plane. From the geometry of lattice 2-polytopes, we know such an Ehrhart polynomial must have half-integral coordinates. Aside from Scott's inequality, we can trivially bound $c_1 \ge 3/2$, since every lattice 2-polytope has at least 3 integral points, namely its vertices. From these considerations, we arrive at Figure 1, which shows regions of possible Ehrhart polynomials of 2-polytopes.

Depicted are (parts of) three lines:

(i) $c_1 = 3/2$
(ii) $c_1 = c_2/2 + 2$
(iii) $c_1 = c_2 + 1$

and the point $(c_2, c_1) = (9/2, 9/2)$. The ray (i) shows the lower bound $c_1 \ge 3/2$. This is a sharp lower bound, in the sense that we can have polygons with exactly three boundary integer points but arbitrarily large area. The ray (ii) illustrates Scott's bound, and the point $(c_2, c_1) = (9/2, 9/2)$ corresponds to the "exceptional" polytope $\operatorname{conv}\{(0,0),(3,0),(0,3)\}$ in Theorem 2.1. The rectangles $\operatorname{conv}\{(0,0),(2,0),(2,x),(0,x)\}$, where x is a positive integer, show that there is a point on (ii) for every half integer. Finally, (iii) corresponds to 2-polytopes which contain no interior lattice point. There is a point on (iii) for every half integer, corresponding to the triangles $\operatorname{conv}\{(0,0),(1,0),(0,x)\}$ for a positive integer x. The rays (i) and (iii) meet in the point $(1/2, 3/2)$, which corresponds to the standard triangle $\operatorname{conv}\{(0,0),(1,0),(0,1)\}$. So the polyhedral complex containing all Ehrhart vectors consists of the polyhedron bounded by (i), (ii), and (iii) (shaded in Figure 1), plus the ray (iii), plus the extra point $(c_2, c_1) = (9/2, 9/2)$. In fact, only points with half-integral coordinates inside the complex are valid Ehrhart vectors. From these constraints, we can locate possible roots of Ehrhart polynomials of lattice 2-polytope fairly precisely.

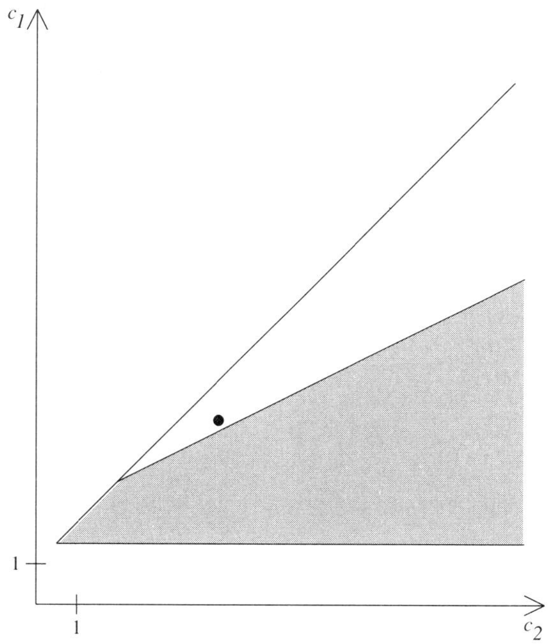

FIGURE 1. Regions in which Ehrhart polynomials of lattice 2-polytopes lie. It consists of 3 half lines, an open region (only points with half-integral coordinates are possible), plus an exceptional point.

THEOREM 2.2. *The roots of the Ehrhart polynomial of any lattice 2-polytope are contained in*

$$\left\{-2,-1,-\frac{2}{3}\right\} \cup \left\{x+iy \in \mathbb{C}:\ -\frac{1}{2} \leq x < 0,\ |y| \leq \frac{\sqrt{15}}{6}\right\}.$$

PROOF. We consider three cases, according to Scott's Theorem 2.1. First, if the lattice 2-polytope P contains no interior lattice point then $i_P(n) = An^2+(A+1)n+1$ (by Pick's Theorem), where A denotes the area of P. The roots of i_P are at -1 and $-1/A$. Note that A is half integral.

The second case is the "exceptional" polytope $P = \operatorname{conv}\{(0,0),(3,0),(0,3)\}$ whose Ehrhart polynomial $i_P(n) = 9/2n^2 + 9/2n + 1$ has roots $-2/3$ and $-1/2$.

This leaves, as the last case, 2-polytopes which contain an interior lattice point and which are not unimodularly equivalent to $\operatorname{conv}\{(0,0),(3,0),(0,3)\}$. The corresponding Ehrhart polynomials $i_P(n) = c_2n^2 + c_1n + 1$ satisfy the Scott inequality $c_1 \leq c_2/2 + 2$. Note that (because P has an interior lattice point) the area of P satisfies $c_2 \geq 3/2$. We have two possibilities:

(A) The discriminant $c_1^2 - 4c_2$ is negative. Then the real part of a root of i_P equals $-\frac{c_1}{2c_2}$ (which is negative). By Pick's Theorem $c_1 = c_2 - I + 1$ where I is the number of interior lattice points, that is, $-\frac{c_1}{2c_2} = -\frac{1}{2} - \frac{1-I}{2c_2}$. For fixed area c_2, this fraction is minimized when I is smallest possible, that is $I = 1$. The imaginary part of a root of i_P is plus or minus

$$\frac{1}{2c_2}\sqrt{4c_2 - c_1^2} \le \frac{1}{2c_2}\sqrt{4c_2 - \frac{9}{4}} = \sqrt{\frac{1}{c_2} - \left(\frac{3}{4c_2}\right)^2} ;$$

here we used $c_1 \ge 3/2$. As a function in c_2, this upper bound is decreasing for $c_2 \ge 1$. Since $c_2 \ge 3/2$, we obtain as an upper bound for the magnitude of the imaginary part of a root

$$\sqrt{\frac{2}{3} - \left(\frac{1}{2}\right)^2} = \frac{\sqrt{15}}{6} .$$

(B) The discriminant $c_1^2 - 4c_2$ is nonnegative. Then the smaller root of i_P is

$$\begin{aligned} -\frac{c_1}{2c_2} - \frac{1}{2c_2}\sqrt{c_1^2 - 4c_2} &\ge -\frac{1}{4} - \frac{1}{c_2} - \frac{1}{2c_2}\sqrt{\left(\frac{c_2}{2} + 2\right)^2 - 4c_2} \\ &= -\frac{1}{4} - \frac{1}{c_2} - \frac{1}{2c_2}\left(\frac{c_2}{2} - 2\right) = -\frac{1}{2} \end{aligned}$$

(Note that in this case $c_2 \ge 4$.)

Finally, the larger root is negative, since all the coefficients of i_P are positive. □

3. Linear inequalities for the coefficients of Ehrhart polynomials

In this section, we prove Theorem 1.1, which bounds the ratio of the k-th and ℓ-th differences of any Ehrhart polynomial solely in terms of d, k, and ℓ. It is perhaps worth observing that most of our arguments are valid for a somewhat larger class of polynomials. To describe this class, we define the *generating function* of the polynomial p as

$$S_p(x) = \sum_{n \ge 0} p(n)\, x^n .$$

It is well known (see, e.g., [**22**, Chapter 4]) that, if p is of degree d, then S_p is a rational function of the form

$$S_p(x) = \frac{f(x)}{(1-x)^{d+1}} , \tag{3.1}$$

where f is a polynomial of degree at most d. Most of our results hold for polynomials p for which the numerator of S_p has only nonnegative coefficients. Ehrhart polynomials are a particular case, as seen from the following theorem of Stanley.

THEOREM 3.1. [**19**, Theorem 2.1] *Suppose P is a convex lattice polytope. Then the generating function $\sum_{n \ge 0} i_P(n)\, x^n$ can be written in the form of (3.1), where $f(x)$ is a polynomial of degree at most d with nonnegative integer coefficients.*

Another well-known (and easy-to-prove) fact about rational generating functions (see, e.g., [**22**, Chapter 4]) is the following.

LEMMA 3.2. *Suppose that* $p \in \mathbb{R}[n]$ *is a polynomial of degree* d *with generating function* $S_p(x) = (a_d x^d + \cdots + a_1 x + a_0)/(1-x)^{d+1}$. *Then* p *can be recovered as*

$$p(n) \quad = \quad \sum_{j=0}^{d} a_j \binom{d+n-j}{d} . \tag{3.2}$$

More generally, we have the identity

$$\Delta^k p(n) \quad = \quad \sum_{j=0}^{d} a_j \binom{d+n-j}{d-k} \qquad \text{for } k \geq 0. \tag{3.3}$$

PROOF. Equation (3.2) follows from expanding $1/(1-x)^{d+1}$ into a binomial series. For (3.3), we proceed by induction on k. For $k=0$, the statement is (3.2), while for $k \geq 1$ we have by the induction hypothesis

$$\begin{aligned} \Delta^k p(n) &= \Delta^{k-1} p(n+1) - \Delta^{k-1} p(n) \\ &= \sum_{j=0}^{d} a_j \left(\binom{d+n+1-j}{d-k+1} - \binom{d+n-j}{d-k+1} \right) \\ &= \sum_{j=0}^{d} a_j \binom{d+n-j}{d-k} . \end{aligned}$$

□

Combining Theorem 3.1 and Lemma 3.2 immediately yields the following fact.

COROLLARY 3.3. *For any lattice polytope* P *and* $k \geq 0$, *we have* $\Delta^k \, i_P(0) \geq 0$.

PROOF. This follows because those binomial coefficients in the final expression for $\Delta^k p(n)$ are either positive or zero. □

PROOF OF THEOREM 1.1. We will use the falling-power notation $d^{\underline{j}} = d(d-1)\cdots(d-j+1)$, along with the obvious relation $k^{\underline{j}} < \ell^{\underline{j}}$ for $j \leq k < \ell$, and the identity

$$\binom{d-j}{d-k} \quad = \quad \binom{d}{k} \frac{k^{\underline{j}}}{d^{\underline{j}}} .$$

The statement now follows from Lemma 3.2 (3.3) by

$$\binom{d}{\ell}\binom{d-j}{d-k} \; = \; \binom{d}{\ell}\binom{d}{k}\frac{k^{\underline{j}}}{d^{\underline{j}}} \; < \; \binom{d}{k}\binom{d}{\ell}\frac{\ell^{\underline{j}}}{d^{\underline{j}}} \; = \; \binom{d}{k}\binom{d-j}{d-\ell} .$$

□

Theorem 1.1 is not the first set of linear inequalities on coefficient vectors of Ehrhart polynomials. Indeed, in 1984, Betke and McMullen [**1**, Theorem 6] obtained the following inequalities.

THEOREM 3.4. *Let P be a lattice d-polytope whose Ehrhart polynomial is $\sum_{i=0}^{d} c_i n^i$. Then*

$$c_r \;\le\; (-1)^{d-r} s(d,r)\, c_d + (-1)^{d-r-1} \frac{s(d,r+1)}{(d-1)!} \qquad \text{for } r = 1, 2, \ldots, d-1,$$

where $s(k,j)$ denote the Stirling numbers of the first kind. □

In that paper, Betke and McMullen sent out a challenge to the community to discover new inequalities for these coefficient vectors. The following theorem sums up the current state of affairs.

THEOREM 3.5. *Let P be a d-dimensional lattice polytope, with Ehrhart polynomial $i_P(n) = \sum_{i=0}^{d} c_i n^i = \sum_{i=0}^{d} a_i \binom{n+d-i}{d}$. Then the following inequalities are valid for $0 \le k < \ell \le d$ and $0 \le i \le d$:*

$$c_r \;\le\; (-1)^{d-r} s(d,r)\, c_d + (-1)^{d-r-1} \frac{s(d,r+1)}{(d-1)!}, \tag{3.4}$$

$$\binom{d}{k} \Delta^{\ell}\, i_P(0) \;\ge\; \binom{d}{\ell} \Delta^{k}\, i_P(0), \tag{3.5}$$

$$\binom{d+1}{2} c_d \;\ge\; c_{d-1}, \tag{3.6}$$

$$i_P(1) \;\ge\; d+1, \tag{3.7}$$

$$\Delta^k\, i_P(0) \;\ge\; \binom{d}{k}, \tag{3.8}$$

$$c_d \;\ge\; c_0/d!, \tag{3.9}$$

$$c_{d-1} \;\ge\; c_0 \frac{d+1}{2(d-1)!}, \tag{3.10}$$

$$\sum_{i=0}^{d} (-1)^{d-i} c_i \;\ge\; 0, \tag{3.11}$$

$$a_i \;\ge\; 0. \tag{3.12}$$

Moreover,

(3.13)

$$a_d + a_{d-1} + \cdots + a_{d-i} \;\le\; a_0 + a_1 + \cdots + a_i + a_{i+1} \quad \text{for all } 0 \le i \le \left\lfloor \frac{d-1}{2} \right\rfloor.$$

Whenever $a_s \neq 0$ but $a_{s+1} = \cdots = a_d = 0$, then

(3.14)

$$a_0 + a_1 + \cdots + a_i \;\le\; a_s + a_{s-1} + \cdots + a_{s-i} \qquad \text{for all } 0 \le i \le s;$$

finally, if $a_d \neq 0$, then

$$a_1 \;\le\; a_i \qquad \text{for all } 2 \le i < d. \tag{3.15}$$

PROOF. The inequalities (3.4) and (3.5) are the contents of Theorems 3.4 and 1.1; while (3.6), (3.7), and (3.8) are the special cases $(k, \ell) = (d-1, d)$, $(k, \ell) = (0, 1)$, and $k = 0$, respectively. (3.9) and (3.10) say that the volume and the normalized surface are at least as big as for a primitive simplex. Inequality (3.11) follows from Ehrhart reciprocity. Inequality (3.12) is the statement of Theorem 3.1. Incidentally, (3.9) also follows from (3.8), and (3.11) follows from (3.12), both by specializing to $i = d$. Inequality (3.14) was proved by Stanley [**21**], and inequalities (3.13) and (3.15) by Hibi [**11, 10**]. □

It is illuminating to compare these inequalities with each other. Since inequality (3.12) was used to prove Theorem 3.4 (by Betke and McMullen) and Theorem 1.1, it seems stronger than the other inequalities. Indeed, the only inequality among (3.4)–(3.12) which does not follow from (3.12) is (3.10). Experimental data for small d shows that neither (3.4) nor (3.5) imply the other.

The set of linear inequalities of Theorem 3.5 describes an unbounded complex of half-open polyhedra in $\mathbb{R}^{d+1}$ inside which all coefficient vectors of Ehrhart polynomials live. From this, we obtain a bounded complex $\mathcal{Q}^d$ by cutting with the normalizing hyperplane $c_d = 1$. By (3.14) each constraint $a_s \neq 0, a_{s+1} = \cdots = a_d = 0$ for $s = 1, 2, \ldots, d$ defines a half-open polytope $E_s \in \mathcal{Q}^d$ of dimension s that is missing one facet; E_0 is a single point.

Here are some particular cases: The bounded complex $\mathcal{Q}^3$ consists of one half-open s-dimensional simplex for each $s = 0, 1, 2, 3$ (Figure 2), and the half-open 3- and 4-dimensional polytopes of $\mathcal{Q}^4$ are shown in Figure 3.

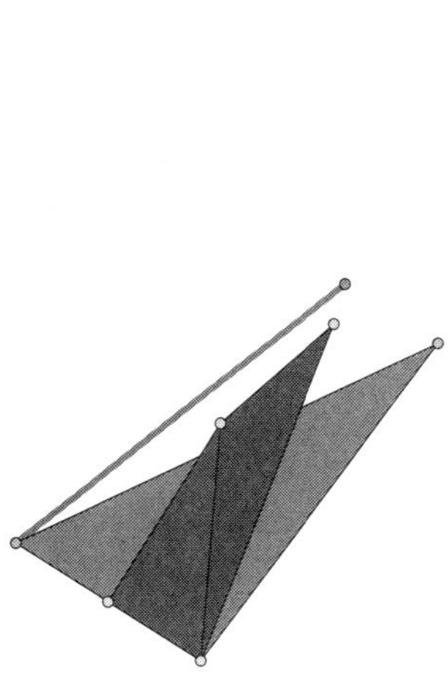

FIGURE 2. The complex $\mathcal{Q}^3$ of half-open polytopes, inside which the possible Ehrhart coefficients of all 3-dimensional polytopes lie. The facets of the tetrahedron corresponding to $a_3 \neq 0$ are $a_0 \geq 0$, $c_2 \geq 1$, (3.14) and (3.15); those of the triangle corresponding to $a_3 = 0$, $a_2 \neq 0$ are $a_0, a_1 \geq 0$ and (3.14); and those of the segment $a_2 = a_3 = 0$, $a_1 \neq 0$ are $a_0 \geq 0$ and (3.14).

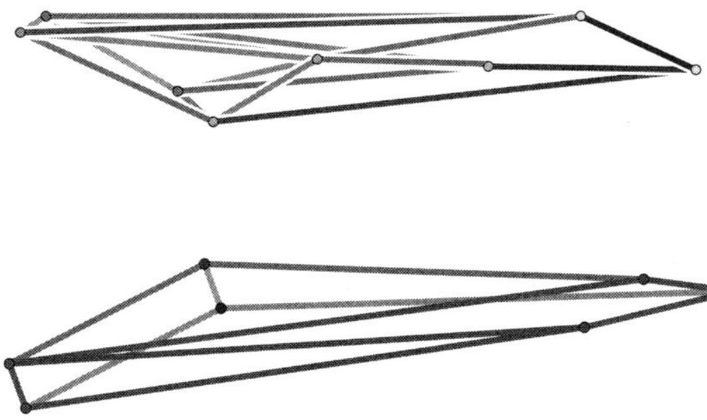

FIGURE 3. The 4-dimensional (top) and 3-dimensional (bottom) member of the complex $\mathcal{Q}^4$. The facets of the 4-dimensional polytope are $a_0 \geq 0$, $c_3 \geq 5/12$, (3.14) for $i = 0$, (3.13) for $i = 0, 1$, and (3.15) for $i = 2, 3$; those of the 3-dimensional one corresponding to $a_4 = 0$ but $a_3 \neq 0$ are $a_0, a_1 \geq 0$, $c_3 \geq 5/12$, (3.13) for $i = 1$, (3.14) for $i = 1$, and (3.15) for $i = 2$; etc.

An important question about any linear inequality is whether or not it defines a facet of $\mathcal{Q}^d$. We rephrase Betke and McMullen's question [**1**]:

Problem. *Are there other linear inequalities for the coefficients of an Ehrhart polynomial aside from those in Theorem 3.5? Do they define facets of the polyhedral complex inside which all coefficient vectors of Ehrhart polynomials live?*

4. The roots of Ehrhart polynomials

When one has a family of polynomials, a natural thing to look at are its roots. What is the general behavior of complex roots of Ehrhart polynomials? As a consequence of the inequalities on its coefficients, we give bounds on the norm of roots of any Ehrhart polynomial in dimension d. The basis $\{\binom{d+n-j}{d} : 0 \leq j \leq d\}$ of the vector space of polynomials of degree d turns out to be much more natural than the basis $\{n^i : 0 \leq i \leq d\}$ for deriving bounds on the roots of Ehrhart polynomials $i_P(n) = \sum_{i=0}^d a_i \binom{n+d-i}{d} = \sum_{i=0}^d c_i n^i$. Also, recall the following classical result of Cauchy (see, for example, [**15**, Chapter VII]).

LEMMA 4.1. *The roots of the polynomial* $p(n) = c_d n^d + c_{d-1} n^{d-1} + \cdots + c_0$ *lie in the open disc*

$$\left\{ z \in \mathbb{C} : \ |z| < 1 + \max_{0 \leq j \leq d} \left| \frac{c_j}{c_d} \right| \right\}.$$

Now we study roots of Ehrhart polynomials in general dimension. We first give an easy proof bounding the norm of all roots.

PROOF OF THEOREM 1.2(A). By Lemma 4.1 and Theorem 3.4, the maximal norm of the roots of i_P is bounded by

$$1 + \max_{0 \le j \le d} \left| \frac{c_j}{c_d} \right| \le 1 + \max_{0 \le j \le d} \left| (-1)^{d-j} s(d,j) + (-1)^{d-j-1} \frac{s(d,j+1)}{(d-1)!\, c_d} \right| \le 1 + d! + d!d = 1 + (d+1)!\,.$$

Here we have used the estimate $s(d,j) \le |s(d,j)| \le d!$ and the fact that $c_d \ge 1/d!$. ☐

While using crude estimates gives us a bound of $1+(d+1)!$, which makes the main point that there exists a bound dependent only on d, the actual bound on the roots can be improved greatly for specific values of d. First of all, for small d, we can compute the inequalities exactly; here the inequalities from Theorem 1.1 are used along with the Betke-McMullen inequalities. This gives appropriate bounds on the ratios of the coefficients of the Ehrhart polynomial. Second of all, Lemma 4.1 is not the best tool to use for specific cases, since calculating the inequalities for small d yields much lower bounds for c_i/c_d when i is large. Instead, we use the following proposition.

PROPOSITION 4.2 (Theorem 27.1 [**15**]). *Let $p(n) = c_d n^d + c_{d-1} n^{d-1} + \cdots + c_0$ be a polynomial. Then the maximal value of the norm of a root of $p(n)$ is the value of the maximal root of $p'(n) = |c_d| n^d - |c_{d-1}| n^{d-1} - |c_{d-2}| n^{d-2} - \cdots - |c_0|$.* ☐

We use this and the exact calculation of the inequalities in question to obtain the following tighter bounds on the roots of Ehrhart polynomials of d-polytopes.

d	2	3	4	5	6	7	8	9
bound	3.6	8.5	15.8	25.7	38.3	53.5	71.4	92.0

The bound appears to grow roughly quadratically. We suspect that there is a bound for the roots of Ehrhart polynomials of d-polytopes which is polynomial in d. For real roots this is certainly the case; we prove next that all real roots of Ehrhart polynomials of d-polytopes lie in the interval $[-d, \lfloor d/2 \rfloor)$. For this, we will use the following well-known bound.

LEMMA 4.3. (Newton Bound) *Let $f \in \mathbb{R}[n]$ be a polynomial of degree d and $B \in \mathbb{R}$ be such that all derivatives of f are positive at B: $f^{(\ell)}(B) > 0$ for $\ell = 0, 1, \ldots, d$. Then all real roots of f are contained in $(-\infty, B)$.* ☐

PROOF OF THEOREM 1.2(B). The lower bound follows from Theorem 3.1 and the simple observation that for (real numbers) $n < -d$ the binomial coefficients in

$$i_P(n) = \sum_{i=0}^{d} a_i \binom{n+d-i}{d}$$

are all positive or all negative, depending on the parity of d.

As for the upper bound, let $B = \lfloor d/2 \rfloor$. We now show that $\alpha < B$ for any real root α of $i_P(n)$. For this, we will make use of the fact that the second highest coefficient of any Ehrhart polynomial measures half the normalized surface area. This coefficient reads

$$c_{d-1} \;=\; \frac{1}{(d-1)!} \sum_{i=0}^{d} a_i\,(d-2i+1)$$

when expressed in terms of the a_i's, so that the following inequality is valid:

$$(d-1)!\,c_{d-1} \;=\; \sum_{i=0}^{d} a_i\,(d-2i+1) \;>\; 0\,. \tag{4.1}$$

Note that the coefficient $s(i) = d - 2i + 1$ of a_i in (4.1) is positive for $0 \le i \le B$ and non-positive for $B+1 \le i \le d$. We now express the ℓ-th derivative of i_P evaluated at $n = B$ as $i_P^{(\ell)}(B) = (\ell!/d!) \sum_{i=0}^{d} a_i\, g_i(B,\ell)$, and claim that for $0 \le \ell \le d$, there exists a $\lambda(\ell) > 0$ with

$$g_i(B,\ell) > \lambda(\ell)\, s(i) \qquad \text{for all } \; 0 \le i \le d.$$

This claim is the statement of Lemma 4.6 below. The proof of Theorem 1.2(b) now follows from this relation, inequality (4.1), $a_0 = 1$ and $a_i \ge 0$ for $1 \le i \le d$ via the following chain of inequalities:

$$\begin{aligned} 0 \;&<\; \sum_{i=0}^{d} \big(g_i(B,\ell) - \lambda(\ell)\, s(i)\big) a_i \\ &<\; \sum_{i=0}^{d} \big(g_i(B,\ell) - \lambda(\ell)\, s(i)\big) a_i + \lambda(\ell) \sum_{i=0}^{d} s(i)\, a_i \\ &=\; \sum_{i=0}^{d} g_i(B,\ell)\, a_i \;=\; i_P^{(\ell)}(B)\,. \end{aligned}$$

□

REMARK 4.4. It is a well-known fact that Ehrhart polynomials of lattice polytopes form a special class of *Hilbert polynomials.* More strongly, they are special examples of Hilbert polynomials of Cohen-Macaulay semi-standard graded k-algebras [**21**] (this is essentially the content of Theorem 3.1). It is then natural to ask whether Ehrhart polynomials are special or whether the bounds proved above hold in more generality. We stress that inequality (4.1), used in previous arguments, comes from geometric information about Ehrhart polynomials $i_P(n)$. Indeed, from the following proposition and Theorem 1.2(b), Ehrhart polynomials are special in their root distribution:

PROPOSITION 4.5. *For degree d Hilbert polynomials associated to arbitrary semi-standard graded k-algebras the negative real roots are arbitrarily small and $d-1$ may appear as a root. In contrast, for fixed degree d, Hilbert polynomials of Cohen-Macaulay semi-standard graded k-algebras have all its real roots in the interval $[-d, d-1)$.*

PROOF. Indeed, it follows from [**2**, Theorem 3.8] that for fixed d and positive integers $a_0, \ldots, a_d$, the polynomial $a_0(x+a_1)(x+a_2)\ldots(x+a_d)$ is the Hilbert polynomial of a semi-standard graded k-algebra. Also, observe that the chromatic polynomial of the complete graph on d vertices has highest root $d-1$, and that chromatic polynomials are known to be Hilbert polynomials of standard graded algebras by a result attributed to Almkvist (see the proof given by Steingrímsson [**23**]). Thus the first statement holds.

Now, in a Cohen-Macaulay semi-standard graded algebra, the Hilbert polynomial can be written as $p(n) = \sum_{i=0}^{d} a_i \binom{n+d-i}{d}$, where $a_i \geq 0$ for $0 \leq i \leq d$. Observe that all the binomial coefficients in $p(n)$ are positive for (real numbers) $n > d-1$, which establishes the upper bound of $d-1$. For the lower bound, observe that for (real numbers) $n < -d$ all the binomial coefficients are positive, respectively negative, depending on the parity of d. □

To complete the proof of Theorem 1.2 (b), we need only to prove the following lemma.

LEMMA 4.6. *Fix $0 \leq \ell \leq d-1$ and consider again the functions $s, g : \{0, 1, \ldots, d\} \to \mathbb{Z}$ defined by $s(i) = d - 2i + 1$ and $g(i) = g_i(B, \ell)$. Moreover, if we set*

$$\lambda(\ell) \;=\; \frac{1}{2}\big(g(B) - g(B+1)\big) \;=\; \frac{d}{2} \sum_{I \in \binom{[d-1]}{d-\ell-1}} \prod_{k \in I} (d-k) \;>\; 0\,, \tag{4.2}$$

then

$$g(i) \;\geq\; \lambda(\ell)\, s(i) \qquad \text{for } \; i = 0, 1, \ldots, d\,. \tag{4.3}$$

For this, we will also need to prove Lemma 4.7 below. We will write $[d-1]_0 = \{0, 1, \ldots, d-1\}$, $[d-1] = \{1, 2, \ldots, d-1\}$, and $\binom{S}{t}$ for the set of all t-element subsets of the finite set S. Now we express $i_P(n)$ as

$$i_P(n) \;=\; \frac{1}{d!} \sum_{i=0}^{d} a_i \prod_{k=0}^{d-1} (n+d-i-k),$$

so that the ℓ-th derivative of i_P is

$$\begin{aligned} i_P^{(\ell)}(n) \;&=\; \frac{\ell!}{d!} \sum_{i=0}^{d} a_i \sum_{I \in \binom{[d-1]_0}{\ell}} \prod_{k \in [d-1]_0 \setminus I} (n+d-i-k) \\ &=\; \frac{\ell!}{d!} \sum_{i=0}^{d} a_i \sum_{I \in \binom{[d-1]_0}{d-\ell}} \prod_{k \in I} (n+d-i-k)\,. \end{aligned}$$

Note that we now have an explicit formula for the coefficient of a_i in $(d!/\ell!)\, i_P^{(\ell)}$:

$$g_i(n, \ell) \;=\; \sum_{I \in \binom{[d-1]_0}{d-\ell}} \prod_{k \in I} (n+d-i-k)\,. \tag{4.4}$$

The following lemma shows that the piece-wise linear function interpolating $g : \{0, 1, \dots, d\} \to \mathbb{Z}$, $g(i) = g_i(B, \ell)$ is positive, and its slope weakly increases in the range $0 \le \ell \le d$ and $0 \le i \le B + 1$. See Figure 4.

LEMMA 4.7. *The following inequalities are satisfied for* $0 \le \ell \le d$ *and* $0 \le i \le B+1$*:*

$$g_i(B,\ell) \;>\; 0, \tag{4.5}$$

$$g_i(B,\ell) - g_{i+1}(B,\ell) \;>\; g_{i+1}(B,\ell) - g_{i+2}(B,\ell). \tag{4.6}$$

PROOF. Equation (4.5) follows because $k \le d-1$ and $i \le B+1$ imply $B+d-i-k \ge 0$. To show (4.6), we abbreviate $m := B + d - i$ and inspect the difference

$$g_i(B,\ell) - g_{i+1}(B,\ell) \;=\; \sum_{I \in \binom{[d-1]_0}{d-\ell}} \prod_{k\in I} (m-k) \;-\; \sum_{J \in \binom{[d-1]_0}{d-\ell}} \prod_{k\in J} (m-k-1)\,.$$

If $0 \notin I$, then the term corresponding to I in the first sum cancels with the term corresponding to $J = \{i - 1 : i \in I\}$ in the second sum:

$$\prod_{k\in I} (m-k) - \prod_{k\in J} (m-k-1) \;=\; \prod_{k\in I} \big((m-k) - (m-(k-1)-1)\big) \;=\; 0,$$

so we are left with summing over the sets $I \in \binom{[d-1]_0}{d-\ell}$ that contain 0 and the sets J that contain $d - 1$. But for such summation sets, the difference simplifies to

$$\begin{aligned} &g_i(B,\ell) - g_{i+1}(B,\ell) \\ &\quad= (m-0) \sum_{I \in \binom{[d-1]}{d-\ell-1}} \prod_{k\in I} (m-k) \;-\; (m-d) \sum_{J \in \binom{[d-2]_0}{d-\ell-1}} \prod_{k\in I} (m-(k+1)) \\ &\quad= d \sum_{I \in \binom{[d-1]}{d-\ell-1}} \prod_{k\in I} (B-d-i-k)\,, \end{aligned}$$

and (4.6) follows by comparing the expressions $g_i(B,\ell) - g_{i+1}(B,\ell)$ and $g_{i+1}(B,\ell) - g_{i+2}(B,\ell)$ term by term. □

In the following, we will use Iverson's notation (see [**9**]): the expression $[S]$ evaluates to 1 resp. 0 according to the truth or falsity of the logical statement S.

PROOF OF LEMMA 4.6. First note that $s(B) = 1$ for even d, so that

$$g(B) - \lambda s(B) \;=\; \tfrac{1}{2} g(B) + \tfrac{1}{2} g(B+1) \;>\; 0\,;$$

for odd d, we have $s(B + 1) = 0$. Now note that the graph of (the piecewise-linear function interpolating) λs is a line, while $g(B+1) > 0$ by (4.5) and the slope of the graph of g is weakly increasing on $[0, B+2]$ by (4.6) (see Figure 4); this proves (4.3) for $0 \le i \le B + [d \text{ odd}]$.

Set $j = i - B$, so that we still need to prove (4.3) for $1 + [d \text{ odd}] \le j \le d - B$. By plugging (4.4) and (4.2) into (4.3) and rearranging, we must show that for these values of j

$$\sum_{I \in \binom{[d-1]_0}{d-\ell}} \prod_{k\in I} (d-j-k) \;+\; \frac{d}{2}\,(2j - [d \text{ odd}] - 1) \sum_{J \in \binom{[d-1]}{d-\ell-1}} \prod_{k\in J} (d-k) \;>\; 0\,. \tag{4.7}$$

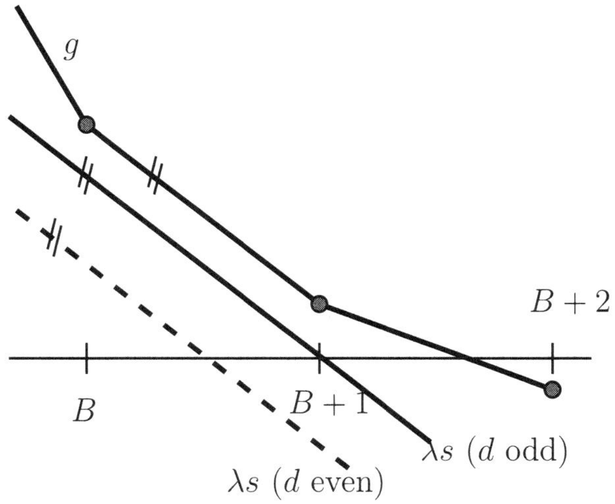

FIGURE 4. The graphs of the functions g and λs (solid for odd d, dashed for even d).

Note that each term in the second sum of (4.7) is positive, and decompose the index sets I in the first sum into disjoint unions $I = I_+ \cup K$ such that $I_+ \subset \{0, 1, \ldots, d-2j\}$ and $K \subset \{d-2j+1, \ldots, d-1\}$, and therefore $d-j-k > 0$ for all $k \in I_+$.

k	0	$\cdots$	$d-2j$	$d-2j+1$	$\cdots$	$d-j$	$\cdots$	$d-1$
value of $d-j-k$	$d-j$	$\cdots$	j	$j-1$	$\cdots$	0	$\cdots$	$-(j-1)$
set in $I = I_+ \cup K$		I_+				K		

If $|K|$ is odd, then the summand $\sigma(K)$ corresponding to $I_+ \cup K$ cancels with the one corresponding to $I_+ \cup (d-j-K)$, so we only need to consider even $|K|$. In that case, $\sigma(K) > 0$ (resp. $\sigma(K) < 0$) if $|K \cap [d-j+1, d-1]|$ is even (resp. odd). In total, there are more than enough positive terms in (4.7) to cancel the negative summands. □

Now we give a very direct and elementary proof ofthe following proposition without using Theorem 1.2(b). It is in fact an improvement on the upper bound of real roots.

PROPOSITION 4.8. *We have $\alpha < 1$ for any real root α of an Ehrhart polynomial i_P of a lattice polytope P of dimension $d \leq 4$.*

PROOF. First note it is enough to prove the statement in dimension 4 because if there was a polytope which counterexample of smaller dimension, then the Cartesian product of the counterexample with a unit standard simplex will be a counterexample of higher dimension. Suppose $f(n) = pn^4 + qn^3 + rn^2 + sn + 1$ is the Ehrhart polynomial of a lattice 4-polytope P. We know $p > 0$ and $q > 0$. Because $f(1)$ counts the lattice points in P, we know that $p+q+r+s+1 \geq 5$. By the reciprocity law, $f(-1) \geq 0$, so $p-q+r-s+1 \geq 0$. The top two coefficients of the shifted polynomial $g(n) = f(n+1) = pn^4 + (4p+q)n^3 + g_2n^2 + g_1n + g_0$ are positive, as is the constant term $g_0 = g(0) = f(1)$. We will show that g_2 and g_1 are nonnegative, and hence, by Descartes' rule of signs, g does not have a positive root. This implies that $f(n) = g(n-1)$ does not have a real root larger than 1. To prove that $g_2 \geq 0$, we add the inequalities $f(1) \geq 5$ and $f(-1) \geq 0$ to obtain $2p + 2r \geq 3$ or $r \geq \frac{3}{2} - p$, whence $g_2 = 6p + 3q + r \geq 5p + 3q + \frac{3}{2} \geq 0$ (because $p, q \geq 0$). A similar reasoning yields

$$g_1 \;=\; 4p+3q+2r+s \;=\; (p+q+r+s)+(3p+2q+r) \;\geq\; 4+2p+2q+\frac{3}{2} \;\geq\; 0\,;$$

here we used the inequality $f(1) \geq 5$ again. □

We now conclude with the proof of Theorem 1.3:

PROOF OF THEOREM 1.3. Given a positive integer d, consider the convex polytope P_d defined by the facet inequalities:

$$0 \leq x_0 \leq x_k \leq 1 \quad \text{for} \quad 1 \leq k \leq d-1.$$

P_d is an *order polytope* in the sense of [**20**] and thus it has 0/1 vertices. We claim that the Ehrhart polynomial of P_d is given by $i_{P_d}(n) = (B_d(n+2) - B_d(0))/d$ where $B_d(x)$ is the d-th Bernoulli polynomial.

Indeed, from the facet-defining inequalities of P_d one sees that $i_{P_d}(n)$ is the number of d-tuples of nonnegative integers $(a_0, a_1, \ldots, a_{d-1})$ such that $a_0 \leq a_k \leq n$. If $a_0 = j$ then there are $n-j+1$ choices for each a_k $(k > 0)$. Hence $i_{P_d}(n) = \sum_{j=0}^{n}(n-j+1)^{d-1} = \sum_{j=1}^{n+1} j^{d-1}$. A classical identity of Bernoulli that says $\sum_{k=0}^{n-1} k^{d-1} = (B_d(n) - B_d(0))/d$. Thus we get $i_{P_d}(n) = (B_d(n+2) - B_d(0))/d$, a polynomial of degree d. Note that when d is odd then $B_d(0) = 0$. Finally, the results of [**24**] imply that the largest real zero of $B_d(n)$ is asymptotically $d/(2\pi e)$. Therefore, as stated, as the degree d grows, the Ehrhart polynomial of P_d has larger and larger real roots. It is worth remarking that since d is the degree of the Ehrhart polynomial of P_d it differs from the upper bound of $\lfloor d/2 \rfloor$ in Theorem 1.2 only by a constant factor. □

5. Special families of polytopes

We begin this section with some charts showing the behavior of roots for hundreds of Ehrhart polynomials computed using `LattE` and `Polymake`. In Figure 5 we show the distribution of roots of a large sample of Ehrhart polynomials of lattice 3-polytopes.

In Table 1 we collected a small sample of Ehrhart polynomials of 0/1 polytopes and cyclic polytopes from the experiments we performed.

5.1. 0/1-polytopes. We computed the Ehrhart polynomials for all 0/1 polytopes of dimension less or equal to 4 (up to symmetry there are 354 different 4-polytopes). In Figure 6 we plotted their roots. In our computations we relied on the on-line data sets of 0/1 polytopes available from `Polymake`'s web page and

name	Ehrhart polynomial $P(s)$
cube	$t^3 + 3t^2 + 3t + 1$
cube minus corner	$5/6\,t^3 + 5/2\,t^2 + 8/3\,t + 1$
prism	$1/2\,t^3 + 2\,t^2 + 5/2\,t + 1$
nameless	$2/3\,t^3 + 2\,t^2 + 7/3\,t + 1$
octahedron	$2/3\,t^3 + 2\,t^2 + 7/3\,t + 1$
square pyramid	$1/3\,t^3 + 3/2\,t^2 + \frac{13}{6}\,t + 1$
bypyramid	$1/2\,t^3 + 3/2\,t^2 + 2\,t + 1$
unimodular tetrahedron	$1/6\,t^3 + t^2 + \frac{11}{6}\,t + 1$
fat tetrahedron	$1/3\,t^3 + t^2 + 5/3\,t + 1$
as:6-18.poly	$\frac{83}{240}\,x^6 + \frac{307}{240}\,x^5 + \frac{41}{16}\,x^4 + \frac{217}{48}\,x^3 + \frac{611}{120}\,x^2 + \frac{16}{5}\,x + 1$
cf:10-11.poly	$\frac{11}{3628800}\,x^{10} + \frac{11}{725760}\,x^9 + \frac{17}{60480}\,x^8 + \frac{121}{24192}\,x^7 + \frac{7643}{172800}\,x^6 + \frac{8591}{34560}\,x^5 + \frac{340873}{362880}\,x^4 + \frac{84095}{36288}\,x^3 + \frac{59071}{16800}\,x^2 + \frac{7381}{2520}\,x + 1$
cf:4-5.poly	$1/12\,x^4 + 1/2\,x^3 + \frac{17}{12}\,x^2 + 2\,x + 1$
cf:9-10.poly	$\frac{1}{120960}\,x^9 + \frac{1}{4480}\,x^8 + \frac{61}{20160}\,x^7 + \frac{79}{2880}\,x^6 + \frac{997}{5760}\,x^5 + \frac{4223}{5760}\,x^4 + \frac{30043}{15120}\,x^3 + \frac{32651}{10080}\,x^2 + \frac{2383}{840}\,x + 1$
cf:8-9.poly	$\frac{11}{40320}\,x^8 + \frac{1}{1120}\,x^7 + \frac{1}{64}\,x^6 + \frac{9}{80}\,x^5 + \frac{1039}{1920}\,x^4 + \frac{267}{160}\,x^3 + \frac{5933}{2016}\,x^2 + \frac{761}{280}\,x + 1$
oa:6-13.poly	$\frac{9}{80}\,x^6 + \frac{43}{80}\,x^5 + \frac{23}{16}\,x^4 + \frac{143}{48}\,x^3 + \frac{79}{20}\,x^2 + \frac{179}{60}\,x + 1$
cut(4)	$\frac{2}{45}\,x^6 + \frac{4}{15}\,x^5 + \frac{7}{9}\,x^4 + 4/3\,x^3 + \frac{98}{45}\,x^2 + \frac{12}{5}\,x + 1$
cyclic01:5-8.poly	$\frac{7}{60}\,x^5 + \frac{5}{12}\,x^4 + 5/4\,x^3 + \frac{31}{12}\,x^2 + \frac{79}{30}\,x + 1$
halfcube(5)	$\frac{13}{15}\,x^5 + 11/3\,x^4 + 16/3\,x^3 + 10/3\,x^2 + 9/5\,x + 1$
Cyclic(2,5)	$10x^2 + 4x + 1$
Cyclic(3,5)	$16x^3 + 10x^2 + 4x + 1$
Cyclic(4,5)	$12x^4 + 16x^3 + 10x^2 + 4x + 1$
Cyclic(2,6)	$20x^2 + 5x + 1$
Cyclic(3,6)	$70x^3 + 20x^2 + 5x + 1$
Cyclic(4,6)	$192x^4 + 70x^3 + 20x^2 + 5x + 1$
Cyclic(5,6)	$288x^5 + 192x^4 + 70x^3 + 20x^2 + 5x + 1$
Cyclic(2,7)	$35x^2 + 6x + 1$
Cyclic(3,7)	$224x^3 + 35x^2 + 6x + 1$
Cyclic(4,7)	$1512x^4 + 224x^3 + 35x^2 + 6x + 1$
Cyclic(2,8)	$56x^2 + 7x + 1$
Cyclic(3,8)	$588x^3 + 56x^2 + 7x + 1$
Cyclic(4,8)	$8064x^4 + 588x^3 + 56x^2 + 7x + 1$

TABLE 1. The Ehrhart polynomials for some well-known lattice polytopes. The choice of coordinates for cyclic polytopes was $t = 1, ..., n$. The rest are listed Ehrhart polynomials comes from 0/1 polytopes selected from Ziegler's list. It includes the Ehrhart polynomials of all 3-dimensional 0/1-polytopes.

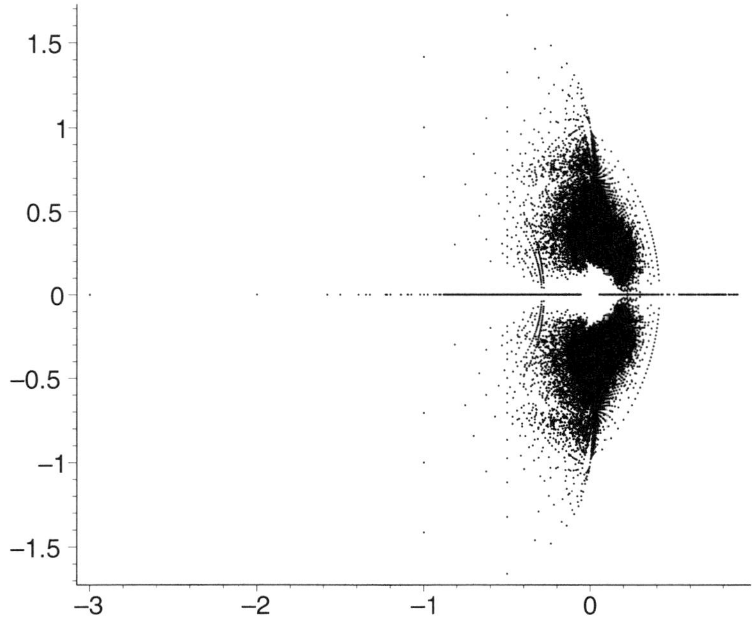

FIGURE 5. The zeros of Ehrhart polynomials corresponding to 100000 random 3-dimensional lattice simplices.

those discussed in Ziegler's lectures on 0/1 polytopes [**13**]. Several phenomena are evident from the data we collected. For example, in Table 1 we see two combinatorially different polytopes that have the same Ehrhart polynomial. These are the so called "nameless" polytope of coordinates $(1,0,0,0)$, $(1,1,0,0)$, $(1,0,1,0)$, $(1,1,1,0)$, $(1,0,1,1)$, $(1,1,0,1)$ and the octahedron. Another example of regular distribution appears also in Figure 6. We show the roots of the Ehrhart polynomials associated to the Birkhoff polytope of doubly stochastic $n \times n$ matrices for $n = 2, \ldots, 9$.

5.2. Cyclic polytopes. Cyclic polytopes form a family whose combinatorial structure (i.e. f-vector, face lattice, etc) is well understood. The canonical choice of coordinates is given using the *moment curve*

$$\nu_d : \left\{ \begin{array}{ccc} \mathbb{R} & \to & \mathbb{R}^d, \\ t & \mapsto & (t^1, t^2, \ldots, t^d). \end{array} \right. \tag{5.1}$$

A cyclic polytope is obtained as the convex hull of n points along the moment curve. Thus we fix $t_1, t_2, \ldots, t_n$ and define $C(n,d) := \operatorname{conv}\{\nu_d(t_1), \nu_d(t_2), \ldots, \nu_d(t_n)\}$. Cyclic polytopes are lattice polytopes exactly when $t_i \in \mathbb{Z}$. There is a natural linear projection connecting these cyclic polytopes.

LEMMA 5.1. *Consider the projection $\pi : \mathbb{R}^d \to \mathbb{R}^{d-1}$ that forgets the last coordinate. The inverse image under π of a lattice point $y \in C(n, d-1) \cap \mathbb{Z}^{d-1}$ is a line that intersects the boundary of $C(n,d)$ in at most two* integral *points.*

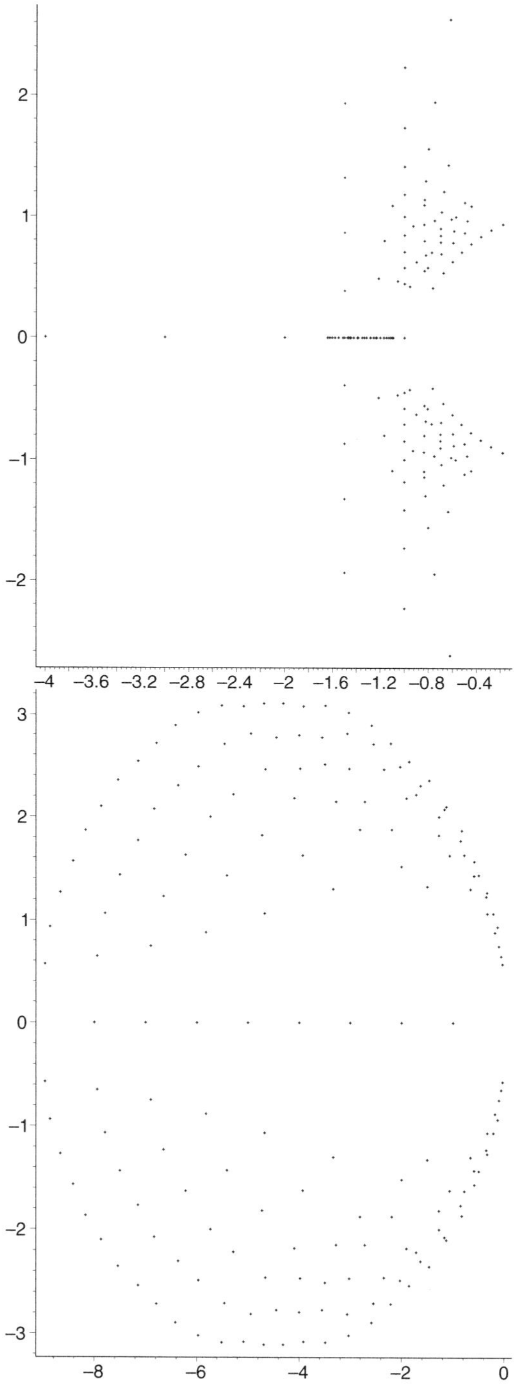

FIGURE 6. *Top:* The zeros of the Ehrhart polynomials of all 3 and 4 dimensional 0/1 polytopes. *Bottom:* The zeros of Ehrhart polynomials for the Birkhoff polytopes up to $n = 9$

PROOF. We need to prove that, given $t_1, t_2, \ldots, t_d \in \mathbb{Z}$ and $\lambda_1, \lambda_2, \ldots, \lambda_d \in \mathbb{R}$,

$$\forall\, 1 \le j < d : \sum_{k=1}^{d} \lambda_k\, t_k^j \in \mathbb{Z} \qquad \Longrightarrow \qquad \sum_{k=1}^{d} \lambda_k\, t_k^d \in \mathbb{Z}\,.$$

For $1 \le j \le d$, let $y_j = \sum_{k=1}^{d} \lambda_k\, t_k^j$; we know that $y_1, y_2, \ldots, y_{d-1} \in \mathbb{Z}$. We need to prove that

$$\mathbf{y} = (y_1, \ldots, y_d) = \sum_{k=1}^{d} \lambda_k\, \nu_d(t_k) \in \mathbb{Z}^d\,.$$

This identity means that $\mathbf{y}$ lies on the hyperplane spanned by $\nu_d(t_1), \ldots, \nu_d(t_d)$, which can be expressed via a determinant:

$$\det \begin{pmatrix} 1 & 1 & \cdots & 1 \\ \mathbf{y} & \nu_d(t_1) & & \nu_d(t_d) \end{pmatrix} = 0\,.$$

Writing this determinant out through the first column and solving for y_d gives

$$\begin{aligned} y_d &= (-1)^d \frac{1}{D} \det \begin{pmatrix} t_1 & \cdots & t_d \\ \vdots & & \vdots \\ t_1^d & \cdots & t_d^d \end{pmatrix} + (-1)^{d-1} \frac{y_1}{D} \det \begin{pmatrix} 1 & \cdots & 1 \\ t_1^2 & \cdots & t_d^2 \\ \vdots & & \vdots \\ t_1^d & \cdots & t_d^d \end{pmatrix} + \cdots \\ &\quad - \frac{y_{d-1}}{D} \det \begin{pmatrix} 1 & \cdots & 1 \\ t_1 & \cdots & t_d \\ \vdots & & \vdots \\ t_1^{d-2} & \cdots & t_d^{d-2} \\ t_1^d & \cdots & t_d^d \end{pmatrix}, \end{aligned}$$

where

$$D = \det \begin{pmatrix} 1 & \cdots & 1 \\ t_1 & \cdots & t_d \\ \vdots & & \vdots \\ t_1^{d-1} & \cdots & t_d^{d-1} \end{pmatrix} = \prod_{1 \le j < k \le d} (t_j - t_k)\,.$$

This expression yields an integer if we can prove that D divides the determinants appearing in the numerators. Equivalently, the substitution $t_j = t_k$ in any of the numerators evaluates the determinant to zero, which is apparent. □

Using the same technique, one can show that Lemma 5.1 also holds for integer dilations of $C(n, d)$. Consequently, Conjecture 1.5 is equivalent to saying that the number of lattice points in a dilation of a cyclic polytope by a positive integer m is equal to its volume plus the number of lattice points in its lower envelope. From the above lemma and Pick's theorem, it follows that Conjecture 1.5 is true for $d = 2$.

Acknowledgements. We thank Christos Athanasiadis, David Eisenbud, Francisco Santos, Bernd Sturmfels, Thomas Zaslavsky and Günter M. Ziegler for helpful discussions and suggestions. This research was supported in part by the Mathematical Sciences Research Institute. Mike Develin was also supported by the American Institute of Mathematics. Jesús De Loera and Richard Stanley were partially supported by NSF grants DMS-0309694 and DMS-9988459 respectively.

References

1. U. Betke and P. McMullen, *Lattice points in lattice polytopes*, Monatsh. Math. **99** (1985), no. 4, 253–265.
2. F. Brenti, *Hilbert polynomials in combinatorics*, J. of Algebraic Combinatorics **7** (1998), 127–156.
3. D. Bump, K.-K. Choi, P. Kurlberg, and J. Vaaler, *A local Riemann hypothesis. I*, Math. Z. **233** (2000), no. 1, 1–19.
4. J. A. De Loera, D. Haws, R. Hemmecke, P. Huggins, J. Tauzer, and R. Yoshida, *A user's guide for* `latte`, (2003), software package `LattE` and manual are available at `http://www.math.ucdavis.edu/~latte/`.
5. J. A. De Loera, R. Hemmecke, J. Tauzer, and R. Yoshida, *Effective lattice point counting in rational convex polytopes*, Journal of Symbolic Computation (2003), to appear, available at `http://www.math.ucdavis.edu/~latte/`.
6. E. Ehrhart, *Sur un problème de géométrie diophantienne linéaire. I. Polyèdres et réseaux*, J. Reine Angew. Math. **226** (1967), 1–29.
7. ———, *Sur un problème de géométrie diophantienne linéaire. II. Systèmes diophantiens linéaires*, J. Reine Angew. Math. **227** (1967), 25–49.
8. ———, *Polynômes arithmétiques et méthode des polyèdres en combinatoire*, Birkhäuser Verlag, Basel, 1977, International Series of Numerical Mathematics, Vol. 35.
9. R. L. Graham, D. E. Knuth, and O. Patashnik, *Concrete mathematics: a foundation for computer science*, 2nd ed., Addison-Wesley, 1994.
10. T. Hibi, *A lower bound theorem for Ehrhart polynomials of convex polytopes*, Advances in Math. **105** (1994), 162–165.
11. ———, *Star-shaped complexes and Ehrhart polynomials*, Proc. Amer. Math. Soc. **123** (1995), 723–726.
12. M. Joswig and E. Gawrilow, `polymake`*: A framework for analyzing convex polytopes*, in Kalai and Ziegler [**13**], pp. 43–73.
13. G. Kalai and G.M. Ziegler (eds.), *Polytopes - Combinatorics and Computation, DMV-Seminar Oberwolfach, Germany, November 1997*, DMV Semin., vol. 29, Birkhäuser, 2000.
14. I. G. Macdonald, *Polynomials associated with finite cell-complexes*, J. London Math. Soc. (2) **4** (1971), 181–192.
15. M. Marden, *Geometry of polynomials*, second ed., Mathematical Surveys, no. 3, American Mathematical Society, Providence, R.I., 1966.
16. F. Rodriguez-Villegas, *On the zeros of certain polynomials*, Proc. Amer. Math. Soc. **130** (2002), no. 8, 2251–2254.
17. P. R. Scott, *On convex lattice polygons*, Bull. Austral. Math. Soc. **15** (1976), no. 3, 395–399.
18. R. P. Stanley, *Combinatorial reciprocity theorems*, Advances in Math. **14** (1974), 194–253.
19. ———, *Decompositions of rational convex polytopes*, Ann. Discrete Math. **6** (1980), 333–342, Combinatorial mathematics, optimal designs and their applications (Proc. Sympos. Combin. Math. and Optimal Design, Colorado State Univ., Fort Collins, Colo., 1978).
20. ———, *Two poset polytopes*, Discrete Comput. Geom. **1** (1986), no. 1, 9–23.
21. ———, *On the Hilbert function of a graded Cohen-Macaulay domain*, J. Pure Appl. Alg. **73** (1991), 307–314.
22. ———, *Enumerative Combinatorics*, 2nd ed., vol. I, Cambridge University Press, 1997.
23. E. Steingrímsson, *The coloring ideal and coloring complex of a graph*, J. Algebraic Combinatorics **14** (2001), 73–84.
24. A.P Veselov and J.P. Ward, *On the real roots of the Bernoulli polynomials and the Hurwitz zeta-function*, preprint available at www.lboro.ac.uk/departments/ma/preprints/papers99/99-35.pdf.
25. G. M. Ziegler, *Lectures on polytopes*, Springer-Verlag, New York, revised edition, 1998.

Department of Mathematics, San Francisco State University, 1600 Holloway Avenue, San Francisco, CA 94132, U.S.A.
E-mail address: beck@math.sfsu.edu

Department of Mathematics, University of California, One Shields Avenue, Davis, CA 95616-8633, USA
E-mail address: deloera@math.ucdavis.edu

American Institute of Mathematics, 360 Portage Ave., Palo Alto, CA 94306-2244, USA
E-mail address: develin@post.harvard.edu

Institut de Matemàtica, Universitat de Barcelona, Gran Via de les Corts Catalanes 585, E-08007 Barcelona, Spain
E-mail address: julian@imub.ub.es

Department of Mathematics 2-375, Massachusetts Institute of Technology, Cambridge, MA 02139-4307, USA
E-mail address: rstan@math.mit.edu

Contemporary Mathematics
Volume **374**, 2005

Ehrhart Polynomials of Lattice Polyhedral Functions

Beifang Chen

ABSTRACT. This paper provides a self-contained introduction to Ehrhart polynomials by elementary methods. We first derive the Ehrhart polynomial for a lattice simplex and prove its reciprocity law. Next, we derive the Ehrhart polynomial for a lattice polytope with polynomial weight; the idea is to reduce the situation to the Ehrhart polynomial for a higher dimensional lattice polytope. Then, using the linearity of the Euler characteristic, we obtain the reciprocity law for arbitrary lattice polyhedral functions with polynomial weight. The procedure is streamlined into a method of transforming automatically a reciprocity law for simplices into a reciprocity law for polyhedral functions. After that, we consider the generating functions of the Ehrhart polynomials for lattice simplices, showing some hyperbolic cotangent formulas. At the end, we demonstrate a proof of the reciprocity law for Dedekind sums, using the constant coefficient formula for the Ehrhart polynomial of a lattice triangle.

1. Introduction

Let Λ be a full dimensional lattice of $\mathbb{R}^N$; and let X be a bounded Λ-lattice polyhedron. For a positive integer m, let $mX = \{m\boldsymbol{x} \mid \boldsymbol{x} \in X\}$. It is shown by Ehrhart [**14, 15**] that the function

$$E(X, \Lambda; m) : = \#(mX \cap \Lambda) = \text{number of points of } mX \cap \Lambda$$

is a polynomial function of degree $\dim X$ in the positive integral variable m; the polynomial is denoted by $L(X, \Lambda; t)$ in the real variable t, called the *Ehrhart polynomial* of X. It was also shown by Ehrhart [**14, 15**] that the Ehrhart polynomials of a Λ-lattice polytope P and its relative interior P^0 satisfy the functional equation

$$L(P, \Lambda; -t) = (-1)^{\dim P} L(P^0, \Lambda; t), \tag{1.1}$$

called the *reciprocity law* of P by Ehrhart. Recently, the polynomial $L(X, \Lambda; t)$ draws a lot of attention [**1, 2, 4, 5, 6, 10, 13, 18, 24, 26, 30**] from different areas because of its rich connection to number theory, combinatorics, topology and geometry. The reciprocity law (1.1) was conjectured by Ehrhart for lattice manifolds with boundaries; and it was proved by Macdonald [**21, 22**], using the

1991 *Mathematics Subject Classification.* 11H06, 51M20, 52B20, 52C07.

Key words and phrases. Lattice points, lattice polyhedra, Ehrhart polynomial, reciprocity law, Dedekind sum.

The work was supported by RGC Competitive Earmarked Research Grants HKUST6119/00P and 600703.

additivity property of the polynomial on lattice polyhedra. Here a lattice manifold means a manifold that is a union of finite number of polytopes whose vertices lie in the given lattice. The reciprocity law was largely extended by Stanley [**28, 29, 30**] in many aspects. Other extensions were recently considered in [**4, 5, 8, 11**].

Let h be a homogeneous polynomial function of degree k on $\mathbb{R}^N$. It was shown by Brion and Vergne [**5**] that the function

$$E(X,h,\Lambda;m) := \sum_{\boldsymbol{\alpha}\in mX\cap\Lambda} h(\boldsymbol{\alpha})$$

is a polynomial function of degree[1] at most $k+\dim X$ in the positive integral variable m; the polynomial is denoted by $L(X,h,\Lambda;t)$ in the real variable t, called the *Ehrhart polynomial of X with the polynomial weight h*. Brion and Vergne [**5**] also showed that the Ehrhart polynomials for a Λ-polytope P and its relative interior P^0 satisfy the reciprocity law:

$$L(P,h,\Lambda;-t) = (-1)^{k+\dim P} L(P^0,h,\Lambda;t). \tag{1.2}$$

Using the additivity of $L(\cdot,h,\Lambda,t)$ on the class $\mathcal{P}_\Lambda^N$ of Λ-lattice polytopes in $\mathbb{R}^N$, it is not hard to extend (1.2) to Λ-lattice manifolds with boundaries. However, it is not obvious to extend either (1.1) or (1.2) to arbitrary lattice polyhedra. In fact, to extend (1.1) and (1.2) to arbitrary lattice polyhedra, it is not enough to consider only lattice polyhedra; we need to consider lattice polyhedra with weights, namely, lattice polyhedral functions.

In this paper we present a streamlined method to obtain automatically the reciprocity law for lattice polyhedral functions, provided that the reciprocity law is proved for simplices. Denote by 1_X the characteristic function of a polyhedron X. For a Λ-lattice polyhedral function $f=\sum a_j 1_{P_j}$, where $P_j\in\mathcal{P}_\Lambda^N$, we consider the polynomial function

$$L(f,h,\Lambda;m) := \sum a_j L(P_j,h,\Lambda;m), \quad m\in\mathbb{Z}.$$

To have a similar reciprocity law for polyhedral functions, we associate with f a *dual polyhedral function* f^*, defined by

$$f^*(\boldsymbol{x}) := \lim_{\delta\to 0}\chi\left(f\cdot 1_{B^0(\boldsymbol{x},\delta)}\right), \quad \boldsymbol{x}\in\mathbb{R}^N,$$

where χ is the Euler characteristic (see [**7, 19**]) and $B^0(\boldsymbol{x},\delta)$ is the open ball of radius δ centered at $\boldsymbol{x}$ in $\mathbb{R}^N$. The main result of the paper is the reciprocity law:

$$L(f,h,\Lambda;-t) = L(f^*,\bar{h},\Lambda;t), \tag{1.3}$$

where h is a polynomial function and $\bar{h}(\boldsymbol{x}) = h(-\boldsymbol{x})$. In particular, if $f=1_M$, where M is a Λ-lattice manifold with boundary ∂M, then $f^* = (-1)^{\dim M}1_{M^0}$, $M^0 = M-\partial M$, and the reciprocity law (1.3) reduces to

$$L(M,h,\Lambda;-t) = (-1)^{\dim M}L(M-\partial M,\bar{h},\Lambda;t).$$

For the sake of smooth and complete exposition of the Ehrhart polynomials, the present paper is somewhat overlapped with the author's previous paper [**10**] on the deriving of the Ehrhart polynomials and their generating functions for a lattice simplex. The main part of the present paper, however, is to count lattice points with weight and to obtain the reciprocity law for a polyhedral weight function. The idea unifies the treatment in literature of the subject on lattice simplices and lattice

[1] It was carelessly misstated in [**5**] that the degree is $k+\dim P$ for any lattice polytope P.

manifolds with boundaries so that arbitrary bounded lattice polyhedra are included, and makes the whole exposition straightforward and natural without much technical effort.

The manuscript is prepared as a self-contained introduction to Ehrhart polynomials and organized as follows: We first derive the Ehrhart polynomial for a lattice simplex of $\mathbb{R}^N$ and prove its reciprocity law in Section 2; then we derive the Ehrhart polynomials with homogeneous polynomial weight for lattice polytopes in Section 3. Using the linearity of the Euler characteristic, we establish the reciprocity law (1.3) in Section 4 for lattice polyhedral functions with polynomial weights. In Section 5, the generating functions of Ehrhart polynomials for lattice simplices are considered; some hyperbolic cotangent formulas are presented by elementary methods. In the final Section 6, we demonstrate a proof of the reciprocity law for Dedekind sums, using a constant coefficient formula for the Ehrhart polynomial of a lattice triangle.

2. Ehrhart polynomials of lattice simplices

We follow the idea of Ehrhart [**14, 15**] to derive Ehrhart polynomials. Let Λ be a full dimensional lattice of $\mathbb{R}^N$. Let σ be an n-simplex with vertices $\boldsymbol{\alpha}_0, \boldsymbol{\alpha}_1, \ldots, \boldsymbol{\alpha}_n \in \Lambda$. We denote by σ^0 the relative interior of σ. For real numbers $t > 0$, we define

$$\begin{aligned} E(\sigma, \Lambda; t) : &= \text{ number of points of } t\sigma \cap \Lambda, \\ E(\sigma^0, \Lambda; t) : &= \text{ number of points of } t\sigma^0 \cap \Lambda. \end{aligned}$$

Let $\boldsymbol{\alpha} \in t\sigma$ and $\boldsymbol{\beta} \in t\sigma^0$. Then $\boldsymbol{\alpha}$ and $\boldsymbol{\beta}$ can be written uniquely as

$$\boldsymbol{\alpha} = \sum_{j=0}^{n} a_j t \boldsymbol{\alpha}_j \quad \text{and} \quad \boldsymbol{\beta} = \sum_{j=0}^{n} b_j t \boldsymbol{\alpha}_j,$$

where $a_j \geq 0$, $b_j > 0$, $\sum_{j=0}^{n} a_j = 1$ and $\sum_{j=0}^{n} b_j = 1$. By the division algorithm, $\boldsymbol{\alpha}$ and $\boldsymbol{\beta}$ can be further written uniquely as

$$\boldsymbol{\alpha} = \sum_{j=0}^{n} u_j \boldsymbol{\alpha}_j + \sum_{j=0}^{n} k_j \boldsymbol{\alpha}_j \quad \text{and} \quad \boldsymbol{\beta} = \sum_{j=0}^{n} v_j \boldsymbol{\alpha}_j + \sum_{j=0}^{n} l_j \boldsymbol{\alpha}_j,$$

where u_j and v_j are real numbers, k_j and l_j are nonnegative integers, satisfying

$$\textstyle\sum_{j=0}^{n} u_j + \sum_{j=0}^{n} k_j = t, \quad 0 \leq u_j < 1, \quad k_j = \lfloor a_j t \rfloor;$$

$$\textstyle\sum_{j=0}^{n} v_j + \sum_{j=0}^{n} l_j = t, \quad 0 < v_j \leq 1, \quad l_j = \lceil b_j t \rceil - 1.$$

It is clear that $\boldsymbol{\alpha}$ is a point of Λ if and only if $\sum_{j=0}^{n} u_j \boldsymbol{\alpha}_j$ is a point of Λ; and that $\boldsymbol{\beta}$ is a point of Λ if and only if $\sum_{j=0}^{n} v_j \boldsymbol{\alpha}_j$ is a point of Λ.

Let $\boldsymbol{\gamma} = \sum_{j=0}^{n} u_j \boldsymbol{\alpha}_j$. We call the real number $|\gamma| = \sum_{j=0}^{n} u_j$ the *level* of $\boldsymbol{\gamma}$. For nonnegative real numbers $\ell \leq n+1$, we define two convex subsets

$$\text{(2.1)} \qquad D(\sigma; \ell) : = \left\{ \gamma = \sum_{j=0}^{n} u_j \boldsymbol{\alpha}_j \;\middle|\; 0 \leq u_j < 1, |\gamma| = \ell \right\};$$

$$\text{(2.2)} \qquad \bar{D}(\sigma; \ell) : = \left\{ \gamma = \sum_{j=0}^{n} u_j \boldsymbol{\alpha}_j \;\middle|\; 0 < u_j \leq 1, |\gamma| = \ell \right\}.$$

If ℓ is a positive integer, the closure of $D(\sigma; \ell)$ is the same as the closure of $\bar{D}(\sigma; \ell)$, and is a Λ-polytope having the vertices $\boldsymbol{\alpha}_I = \sum_{i \in I} \boldsymbol{\alpha}_i$, where $I \subset \{0, 1, \ldots, n\}$ and

$|I| = \ell$. Let $\mathcal{I}(\sigma;\ell)$ denote the set of all subsets I of $\{0, 1, \ldots, n\}$ such that $|I| = \ell$. Obviously, $D(\sigma;\ell)$ consists of the points $\sum_{I\in\mathcal{I}(\sigma;\ell)} u_I \boldsymbol{\alpha}_I$ such that

$$u_I \geq 0, \quad \sum_{I\in\mathcal{I}(\sigma;\ell)} u_I = 1, \quad \sum_{i\in I\in\mathcal{I}(\sigma;\ell)} u_I < 1$$

for all $0 \leq i \leq n$; and $\bar{D}(\sigma;\ell)$ consists of the points $\sum_{I\in\mathcal{I}(\sigma;\ell)} u_I \boldsymbol{\alpha}_I$ such that

$$u_I \geq 0, \quad \sum_{I\in\mathcal{I}(\sigma;\ell)} u_I = 1, \quad \sum_{i\in I\in\mathcal{I}(\sigma;\ell)} u_I > 0$$

for all $0 \leq i \leq n$. We call the subsets

$$D(\sigma, \Lambda; \ell) := D(\sigma;\ell) \cap \Lambda \quad \text{and} \quad \bar{D}(\sigma, \Lambda; \ell) := \bar{D}(\sigma;\ell) \cap \Lambda \tag{2.3}$$

the *determined sets of σ at level ℓ*. There is no simple relation between $D(\sigma, \Lambda; \ell)$ (or $\bar{D}(\sigma, \Lambda; \ell)$) and the fundamental region of the lattice Λ. If $\boldsymbol{\alpha}_0, \boldsymbol{\alpha}_1, \ldots, \boldsymbol{\alpha}_n$ are the vertices of a fundamental region of Λ, then $D(\sigma, \Lambda; \ell) = \emptyset$ for $1 \leq \ell \leq n+1$ and $\bar{D}(\sigma, \Lambda; \ell) = \emptyset$ for $0 \leq \ell \leq n$; $D(\sigma, \Lambda; 0)$ and $\bar{D}(\sigma, \Lambda; n+1)$ contain one single point.

Now, for each $\boldsymbol{\gamma} \in D(\sigma, \Lambda; \ell)$, since $0 \leq |\boldsymbol{\gamma}| < n+1$, the integer $k = t - |\boldsymbol{\gamma}|$ belongs to the half-open and half-closed interval $(t-n-1, t]$; while for each $\boldsymbol{\gamma} \in \bar{D}(\sigma, \Lambda; \ell)$, since $0 < |\boldsymbol{\gamma}| \leq n+1$, the integer $k = t - |\boldsymbol{\gamma}|$ belongs to the interval $[t-n-1, t)$. Note that when $|\boldsymbol{\gamma}| = t - k$ is fixed, the number of tuples $(k_0, k_1, \ldots, k_n)$ of nonnegative integers satisfying the equation $\sum_{j=0}^{n} k_j = k$ is the binomial coefficient $\binom{k+n}{n}$. We thus have proved the following proposition.

PROPOSITION 2.1. *Let σ be a Λ-simplex of dimension n. Then for real numbers $t > 0$,*

$$E(\sigma, \Lambda; t) = \sum_{k\in(t-n-1,t]\cap\mathbb{Z}} \#(D(\sigma, \Lambda; t-k)) \binom{k+n}{n}; \tag{2.4}$$

$$E(\sigma^0, \Lambda; t) = \sum_{k\in[t-n-1,t)\cap\mathbb{Z}} \#(\bar{D}(\sigma, \Lambda; t-k)) \binom{k+n}{n}. \tag{2.5}$$

Since the right-hand sides of (2.4) and (2.5) are well defined for real numbers $t \leq 0$, we then define $E(\sigma, \Lambda; t)$ and $E(\sigma^0, \Lambda; t)$ for the real variable t by (2.4) and (2.5), respectively.

PROPOSITION 2.2 (Ehrhart Reciprocity Law). *Let σ be a Λ-simplex of dimension n. Then*

$$E(\sigma^0, \Lambda; -t) = (-1)^n E(\sigma, \Lambda; t). \tag{2.6}$$

PROOF. Let r and s be nonnegative real numbers such that $r+s = n+1$. There is a bijection $D(\sigma, \Lambda; r) \to \bar{D}(\sigma, \Lambda; s)$, sending $\sum_{j=0}^{n} u_j \boldsymbol{\alpha}_j$ to $\sum_{j=0}^{n}(1-u_j)\boldsymbol{\alpha}_j$. Write $s = t-k$; the determined set $\bar{D}(\sigma, \Lambda; t-k)$ in (2.5) can be replaced by the determined set $D(\sigma, \Lambda; r)$ with $r = n+1-t+k$. We then have

$$E(\sigma^0, \Lambda; t) = \sum_{k\in[t-n-1,t)\cap\mathbb{Z}} \#(D(\sigma, \Lambda; n+1-t+k)) \binom{k+n}{n}.$$

Write $n+1-t+k = -t-l$ for some l; that is, $l = -k-n-1$. Then $k \in [t-n-1, t)$ if and only if $l \in (-t-n-1, -t]$. Applying the reciprocity property of binomial coefficients, we have

$$\binom{k+n}{n} = \binom{-l-1}{n} = (-1)^n \binom{l+n}{n}.$$

Thus

$$\begin{aligned} E(\sigma^0, \Lambda; t) &= \sum_{l \in (-t-n-1, -t] \cap \mathbb{Z}} \#(D(\sigma, \Lambda; -t-l))(-1)^n \binom{l+n}{n} \\ &= (-1)^n E(\sigma, \Lambda; -t). \end{aligned}$$

□

Notice that when the real variable t takes integers m, the functions $E(\sigma, \Lambda; m)$ and $E(\sigma^0, \Lambda; m)$ are polynomial functions of the integral variable m. We denote such polynomials by $L(\sigma, \Lambda; t)$ and $L(\sigma^0, \Lambda; t)$ in the real variable t, respectively; they are called the *Ehrhart polynomials of* σ. Of course, the polynomials $L(\sigma, \Lambda; t)$ and $L(\sigma^0, \Lambda; t)$ are different from the integer-valued functions $E(\sigma, \Lambda; t)$ and $E(\sigma^0, \Lambda; t)$. In fact, $E(\sigma, \Lambda; t) = L(\sigma, \Lambda; \lfloor t \rfloor)$ and $E(\sigma^0, \Lambda; t) = L(\sigma^0, \Lambda; \lceil t \rceil)$. By changing the summation indices in (2.4) and (2.5), the Ehrhart polynomials $L(\sigma, \Lambda; m)$ and $L(\sigma^0, \Lambda; m)$ can be explicitly written as

$$L(\sigma, \Lambda; m) = \sum_{k=0}^{n} \#(D(\sigma, \Lambda; k)) \binom{m+n-k}{n}; \tag{2.7}$$

$$L(\sigma^0, \Lambda; m) = \sum_{k=0}^{n} \#(D(\sigma, \Lambda; k)) \binom{m+k-1}{n}. \tag{2.8}$$

Let s_j denote the jth elementary symmetric polynomial of n variables; that is,

$$s_j(x_1, x_2, \ldots, x_n) = \sum_{k_1 < k_2 < \cdots < k_j} x_{k_1} x_{k_2} \cdots x_{k_j}.$$

With the convention $s_0 \equiv 1$, the binomial coefficient in (2.8) can be written as

$$\begin{aligned} \binom{m+k-1}{n} &= \frac{1}{n!}(m+k-1)(m+k-2)\cdots(m+k-n) \\ &= \frac{1}{n!}\sum_{j=0}^{n} m^j s_{n-j}(k-1, k-2, \ldots, k-n). \end{aligned}$$

This immediately gives rise to the following simple formulas on the jth coefficient $c_j(\sigma^0, \Lambda)$ of $L(\sigma^0, \Lambda; t)$ and on the jth coefficient $c_j(\sigma, \Lambda)$ of $L(\sigma, \Lambda; t)$:

$$c_j(\sigma^0, \Lambda) = \frac{1}{n!}\sum_{k=0}^{n} \#(D(\sigma, \Lambda; k)) s_{n-j}(k-1, k-2, \ldots, k-n); \tag{2.9}$$

$$c_j(\sigma, \Lambda) = \frac{1}{n!}\sum_{k=0}^{n} \#(D(\sigma, \Lambda; k)) s_{n-j}(1-k, 2-k, \ldots, n-k). \tag{2.10}$$

Notice that counting lattice points is invariant under integral translations. Let σ' be the simplex whose vertices are $\boldsymbol{\alpha}_1 - \boldsymbol{\alpha}_0, \ldots, \boldsymbol{\alpha}_n - \boldsymbol{\alpha}_0$ and the origin. Then the number of lattice points in $m\sigma$ is the same as the number of lattice points in $m\sigma'$.

Following the notation of [**30**], we define the *determined sets of σ at the vertex $\boldsymbol{\alpha}_0$* to be the sets

$$\begin{aligned} D(\sigma, \boldsymbol{\alpha}_0; \Lambda): &= \left\{ \sum_{j=1}^{n} u_j(\boldsymbol{\alpha}_j - \boldsymbol{\alpha}_0) \in \Lambda \;\middle|\; 0 \le u_j < 1 \right\}; \\ \bar{D}(\sigma, \boldsymbol{\alpha}_0; \Lambda): &= \left\{ \sum_{j=1}^{n} v_j(\boldsymbol{\alpha}_j - \boldsymbol{\alpha}_0) \in \Lambda \;\middle|\; 0 < v_j \le 1 \right\}. \end{aligned}$$

Similarly, for $\boldsymbol{\gamma} = \sum_{j=1}^{n} u_j(\boldsymbol{\alpha}_j - \boldsymbol{\alpha}_0) \in D(\sigma, \boldsymbol{\alpha}_0; \Lambda) \cup \bar{D}(\sigma, \boldsymbol{\alpha}; \Lambda)$, we write $|\boldsymbol{\gamma}| = \sum_{j=1}^{n} u_j$.

THEOREM 2.3 (Ehrhart). *Let σ be a Λ-simplex of dimension n. Then*

$$L(\sigma, \Lambda; m) = \sum_{\boldsymbol{\gamma} \in D(\sigma, \boldsymbol{\alpha}_0; \Lambda)} \binom{m + n - \lceil |\boldsymbol{\gamma}| \rceil}{n}; \tag{2.11}$$

$$L(\sigma^0, \Lambda; m) = \sum_{\boldsymbol{\gamma} \in \bar{D}(\sigma, \boldsymbol{\alpha}_0; \Lambda)} \binom{m + n - 1 - \lfloor |\boldsymbol{\gamma}| \rfloor}{n} \tag{2.12}$$

$$= \sum_{\boldsymbol{\gamma} \in D(\sigma, \boldsymbol{\alpha}_0; \Lambda)} \binom{m + \lceil |\boldsymbol{\gamma}| \rceil - 1}{n}. \tag{2.13}$$

Moreover, the coefficients of $L(\sigma^0, m)$ and $L(\sigma, m)$ are respectively given by

$$c_j(\sigma^0, \Lambda) = \frac{1}{n!} \sum_{\boldsymbol{\gamma} \in D(\sigma, \boldsymbol{\alpha}_0; \Lambda)} s_{n-j}(\lceil |\boldsymbol{\gamma}| \rceil - 1, \lceil |\boldsymbol{\gamma}| \rceil - 2, \ldots, \lceil |\boldsymbol{\gamma}| \rceil - n); \tag{2.14}$$

$$c_j(\sigma, \Lambda) = \frac{1}{n!} \sum_{\boldsymbol{\gamma} \in D(\sigma, \boldsymbol{\alpha}_0; \Lambda)} s_{n-j}(1 - \lceil |\boldsymbol{\gamma}| \rceil, 2 - \lceil |\boldsymbol{\gamma}| \rceil, \ldots, n - \lceil |\boldsymbol{\gamma}| \rceil). \tag{2.15}$$

PROOF. Every $\boldsymbol{\alpha} \in m\sigma'$ and every $\boldsymbol{\beta} \in m\sigma'^0$ can be written uniquely as

$$\boldsymbol{\alpha} = \sum_{j=1}^{n} a_j(\boldsymbol{\alpha}_j - \boldsymbol{\alpha}_0), \quad \text{where } \textstyle\sum_{j=1}^{n} a_j \le m, \quad a_j \ge 0;$$

$$\boldsymbol{\beta} = \sum_{j=1}^{n} b_j(\boldsymbol{\alpha}_j - \boldsymbol{\alpha}_0), \quad \text{where } \textstyle\sum_{j=1}^{n} b_j < m, \quad b_j > 0.$$

Then, $\boldsymbol{\alpha}$ and $\boldsymbol{\beta}$ can be further written uniquely as

$$\boldsymbol{\alpha} = \sum_{j=1}^{n} u_j(\boldsymbol{\alpha}_j - \boldsymbol{\alpha}_0) + \sum_{j=1}^{n} k_j(\boldsymbol{\alpha}_j - \boldsymbol{\alpha}_0),$$

$$\boldsymbol{\beta} = \sum_{j=1}^{n} v_j(\boldsymbol{\alpha}_j - \boldsymbol{\alpha}_0) + \sum_{j=1}^{n} l_j(\boldsymbol{\alpha}_j - \boldsymbol{\alpha}_0),$$

where u_j and v_j are real numbers, k_j and l_j are nonnegative integers, satisfying

$$\textstyle\sum_{j=1}^{n} u_j + \sum_{i=1}^{n} k_j \le m, \quad 0 \le u_j < 1;$$

$$\textstyle\sum_{j=1}^{n} v_j + \sum_{j=1}^{n} l_j < m, \quad 0 < v_j \le 1.$$

Notice that $\boldsymbol{\alpha}$ is a point of Λ if and only if $\sum_{j=1}^{n} u_j(\boldsymbol{\alpha}_j - \boldsymbol{\alpha}_0)$ is a point of Λ; and that $\boldsymbol{\beta}$ is a point of Λ if and only if $\sum_{j=1}^{n} v_j(\boldsymbol{\alpha}_j - \boldsymbol{\alpha}_0)$ is a point of Λ. Also notice that

$$\sum_{j=1}^{n} k_j \le m - |\gamma| \quad \Leftrightarrow \quad \sum_{j=1}^{n} k_j \le m - \lceil |\gamma| \rceil; \tag{2.16}$$

$$\sum_{j=1}^{n} l_j < m - |\gamma| \quad \Leftrightarrow \quad \sum_{j=1}^{n} l_j \le m - \lfloor |\gamma| \rfloor - 1. \tag{2.17}$$

We thus have

$$\boldsymbol{\alpha} \in m\sigma' \cap \Lambda \quad \Leftrightarrow \quad \sum_{j=1}^{n} u_j(\boldsymbol{\alpha}_j - \boldsymbol{\alpha}_0) \in D(\sigma, \boldsymbol{\alpha}_0; \Lambda),\ \textstyle\sum_{j=1}^{n} k_j \le m - \lceil |\gamma| \rceil;$$

$$\boldsymbol{\beta} \in D(\sigma, \boldsymbol{\alpha}_0; \Lambda) \quad \Leftrightarrow \quad \sum_{j=1}^{n} v_j(\boldsymbol{\alpha}_j - \boldsymbol{\alpha}_0) \in \bar{D}(\sigma, \boldsymbol{\alpha}_0; \Lambda),\ \textstyle\sum_{j=1}^{n} l_j \le m - \lfloor |\gamma| \rfloor - 1.$$

Notice that the number of nonnegative integer solutions for (2.16) is the binomial coefficient $\binom{m+n-\lceil |\gamma| \rceil}{n}$; and that the number of nonnegative integer solutions for (2.17) is $\binom{m+n-\lfloor |\gamma| \rfloor -1}{n}$. We then obtain (2.11) and (2.12). Using the bijection

$$D(\sigma, \boldsymbol{\alpha}_0; \Lambda) \to \bar{D}(\sigma, \boldsymbol{\alpha}_0, \Lambda), \quad \sum_{j=1}^{n} u_j(\boldsymbol{\alpha}_j - \boldsymbol{\alpha}_0) \mapsto \sum_{j=1}^{n} (1 - u_j)(\boldsymbol{\alpha}_j - \boldsymbol{\alpha}_0)$$

and the reciprocity property of binomial coefficients, we further obtain (2.13). □

3. Ehrhart polynomials with polynomial weight

In this section we count lattice points evaluated by a polynomial function inside a lattice polytope. Let Λ be a full dimensional lattice of $\mathbb{R}^N$ with a lattice basis $\{\boldsymbol{\xi}_i\}_{i=1}^{N}$. Let X be a bounded polyhedron of $\mathbb{R}^N$. Let $h : \mathbb{R}^N \to \mathbb{R}$ be a polynomial function relative to $\{\boldsymbol{\xi}_i\}_{i=1}^{N}$; that is, for $\boldsymbol{x} = x_1\boldsymbol{\xi}_1 + \cdots + x_N\boldsymbol{\xi}_N$, $h(\boldsymbol{x}) = \sum a_{r_1,\ldots,r_N} x_1^{r_1} \cdots x_N^{r_N}$. For integers $m > 0$, we define

$$E(X, h, \Lambda; m) := \sum_{\boldsymbol{\alpha} \in mX \cap \Lambda} h(\boldsymbol{\alpha}). \tag{3.1}$$

If X is a Λ-lattice polyhedron and h is of degree k, then it is shown (indirectly) that $E(X, h, \Lambda; m)$ is a polynomial function of positive integers m, having degree at most $k + \dim X$; see [**5**]. (However, it was carelessly stated in [**5**] that the Ehrhart polynomials $E(P, h, \Lambda; m)$ for Λ-lattice polytopes P have degree $k + \dim P$; counterexamples can be easily found.) We call $E(X, h, \Lambda; m)$ the *Ehrhart polynomial of X with the polynomial weight h*.

Let $\boldsymbol{r} = (r_1, \ldots, r_N)$, $\boldsymbol{s} = (s_1, \ldots, s_N) \in \mathbb{Z}^N$ with nonnegative entries. If each entry of $\boldsymbol{s}$ is less than or equal to the corresponding entry of $\boldsymbol{r}$, we write $\boldsymbol{s} \le \boldsymbol{r}$ and define

$$\binom{\boldsymbol{r}}{\boldsymbol{s}} := \binom{r_1}{s_1} \cdots \binom{r_N}{s_N}.$$

We write $k = |\boldsymbol{r}| := r_1 + \cdots + r_N$. For $\boldsymbol{x} = x_1\boldsymbol{\xi}_1 + \cdots + x_N\boldsymbol{\xi}_N$, we define polynomials

$$\begin{aligned} h_{\boldsymbol{r}}(\boldsymbol{x}) : &= \boldsymbol{x}^{\boldsymbol{r}} = x_1^{r_1} \cdots x_N^{r_N}, \\ h_{\boldsymbol{r}}^{+}(\boldsymbol{x}) : &= (\boldsymbol{x} + \mathbf{1})^{\boldsymbol{r}} = (x_1 + 1)^{r_1} \cdots (x_N + 1)^{r_N}, \\ h_{\boldsymbol{r}}^{-}(\boldsymbol{x}) : &= (\boldsymbol{x} - \mathbf{1})^{\boldsymbol{r}} = (x_1 - 1)^{r_1} \cdots (x_N - 1)^{r_N}. \end{aligned}$$

For each $\boldsymbol{\alpha} = a_1\boldsymbol{\xi}_1 + \cdots + a_N\boldsymbol{\xi}_N \in \mathbb{R}^N$ and $1 \le i \le N$, we associate vectors $\boldsymbol{\alpha}_{ij} \in \mathbb{R}^{r_i}$ whose jth entry is a_i and all other entries are 0, $1 \le j \le r_i$. For a polytope P in $\mathbb{R}^N$, we associate a polytope $\tilde{P}$ in $\mathbb{R}^{N+k}$, defined as the convex hull of the vectors

$$\left(\boldsymbol{\alpha}, \textstyle\sum_{j=1}^{r_1} \varepsilon_{1j}\boldsymbol{\alpha}_{1j}, \ldots, \sum_{j=1}^{r_N} \varepsilon_{Nj}\boldsymbol{\alpha}_{Nj}\right) \in \mathbb{R}^{N+k}, \quad \text{where } \boldsymbol{\alpha} \in P,\ \varepsilon_{ij} = 0 \text{ or } 1.$$

Identifying $\mathbb{R}^N$ with the subspace $\mathbb{R}^N \times \mathbf{0} \subset \mathbb{R}^{N+k}$, we have a lattice $\widetilde{\Lambda} = \Lambda \times \mathbb{Z}^k$ of $\mathbb{R}^{N+k}$. If P is a Λ-lattice polytope, then $\tilde{P}$ is a $\tilde{\Lambda}$-lattice polytope.

LEMMA 3.1. *Let P be a Λ-lattice polytope such that every vector $\boldsymbol{\alpha} \in P$ has positive coordinates relative to the lattice basis $\{\boldsymbol{\xi}_i\}_{i=1}^N$. Then $E(P, h_{\boldsymbol{r}}^+, \Lambda; m)$ and $E(P^0, h_{\boldsymbol{r}}^-, \Lambda; m)$ are polynomial functions of degree $|\boldsymbol{r}| + \dim P$ in the integral variable $m > 0$. Moreover, as polynomials, they satisfy the reciprocity law:*

$$E(P, h_{\boldsymbol{r}}^+, \Lambda; -m) = (-1)^{|\boldsymbol{r}|+\dim P} E(P^0, h_{\boldsymbol{r}}^-, \Lambda; m).$$

PROOF. Let $\pi : \mathbb{R}^{N+k} \to \mathbb{R}^N$ be the obvious projection. Then $\pi(\widetilde{mP}) = mP$ and $\pi(\widetilde{mP}^0) = mP^0$. If $\boldsymbol{\alpha} = a_1\boldsymbol{\xi}_1 + \cdots + a_N\boldsymbol{\xi}_N \in mP \cap \Lambda$, then the fiber $\pi^{-1}(\boldsymbol{\alpha}) \cap \widetilde{mP}$ is a closed coordinate rectangular box in $\boldsymbol{\alpha} \times \mathbb{R}^k$ with the vertices $\left(\boldsymbol{\alpha}, \sum_{j=1}^{r_1} \varepsilon_{1j}\boldsymbol{\alpha}_{1j}, \ldots, \sum_{j=1}^{r_N} \varepsilon_{Nj}\boldsymbol{\alpha}_{Nj}\right)$, where $\varepsilon_{ij} = 0$ or 1, and side lengths

$$\underbrace{a_1, \ldots, a_1}_{r_1}, \ldots, \underbrace{a_N, \ldots, a_N}_{r_N}.$$

If $\boldsymbol{\alpha} \in mP^0 \cap \Lambda$, then the fiber $\pi^{-1}(\boldsymbol{\alpha}) \cap \widetilde{mP}^0$ is the relative interior of the coordinate rectangular box $\pi^{-1}(\boldsymbol{\alpha}) \cap \widetilde{mP}$. Thus, for $\boldsymbol{\alpha} \in mP$,

$$\#(\pi^{-1}(\boldsymbol{\alpha}) \cap \widetilde{mP} \cap \widetilde{\Lambda}) = (a_1 + 1)^{r_1} \cdots (a_N + 1)^{r_N};$$

and for $\boldsymbol{\alpha} \in mP^0$,

$$\#(\pi^{-1}(\boldsymbol{\alpha}) \cap \widetilde{mP}^0 \cap \widetilde{\Lambda}) = (a_1 - 1)^{r_1} \cdots (a_N - 1)^{r_N}.$$

Notice that $\widetilde{mP} = m\widetilde{P}$ and $\widetilde{mP}^0 = m\widetilde{P}^0$. We have

$$\begin{aligned} E(P, h_{\boldsymbol{r}}^+, \Lambda; m) &= \sum_{\boldsymbol{\alpha} = a_1\boldsymbol{\xi}_1 + \cdots + a_N\boldsymbol{\xi}_N \in mP \cap \Lambda} (a_1 + 1)^{r_1} \cdots (a_N + 1)^{r_N} \\ &= \sum_{\boldsymbol{\alpha} \in mP \cap \Lambda} \#\left(\pi^{-1}(\boldsymbol{\alpha}) \cap \widetilde{mP} \cap \widetilde{\Lambda}\right) = \#\left(\widetilde{mP} \cap \widetilde{\Lambda}\right) \\ &= \#\left(m\widetilde{P} \cap \widetilde{\Lambda}\right) = L(\widetilde{P}, \widetilde{\Lambda}; m); \end{aligned}$$

$$\begin{aligned} E(P^0, h_{\boldsymbol r}^-, \Lambda; m) &= \sum_{\boldsymbol\alpha = a_1\boldsymbol\xi_1 + \cdots + a_N\boldsymbol\xi_N \in mP^0\cap\Lambda} (a_1-1)^{r_1}\cdots(a_N-1)^{r_N} \\ &= \sum_{\boldsymbol\alpha\in mP^0\cap\Lambda} \#\left(\pi^{-1}(\boldsymbol\alpha)\cap \widetilde{mP}^0 \cap \widetilde\Lambda\right) = \#\left(\widetilde{mP}^0\cap\widetilde\Lambda\right) \\ &= \#\left(m\widetilde P^0 \cap \widetilde\Lambda\right) = L(\widetilde P^0, \widetilde\Lambda; m). \end{aligned}$$

Therefore $E(P, h_{\boldsymbol r}^+, \Lambda; m)$ and $E(P^0, h_{\boldsymbol r}^-, \Lambda; m)$ are polynomial functions of m of degree $\dim \widetilde P = |\boldsymbol r| + \dim P$. The reciprocity law follows from the reciprocity law about the Ehrhart polynomials of the polytope $\widetilde P$ and its relative interior $\widetilde P^0$. □

LEMMA 3.2. *Let P be a Λ-lattice polytope such that every $\boldsymbol\alpha \in P$ has positive coordinates relative to the lattice basis $\{\boldsymbol\xi_i\}_{i=1}^N$. Then $E(P, h_{\boldsymbol r}, \Lambda; m)$ and $E(P^0, h_{\boldsymbol r}, \Lambda; m)$ are polynomial functions of degree $|\boldsymbol r| + \dim P$ in the integral variable $m > 0$. Moreover, as polynomials, they satisfy the reciprocity law:*

$$E(P, h_{\boldsymbol r}, \Lambda; -m) = (-1)^{|\boldsymbol r| + \dim P} E(P^0, h_{\boldsymbol r}, \Lambda; m).$$

PROOF. We proceed by induction on $\boldsymbol r$. Obviously, it is true for $\boldsymbol r = \boldsymbol 0$. Assume that it is true for all $\boldsymbol s$ such that $\boldsymbol 0 \le \boldsymbol s < \boldsymbol r$. Notice that

$$\begin{aligned} h_{\boldsymbol r}^+(\boldsymbol x) &= \sum_{\boldsymbol 0\le \boldsymbol s\le \boldsymbol r} \binom{\boldsymbol r}{\boldsymbol s} \boldsymbol x^{\boldsymbol s} \\ &= h_{\boldsymbol r}(\boldsymbol x) + \sum_{\boldsymbol 0\le\boldsymbol s<\boldsymbol r} \binom{\boldsymbol r}{\boldsymbol s} h_{\boldsymbol s}(\boldsymbol x); \\ h_{\boldsymbol r}^-(\boldsymbol x) &= \sum_{\boldsymbol 0\le \boldsymbol s\le \boldsymbol r} (-1)^{|\boldsymbol r - \boldsymbol s|}\binom{\boldsymbol r}{\boldsymbol s} \boldsymbol x^{\boldsymbol s} \\ &= h_{\boldsymbol r}(\boldsymbol x) + \sum_{\boldsymbol 0\le\boldsymbol s<\boldsymbol r} (-1)^{|\boldsymbol r - \boldsymbol s|}\binom{\boldsymbol r}{\boldsymbol s} h_{\boldsymbol s}(\boldsymbol x). \end{aligned}$$

Then

$$h_{\boldsymbol r} = h_{\boldsymbol r}^+ - \sum_{\boldsymbol 0\le\boldsymbol s<\boldsymbol r}\binom{\boldsymbol r}{\boldsymbol s} h_{\boldsymbol s} = h_{\boldsymbol r}^- - \sum_{\boldsymbol 0\le\boldsymbol s<\boldsymbol r} (-1)^{|\boldsymbol r-\boldsymbol s|}\binom{\boldsymbol r}{\boldsymbol s} h_{\boldsymbol s}.$$

By Lemma 3.1, we see that both $E(P, h_{\boldsymbol r}, \Lambda; m)$ and $E(P^0, h_{\boldsymbol r}, \Lambda; m)$ are polynomial functions of degree $|\boldsymbol r| + \dim P$ in the positive integral variable m. Moreover, as polynomials,

$$\begin{aligned} E(P, h_{\boldsymbol r}, \Lambda; -m) &= E(P, h_{\boldsymbol r}^+, \Lambda; -m) - \sum_{\boldsymbol 0\le\boldsymbol s<\boldsymbol r}\binom{\boldsymbol r}{\boldsymbol s} E(P, h_{\boldsymbol s}, \Lambda; -m) \\ &= (-1)^{|\boldsymbol r|+\dim P} E(P^0, h_{\boldsymbol r}^-, \Lambda; m) \\ &\quad - \sum_{\boldsymbol 0\le\boldsymbol s<\boldsymbol r} (-1)^{|\boldsymbol s|+\dim P}\binom{\boldsymbol r}{\boldsymbol s} E(P^0, h_{\boldsymbol s}, \Lambda; m) \\ &= (-1)^{|\boldsymbol r|+\dim P} E(P^0, h_{\boldsymbol r}, \Lambda; m). \end{aligned}$$

□

THEOREM 3.3. *Let h be a polynomial function of degree k, relative to a Λ-lattice basis $\{\boldsymbol\xi_i\}_{i=1}^N$; and $\bar h(\boldsymbol x) = h(-\boldsymbol x)$. Let P be a Λ-lattice polytope. Then $E(P, h, \Lambda; m)$ and $E(P^0, h, \Lambda; m)$ are polynomial functions of degree at most $k + \dim P$ in the* *positive integral variable m. Moreover, as polynomials, they satisfy the reciprocity law:*

$$E(P,h,\Lambda;-m)=(-1)^{\dim P}E(P^0,\bar{h},\Lambda;m).$$

PROOF. It suffices to prove the statement only for the homogenous polynomial $h_{\boldsymbol{r}}(\boldsymbol{x})=\boldsymbol{x}^{\boldsymbol{r}}$. Let $\boldsymbol{\gamma}=c_1\boldsymbol{\xi}_1+\cdots+c_N\boldsymbol{\xi}_N\in\Lambda$ be such that $\boldsymbol{\alpha}+\boldsymbol{\gamma}$ have positive coordinates for all $\boldsymbol{\alpha}\in P$ relative to the lattice basis $\{\boldsymbol{\xi}\}_{i=1}^N$. Then for positive integers m,

$$\begin{aligned}
E(P,h_{\boldsymbol{r}},\Lambda;m) &= \sum_{\boldsymbol{\alpha}\in m(\boldsymbol{\gamma}+P)\cap\Lambda}(\boldsymbol{\alpha}-m\boldsymbol{\gamma})^{\boldsymbol{r}}\\
&= \sum_{\boldsymbol{\alpha}\in m(\boldsymbol{\gamma}+P)\cap\Lambda}\sum_{\boldsymbol{s}=\boldsymbol{0}}^{\boldsymbol{r}}\binom{\boldsymbol{r}}{\boldsymbol{s}}\boldsymbol{\alpha}^{\boldsymbol{s}}(-m\boldsymbol{\gamma})^{\boldsymbol{r}-\boldsymbol{s}}\\
&= \sum_{\boldsymbol{s}=\boldsymbol{0}}^{\boldsymbol{r}}\binom{\boldsymbol{r}}{\boldsymbol{s}}(-m\boldsymbol{\gamma})^{\boldsymbol{r}-\boldsymbol{s}}\sum_{\boldsymbol{\alpha}\in m(\boldsymbol{\gamma}+P)\cap\Lambda}\boldsymbol{\alpha}^{\boldsymbol{s}}\\
&= \sum_{\boldsymbol{s}=\boldsymbol{0}}^{\boldsymbol{r}}\binom{\boldsymbol{r}}{\boldsymbol{s}}(-m\boldsymbol{\gamma})^{\boldsymbol{r}-\boldsymbol{s}}E(\boldsymbol{\gamma}+P,h_{\boldsymbol{s}},\Lambda;m).
\end{aligned}$$

By Lemma 3.2, $E(P,h_{\boldsymbol{r}},\Lambda;m)$ is a polynomial function of the positive integral variable m, having degree at most $|\boldsymbol{r}|+\dim P$. (Terms of degree $|\boldsymbol{r}|+\dim P$ may be cancelled out.) Similarly,

$$E(P^0,h_{\boldsymbol{r}},\Lambda;m)=\sum_{\boldsymbol{s}=\boldsymbol{0}}^{\boldsymbol{r}}\binom{\boldsymbol{r}}{\boldsymbol{s}}(-m\boldsymbol{\gamma})^{\boldsymbol{r}-\boldsymbol{s}}E(\boldsymbol{\gamma}+P^0,h_{\boldsymbol{s}},\Lambda;m)$$

is a polynomial function of the positive integral variable m, having degree at most $|\boldsymbol{r}|+\dim P$. Moreover, as polynomials,

$$\begin{aligned}
E(P,h_{\boldsymbol{r}},\Lambda;-m) &= \sum_{\boldsymbol{s}=\boldsymbol{0}}^{\boldsymbol{r}}\binom{\boldsymbol{r}}{\boldsymbol{s}}(m\boldsymbol{\gamma})^{\boldsymbol{r}-\boldsymbol{s}}E(\boldsymbol{\gamma}+P,h_{\boldsymbol{s}},\Lambda;-m)\\
&= \sum_{\boldsymbol{s}=\boldsymbol{0}}^{\boldsymbol{r}}\binom{\boldsymbol{r}}{\boldsymbol{s}}(m\boldsymbol{\gamma})^{\boldsymbol{r}-\boldsymbol{s}}(-1)^{|\boldsymbol{s}|+\dim P}E(\boldsymbol{\gamma}+P^0,h_{\boldsymbol{s}},\Lambda;m)\\
&= (-1)^{|\boldsymbol{r}|+\dim P}\sum_{\boldsymbol{s}=\boldsymbol{0}}^{\boldsymbol{r}}\binom{\boldsymbol{r}}{\boldsymbol{s}}(-m\boldsymbol{\gamma})^{\boldsymbol{r}-\boldsymbol{s}}E(\boldsymbol{\gamma}+P^0,h_{\boldsymbol{s}},\Lambda;m)\\
&= (-1)^{|\boldsymbol{r}|+\dim P}E(P^0,h_{\boldsymbol{r}},\Lambda;m)=(-1)^{\dim P}E(P^0,\bar{h}_{\boldsymbol{r}},\Lambda;m).
\end{aligned}$$

□

4. Reciprocity law for lattice polyhedral functions

The reciprocity law for a lattice simplex can be extended to more general lattice spaces. Ehrhart [**14, 15**] conjectured that the reciprocity law is true for any lattice manifold with boundary; this was proved by Macdonald [**22**]. Recently, the reciprocity law was extended to arbitrary lattice polyhedra, using weight and boundary weight functions; see [**8**]. This section is to further extend the reciprocity law to lattice polyhedral functions without using weight and boundary weight functions.

Let $\mathcal{P}^N$ denote the class of polytopes in $\mathbb{R}^N$. We assume that the empty set $\emptyset$ is a polytope. The *indicator function* of a polyhedron $X \subset \mathbb{R}^N$ is the characteristic function

$$1_X(\boldsymbol{x}) = \begin{cases} 1 & \text{for} \quad \boldsymbol{x} \in X \\ 0 & \text{for} \quad \boldsymbol{x} \notin X. \end{cases}$$

A *polyhedral function with bounded support* on $\mathbb{R}^N$ is a linear combination $f = \sum_{j=1}^k a_j 1_{P_j}$ of indicator functions of some polytopes P_j. Notice that every polyhedral function with bounded support can be also written as a linear combination of indicator functions of some simplices. Bounded polyhedral functions can be classified as functions $f : \mathbb{R}^N \to \mathbb{R}$ such that, (i) $f(\mathbb{R}^N)$ is a finite set; and (ii) for any nonzero $a \in \mathbb{R}$ the inverse image $f^{-1}(a)$ is a bounded polyhedron. We denote by $\mathbb{R}(\mathcal{P}^N)$ the vector space of polyhedral functions with bounded supports on $\mathbb{R}^N$. The *support* of $f \in \mathbb{R}(\mathcal{P}^N)$ is the closure of the set $\{\boldsymbol{x} \in \mathbb{R}^n \mid f(\boldsymbol{x}) \neq 0\}$, denoted $\operatorname{supp}(f)$.

A set function $\mu : \mathcal{P}^N \to \mathbb{R}$ is called a *valuation* if (i) $\mu(\emptyset) = 0$ and (ii) for polytopes P and Q, if $P \cup Q$ is a polytope, then

$$\mu(P \cup Q) = \mu(P) + \mu(Q) - \mu(P \cap Q).$$

It is well known that any valuation μ on $\mathcal{P}^N$ can be extended uniquely to a linear functional on the vector space $\mathbb{R}(\mathcal{P}^N)$ such that $\mu(1_P) = \mu(P)$ for all $P \in \mathcal{P}^N$; see [**16**].

An important valuation on $\mathcal{P}^N$ is the *Euler characteristic* $\chi : \mathcal{P}^N \to \mathbb{R}$, defined by $\chi(P) = 1$ for nonempty $P \in \mathcal{P}^N$. Viewing χ as a linear functional on $\mathbb{R}(\mathcal{P}^N)$, we have $\chi(f) = \sum_{j=1}^k a_j$ for any polyhedral function $f = \sum_{j=1}^k a_j 1_{P_j}$, $P_j \in \mathcal{P}^N$. An easy and efficient way to introduce the Euler characteristic for polyhedral functions can be found in [**7, 19**].

Let $\phi : \mathbb{R}^N \to \mathbb{R}^M$ be an affine linear map. There are push-forward linear maps ϕ_* and $\bar{\phi}_*$ from $\mathbb{R}(\mathcal{P}^N)$ to $\mathbb{R}(\mathcal{P}^M)$, defined by

$$\begin{aligned} (\phi_* f)(\boldsymbol{x}) : &= \lim_{\delta \to 0} \chi\left(f \cdot 1_{\phi^{-1}(B(\boldsymbol{x},\delta))}\right), \\ (\bar{\phi}_* f)(\boldsymbol{x}) : &= \lim_{\delta \to 0} \chi\left(f \cdot 1_{\phi^{-1}(B^0(\boldsymbol{x},\delta))}\right), \end{aligned}$$

where $f \in \mathbb{R}(\mathcal{P}^N)$, $B(\boldsymbol{x}, \delta)$ is the closed ball of radius δ centered at $\boldsymbol{x}$ in $\mathbb{R}^M$, and $B^0(\boldsymbol{x}, \delta)$ is the relative interior of $B(\boldsymbol{x}, \delta)$; see [**9**] for detailed properties of ϕ_* and $\bar{\phi}_*$. In particular, for the identity map $\mathrm{id} : \mathbb{R}^N \to \mathbb{R}^N$, we have induced linear maps id_* and $\bar{\mathrm{id}}_*$ from $\mathbb{R}(\mathcal{P}^N)$ to itself. The map id_* is just the identity map, while $\bar{\mathrm{id}}_*$ is an involution on $\mathbb{R}(\mathcal{P}^N)$. For any polytope P, we have

$$\begin{aligned} \bar{\mathrm{id}}_*(1_P) &= (-1)^{\dim P} 1_{P^0}, \\ \bar{\mathrm{id}}_*(1_{P^0}) &= (-1)^{\dim P} 1_P. \end{aligned}$$

Let t be a real variable. We denote by ϕ_t the dilation $\phi_t(\boldsymbol{x}) = t\boldsymbol{x}$, $\boldsymbol{x} \in \mathbb{R}^N$. It is straightforward that for any simplex σ,

$$\begin{aligned} (\phi_t)_*(1_\sigma) &= 1_{t\sigma}; \\ (\bar{\phi}_t)_*(1_\sigma) &= \begin{cases} 1_{\{\boldsymbol{o}\}} & \text{if } t = 0 \\ (-1)^{\dim \sigma} 1_{t\sigma^0} & \text{if } t \neq 0. \end{cases} \end{aligned}$$

For simplicity, we write

$$t * 1_\sigma := \begin{cases} (\phi_t)_*(1_\sigma) = 1_{t\sigma} & \text{if } t \geq 0 \\ (\bar{\phi}_t)_*(1_\sigma) = (-1)^{\dim \sigma} 1_{t\sigma^0} & \text{if } t < 0. \end{cases}$$

Then

$$\begin{aligned} E(1_\sigma, \Lambda; t) : &= \sum_{\boldsymbol{\alpha} \in \Lambda} (t * 1_\sigma)(\boldsymbol{\alpha}) \\ &= \begin{cases} \#(t\sigma \cap \Lambda) & \text{if } t \geq 0 \\ (-1)^{\dim \sigma} \#(t\sigma^0 \cap \Lambda) & \text{if } t < 0 \end{cases} \\ &= \begin{cases} E(\sigma, \Lambda; t) & \text{if } t \geq 0 \\ (-1)^{\dim \sigma} E(\sigma^0, \Lambda; -t) & \text{if } t < 0 \end{cases} \\ &= E(\sigma, \Lambda; t). \end{aligned}$$

Similarly, for a polyhedral function $f \in \mathbb{R}(\mathcal{P}^N)$ with bounded support, we define

$$t * f : = \begin{cases} (\phi_t)_*(f) & \text{if } t \geq 0 \\ (\bar{\phi}_t)_*(f) & \text{if } t < 0. \end{cases}$$

The *dual* of f is the polyhedral function with bounded support

$$f^* := \bar{\mathrm{id}}_* f.$$

Since $(\phi_t)_*$ and $(\bar{\phi}_t)_*$ are linear operators, each real number t can be viewed as two linear operators on $\mathbb{R}(\mathcal{P}^N)$. If f is a Λ-lattice polyhedral function, then $f = \sum_{j=1}^k a_j 1_{\sigma_j}$ for some Λ-simplices σ_j. Thus, $t * f = \sum_{j=1}^k a_j (t * 1_{\sigma_j})$ and

$$E(f, \Lambda; t) := \sum_{\boldsymbol{\alpha} \in \Lambda} (t * f)(\boldsymbol{\alpha}) = \sum_{j=1}^k a_j E(\sigma_j, \Lambda; t). \tag{4.1}$$

This means that $E(f, \Lambda; m)$ is a polynomial function of the integral variable m. We denote the polynomial by $L(f, \Lambda; t)$ in the real variable t, called the *Ehrhart polynomial of f*.

THEOREM 4.1 (Reciprocity Law for Polyhedral Functions). *For any Λ-lattice polyhedral function f with bounded support,*

$$E(f, \Lambda; -t) = E(f^*, \Lambda; t). \tag{4.2}$$

PROOF. Let $f = \sum_{j=1}^{k} a_j 1_{\sigma_j}$ for some Λ-simplices σ_j. Then

$$\begin{aligned} E(f, \Lambda; -t) &= \sum_{j=1}^{k} a_j E(\sigma_j, \Lambda; -t) \\ &= \sum_{j=1}^{k} a_j (-1)^{\dim \sigma_j} E\left(\sigma_j^0, \Lambda; t\right) \\ &= \sum_{j=1}^{k} a_j E\left((-1)^{\dim \sigma_j} 1_{\sigma_j^0}, \Lambda; t\right) \\ &= \sum_{j=1}^{k} a_j E\left(\left(1_{\sigma_j}\right)^*, \Lambda; t\right) \\ &= E\left(\sum_{j=1}^{k} a_j \left(1_{\sigma_j}\right)^*, \Lambda; t\right) \\ &= E(f^*, \Lambda; t). \end{aligned}$$

□

Let μ be a valuation on $\mathcal{P}^N$. Then μ is extended to a linear functional on $\mathbb{R}(\mathcal{P}^N)$. For any polyhedral function $f \in \mathbb{R}(\mathcal{P}^N)$ with bounded support, we define

$$\mu(f, t) := \mu(t * f). \tag{4.3}$$

Then, for a fixed real number t, $\mu(\cdot\,, t)$ is a linear functional on $\mathbb{R}(\mathcal{P}^N)$. In particular, since $0 * f = \chi(f) 1_{\{o\}} = \chi(f^*) 1_{\{o\}} = 0 * f^*$, we have

$$\mu(f, 0) = \chi(f)\mu(\{o\}) = \chi(f^*)\mu(\{o\}) = \mu(f^*, 0).$$

Let σ be a simplex of $\mathbb{R}^N$. Then, for real numbers $t > 0$,

$$\begin{aligned} (-t) * 1_\sigma &= (-1)^{\dim \sigma} 1_{-t\sigma^0} = (-1)^{\dim \sigma} 1_{t(-\sigma^0)} \\ &= t * (-1)^{\dim \sigma} 1_{-\sigma^0} = t * [(-1) * 1_\sigma]; \end{aligned}$$

and for real numbers $t < 0$,

$$(-t) * 1_\sigma = 1_{-t\sigma} = 1_{t(-\sigma)} = t * (-1)^{\dim \sigma} 1_{-\sigma^0} = t * [(-1) * 1_\sigma].$$

Thus, by the linearity of the operators $t*$ and $(-t)*$, we have

$$(-t) * f = t * f^* = t * [(-1) * f].$$

We obtain the following theorem.

THEOREM 4.2. (Reciprocity Law for Valuations) *Let μ be a valuation on $\mathcal{P}^N$. Then for any $f \in \mathbb{R}(\mathcal{P}^N)$,*

$$\mu(f, -t) = \mu((-1) * f, t). \tag{4.4}$$

For the indicator function 1_X of a bounded polyhedron X, we write $\mu(X, t)$ for $\mu(1_X, t)$. Let M be a polyhedral manifold M with boundary ∂M. Then

$$(-1) * 1_M = (-1)^{\dim M} 1_{-M^0},$$

where $M^0 = M - \partial M$ and $-M^0$ is the reflection of M^0 about the origin. We thus have the following corollary.

COROLLARY 4.3. *Let μ be a valuation on $\mathcal{P}^N$. Let M be a polyhedral manifold with boundary ∂M and $M^0 = M - \partial M$. Then*

$$\mu(M, -t) = (-1)^{\dim M} \mu(-M^0, t). \tag{4.5}$$

Let $h : \mathbb{R}^N \to \mathbb{R}$ be a polynomial function of degree k, relative to a lattice basis of Λ. We define $\bar{h} : \mathbb{R}^N \to \mathbb{R}$ by $\bar{h}(\boldsymbol{x}) = h(-\boldsymbol{x})$. There is a valuation $E(\cdot\,, h, \Lambda)$ on $\mathcal{P}^N$, defined by

$$E(P, h, \Lambda) := \sum_{\boldsymbol{\alpha} \in P \cap \Lambda} h(\boldsymbol{\alpha}), \quad P \in \mathcal{P}^N. \tag{4.6}$$

This valuation extends to a linear functional $E(\cdot\,, h, \Lambda)$ on $\mathbb{R}(\mathcal{P}^N)$; thus we can define

$$E(f, h, \Lambda; t) := E(t * f, h, \Lambda), \quad t \in \mathbb{R}.$$

More explicitly, the function $E(f, h, \Lambda; t)$ can be written as

$$E(f, h, \Lambda; t) = \sum_{\boldsymbol{\alpha} \in \Lambda} (t * f)(\boldsymbol{\alpha}) h(\boldsymbol{\alpha}).$$

Let σ be a Λ-simplex and let m be integers. Then by Theorem 3.3,

$$\begin{aligned}
E(1_\sigma, h, \Lambda; m) &= \begin{cases} \sum_{\boldsymbol{\alpha} \in m\sigma \cap \Lambda} h(\boldsymbol{\alpha}) & \text{if } m \geq 0 \\ \sum_{\boldsymbol{\alpha} \in m\sigma^0 \cap \Lambda} (-1)^{\dim \sigma} h(\boldsymbol{\alpha}) & \text{if } m < 0 \end{cases} \\
&= \begin{cases} \sum_{\boldsymbol{\alpha} \in m\sigma \cap \Lambda} h(\boldsymbol{\alpha}) & \text{if } m \geq 0 \\ (-1)^{k + \dim \sigma} \sum_{\boldsymbol{\alpha} \in -m\sigma^0 \cap \Lambda} h(\boldsymbol{\alpha}) & \text{if } m < 0 \end{cases} \\
&= \begin{cases} E(\sigma, h, \Lambda; m) & \text{if } m \geq 0 \\ (-1)^{k + \dim \sigma} E(\sigma^0, h, \Lambda; -m) & \text{if } m < 0 \end{cases} \\
&= E(\sigma, h, \Lambda; m).
\end{aligned}$$

Similarly, $E\left(1_{\sigma^0}, h, \Lambda; m\right) = E\left(\sigma^0, h, \Lambda; m\right)$. Then $E\left(1_\sigma, h, \Lambda; m\right)$ and $E\left(1_{\sigma^0}, h, \Lambda; m\right)$ are polynomial functions of degree at most $k + \dim \sigma$ in the integral variable m, and

$$E\left(1_\sigma, h, \Lambda; -m\right) = (-1)^{\dim \sigma} E\left(1_{\sigma^0}, \bar{h}, \Lambda; m\right) = E\left((1_\sigma)^*, \bar{h}, \Lambda; m)\right).$$

If f is a Λ-lattice polyhedral function with bounded support, then $f = \sum a_j 1_{\sigma_j}$ for some Λ-simplices σ_j. Hence

$$E(f, h, \Lambda; m) = \sum a_j E(1_{\sigma_j}, h, \Lambda; m) = \sum a_j E(\sigma_j, h, \Lambda; m), \quad m \in \mathbb{Z},$$

is a polynomial function of degree at most $k + \dim \operatorname{supp}(f)$ in the integral variable m; we denote the polynomial by $L(f, h, \Lambda; t)$ in the real variable t. If X is a bounded Λ-polyhedron, we have $L(1_X, h, \Lambda; t) = L(X, h, \Lambda; t)$.

THEOREM 4.4. *Let $h : \mathbb{R}^N \to \mathbb{R}$ be a polynomial function of degree k, relative to a Λ-lattice basis. Then, for any Λ-lattice polyhedral function f with bounded support, $E(f, h, \Lambda; m)$ is a polynomial function of degree at most $k + \dim \operatorname{supp}(f)$ in the integral variable m. Denote the polynomial by $L(f, h, \Lambda; t)$ in the real variable t; we have the reciprocity law:*

$$L(f, h, \Lambda; -t) = L(f^*, \bar{h}, \Lambda; t). \tag{4.7}$$

PROOF. The reciprocity law follows from Theorem 4.2 or directly from the additivity of the valuation $E(\cdot\,, h, \Lambda; m)$ and the linearity of the dual operator *. $\square$

COROLLARY 4.5. *Let M be a Λ-lattice polyhedral manifold with boundary ∂M, and let $h:\mathbb{R}^N\to\mathbb{R}$ be a homogeneous polynomial function of degree k, relative to a Λ-lattice basis. Then*

$$L(M,h,\Lambda;-t)=(-1)^{k+\dim M}L(M-\partial M,h,\Lambda;t). \tag{4.8}$$

PROOF. Notice that the set function $E(\cdot\,,h,\Lambda)$ defined on $\mathcal{P}^N$ by (4.6) is a valuation. Then by Corollary 4.3, $L(M,h,\Lambda;-t)=(-1)^{\dim M}L(-M^0,h,\Lambda;t)$. Since h is homogeneous of degree k, we have

$$L(-M^0,h,\Lambda;t)=(-1)^kL(M^0,h,\Lambda;t).$$

The reciprocity law (4.8) follows immediately. □

The dual operator * is closely related to the boundary operator ∂ that is used to define boundary weight functions introduced in [**8, 12**]. Let X be a compact subspace of a manifold without boundary. By an *Eulerian stratification* of X we mean a finite collection of disjoint subspaces X_a, indexed by a poset Γ, such that (i) $X=\bigcup_{a\in\Gamma}X_a$; (ii) $\bar{X}_b=\bigcup_{a\le b}X_a$; and (iii) if $a\le b$, then $\chi(\mathrm{lk}\,(x,X_b))=\chi(\mathrm{lk}\,(y,X_b))$ for $x,y\in X_a$. We call each subspace X_a in the collection a *pure stratum* and its closure $\bar{X}_a$ a *closed stratum* of the stratification. The *dimension* of X is the highest dimension of its strata. We define

$$\chi(a,b):=\chi(\mathrm{lk}\,(x,X_b)),\quad \text{where } x\in X_a,\ a\le b.$$

There are various versions of stratifications considered in topology such as geometric stratification, Whitney stratification, PL-linear stratification, intrinsic stratification, and homotopy stratification, etc. These stratifications share the homogeneity property: for $x,y\in X_a$ and $a\le b$, there are homeomorphic neighborhoods of x and y in X_b; so all of them are Eulerian stratifications. In particular, with the intrinsic stratifications, the spaces such as polyhedra, algebraic varieties, semialgebraic sets, subanalytic sets, are Eulerian stratified if they are compact; see [**12**].

A function $w:X\to\mathbb{R}$ is called a *weight function* with respect to a stratification $\{X_a\,|\,a\in\Gamma\}$ if w is constant on each pure stratum. So w can be viewed as a function on the poset Γ. The *nth boundary* of w is a function $\partial_n w:\Gamma\to\mathbb{R}$, defined by

$$\partial_n w(a):=(1-(-1)^n)\,w(a)+(-1)^n\sum_{a\le b}\chi(a,b)w(b). \tag{4.9}$$

A triangulation Δ of X is called *stratified* with respect to a stratification Γ if each pure stratum X_a is a union of some open simplices in Δ. We assume that all simplices in Δ are relatively open simplices. A weight function is automatically viewed as a function on any stratified triangulation.

LEMMA 4.6. *Let Δ be a stratified triangulation of X with respect to a stratification Γ. Then for any $w:\Gamma\to\mathbb{R}$ and integer n,*

$$\partial_n w(\sigma)=w(\sigma)-(-1)^n\sum_{\sigma\le\tau\in\Delta}(-1)^{\dim\tau}w(\tau),\quad \sigma\in\Delta.$$

Moreover, when w and $\partial_n w$ are considered as polyhedral functions with bounded supports on $\mathbb{R}^N$, vanishing on the complement of X, we have

$$\partial_n w=w-(-1)^n w^*.$$

PROOF. Let $\sigma \in \Delta$ and $\sigma \subset X_a$. Fix a point $x \in \sigma$. Let Δ_x be the triangulation of $\bigcup_{\sigma \leq \tau \in \Delta} \tau$ by taking joint from x to the open simplices ρ, where $\rho < \tau$ for some $\tau \geq \sigma$ in Δ. Then

$$\begin{aligned}
\partial_n w(\sigma) &= (1-(-1)^n)w(a) + (-1)^n \sum_{a \leq b} \chi(\mathrm{lk}\,(x, X_b))w(b) \\
&= (1-(-1)^n)w(\sigma) + (-1)^n \sum_{x < \rho \in \Delta_x} \chi(\mathrm{lk}\,(x, \rho))w(\rho) \\
&= (1-(-1)^n)w(\sigma) - (-1)^n \sum_{x < \rho \in \Delta_x} (-1)^{\dim \rho} w(\rho) \\
&= w(\sigma) - (-1)^n \sum_{x \leq \rho \in \Delta_x} (-1)^{\dim \rho} w(\rho) \\
&= w(\sigma) - (-1)^n \sum_{\sigma \leq \tau \in \Delta} w(\tau) \sum_{\substack{\rho \subset \tau \\ x \leq \rho \in \Delta_x}} (-1)^{\dim \rho} \\
&= w(\sigma) - (-1)^n \sum_{\sigma \leq \tau \in \Delta} (-1)^{\dim \tau} w(\tau).
\end{aligned}$$

Since $(\bar{\mathrm{id}}_* w)(x) = \sum_{x \leq \rho \in \Delta_x} (-1)^{\dim \rho} w(\rho)$, we have $\partial_n w = w - (-1)^n \bar{\mathrm{id}}_* w$. □

Let $h : \mathbb{R}^N \to \mathbb{R}$ be a polynomial function of degree k. For integers $m > 0$, let

$$L(X, w, h, \Lambda; m) := \sum_{\boldsymbol{\alpha} \in X \cap m^{-1}\Lambda} w(\boldsymbol{\alpha}) h(m\boldsymbol{\alpha}). \tag{4.10}$$

If Δ is a stratified triangulation of X, then

$$L(X, w, h, \Lambda; m) = \sum_{\sigma \in \Delta} w(\sigma) L(\sigma, h, \Lambda; m).$$

This means that $L(X, w, h, \Lambda; m)$ is a polynomial function of degree at most $k + \dim \mathrm{supp}\,(w)$ in the integral variable $m > 0$.

THEOREM 4.7 (Stratified Reciprocity Law). *Let X be a compact Λ-polyhedron of dimension n and let w be a weight function on X with respect to a stratification. Then, for real variables s and t,*

$$L(X, w - s\partial_n w, h, \Lambda; -t) = (-1)^n L(X, w - (1-s)\partial_n w, \bar{h}, \Lambda; t). \tag{4.11}$$

PROOF. Let Δ be a stratified Λ-triangulation of X with respect to the stratification; that is, each stratum is a union of some open simplices in Δ. Then

$$\begin{aligned}
L(X, w - s\partial_n w, h, \Lambda; -t) &= \sum_{\sigma \in \Delta} (w - s\partial_n w)(\sigma) L(\sigma, h, \Lambda; -t) \\
&= \sum_{\sigma \in \Delta} (w - s\partial_n w)(\sigma)(-1)^{\dim \sigma} L(\bar{\sigma}, \bar{h}, \Lambda; t) \\
&= \sum_{\sigma \in \Delta} (w - s\partial_n w)(\sigma)(-1)^{\dim \sigma} \sum_{\rho \leq \sigma} L(\rho, \bar{h}, \Lambda; t) \\
&= \sum_{\rho \in \Delta} L(\rho, \bar{h}, \Lambda; t) \sum_{\rho \leq \sigma} (-1)^{\dim \sigma} (w - s\partial_n w)(\sigma).
\end{aligned}$$

Now, for each fixed $\rho \in \Delta$, by Lemma 4.6, we have

$$
\begin{aligned}
w'(\rho) : &= \sum_{\rho \le \sigma \in \Delta} (-1)^{\dim \sigma} (w - s\partial_n w)(\sigma) \\
&= \sum_{\rho \le \sigma \in \Delta} (-1)^{\dim \sigma} \left(w(\sigma) - sw(\sigma) + s(-1)^n \sum_{\sigma \le \tau \in \Delta} (-1)^{\dim \tau} w(\tau) \right) \\
&= (1-s) \sum_{\rho \le \sigma \in \Delta} (-1)^{\dim \sigma} w(\sigma) + s(-1)^n \sum_{\substack{\rho \le \sigma \le \tau \\ \sigma, \tau \in \Delta}} (-1)^{\dim \sigma + \dim \tau} w(\tau) \\
&= (1-s) \sum_{\rho \le \sigma \in \Delta} (-1)^{\dim \sigma} w(\sigma) + s(-1)^n w(\rho) \\
&= (-1)^n \left[w(\rho) - (1-s) \left(w(\rho) - (-1)^n \sum_{\rho \le \sigma \in \Delta} (-1)^{\dim \sigma} w(\sigma) \right) \right] \\
&= (-1)^n [w - (1-s)\partial_n w](\rho).
\end{aligned}
$$

This completes the proof of the reciprocity law. □

THEOREM 4.8. *Let f be a polyhedral function with bounded support, $n =$ $\dim \operatorname{supp}(f)$. Then*

$$f - s\partial_n f = (-1)^n [f - (1-s)\partial_n f]^*.$$

If μ is a valuation on $\mathcal{P}^N$ invariant under the reflection about the origin, then

$$\mu(f - s\partial_n f, -t) = (-1)^n \mu(f - (1-s)\partial_n f, t).$$

PROOF. Since $\partial_n f = f - (-1)^n f^*$, we have

$$
\begin{aligned}
(-1)^n [f - (1-s)\partial_n f]^* &= (-1)^n [sf + (1-s)(-1)^n f^*]^* \\
&= s(-1)^n f^* + (1-s) f \\
&= f - s[f - (-1)^n f^*] \\
&= f - s\partial_n f.
\end{aligned}
$$

Notice that $\mu(f, t) = \mu(t * f)$. For $t > 0$, we have

$$
\begin{aligned}
\mu(f - s\partial_n f, -t) &= (-1)^n \mu\left([f - (1-s)\partial_n f]^*, -t\right) \\
&= (-1)^n \mu\left((\bar{\phi}_{-t})_* \bar{\mathrm{id}}_* [f - (1-s)\partial_n f]\right) \\
&= (-1)^n \mu((\phi_{-t})_* (f - (1-s)\partial_n f)) \\
&= (-1)^n \mu((\phi_{-1})_* (\phi_t)_* (f - (1-s)\partial_n f)) \\
&= (-1)^n \mu((\phi_t)_* (f - (1-s)\partial_n f)) \\
&= (-1)^n \mu(f - (1-s)\partial_n f, t).
\end{aligned}
$$

The situation for $t < 0$ is similar. For $t = 0$, it suffices to show that $\chi(f - s\partial_n f) =$ $(-1)^n \chi(f - (1-s)\partial_n f)$. This follows from the linearity of χ and $\chi(1_\sigma - s\partial_n 1_\sigma) =$ $(-1)^n \chi(1_\sigma - (1-s)\partial_n 1_\sigma)$ for any simplex σ. □

5. Generating functions of Ehrhart polynomials

Let Λ be a full dimensional lattice of $\mathbb{R}^N$ with a lattice basis $\{\boldsymbol{\xi}_i\}_{i=1}^N$. Let $\boldsymbol{x} = (x_1, \ldots, x_N)$ be a multi-indeterminant. We define Laurent monomials

$$\boldsymbol{x}^{\boldsymbol{\alpha}} := x_1^{a_1} \cdots x_N^{a_N}, \quad \boldsymbol{\alpha} = a_1\boldsymbol{\xi}_1 + \cdots + a_N\boldsymbol{\xi}_N \in \Lambda.$$

For a Λ-polyhedron P, we define the formal power series

$$F(P, \Lambda; \boldsymbol{x}) := \sum_{\boldsymbol{\alpha} \in P \cap \Lambda} \boldsymbol{x}^{\boldsymbol{\alpha}} \in \mathbf{C}[[\boldsymbol{x}, \boldsymbol{x}^{-1}]];$$

and for a polyhedral function $f \in \mathbb{R}(\mathcal{P}^N)$, define

$$F(f, \Lambda; \boldsymbol{x}) := \sum_{\boldsymbol{\alpha} \in \Lambda} f(\boldsymbol{\alpha})\boldsymbol{x}^{\boldsymbol{\alpha}} \in \mathbf{C}[[\boldsymbol{x}, \boldsymbol{x}^{-1}]].$$

As usual rational functions of $\boldsymbol{x}$ are identified as elements of $\mathbf{C}[[\boldsymbol{x}, \boldsymbol{x}^{-1}]]$.

Let K be a convex cone generated by independent vectors $\boldsymbol{\alpha}_1, \ldots, \boldsymbol{\alpha}_n$ of Λ; that is, $K = \bigoplus_{j=1}^n \mathbb{R}_{\geq 0}\boldsymbol{\alpha}_j$. Let $K^0 = K - \partial K$ be the relative interior of K. We define *determined sets* of K to be the sets

$$D(K, \Lambda) : = \left\{ \sum_{j=1}^n u_j \boldsymbol{\alpha}_j \in \Lambda \;\middle|\; 0 \leq u_j < 1 \right\};$$

$$\bar{D}(K, \Lambda) : = \left\{ \sum_{j=1}^n v_j \boldsymbol{\alpha}_j \in \Lambda \;\middle|\; 0 < v_j \leq 1 \right\}.$$

Notice that every point $\boldsymbol{\alpha} \in K \cap \Lambda$ (or $\boldsymbol{\alpha} \in K^0 \cap \Lambda$) can be written uniquely as $\boldsymbol{\alpha} = \boldsymbol{\gamma} + \sum_{j=1}^n k_j \boldsymbol{\alpha}_j$, where $\boldsymbol{\gamma} \in D(K, \Lambda)$ (or $\boldsymbol{\gamma} \in \bar{D}(K, \Lambda)$) and k_j are nonnegative integers. Then

$$F(K, \Lambda; \boldsymbol{x}) = \sum_{\boldsymbol{\gamma} \in D(K,\Lambda)} \sum_{k_1 \geq 0, \ldots, k_n \geq 0} \boldsymbol{x}^{\boldsymbol{\gamma} + k_1\boldsymbol{\alpha}_1 + \cdots + k_n\boldsymbol{\alpha}_n}$$

$$(5.1) \qquad = \left(\sum_{\boldsymbol{\gamma} \in D(K,\Lambda)} \boldsymbol{x}^{\boldsymbol{\gamma}} \right) \prod_{j=1}^n \frac{1}{1 - \boldsymbol{x}^{\boldsymbol{\alpha}_j}};$$

$$(5.2) \qquad F(K^0, \Lambda; \boldsymbol{x}) = \left(\sum_{\boldsymbol{\gamma} \in \bar{D}(K,\Lambda)} \boldsymbol{x}^{\boldsymbol{\gamma}} \right) \prod_{j=1}^n \frac{1}{1 - \boldsymbol{x}^{\boldsymbol{\alpha}_j}}$$

$$(5.3) \qquad = \left(\sum_{\boldsymbol{\gamma} \in D(K,\Lambda)} \boldsymbol{x}^{-\boldsymbol{\gamma}} \right) \prod_{j=1}^n \frac{\boldsymbol{x}^{\boldsymbol{\alpha}_j}}{1 - \boldsymbol{x}^{\boldsymbol{\alpha}_j}}$$

$$(5.4) \qquad = (-1)^n F(K, \Lambda; \boldsymbol{x}^{-1}).$$

By a Λ-lattice *cone function* we mean a linear combination $f = \sum_{j=1}^k a_j 1_{K_j}$, where K_j are Λ-lattice polyhedral cones. Since

$$\begin{aligned} F\left(1_{K_j}, \Lambda; \boldsymbol{x}^{-1}\right) &= F\left(K_j, \Lambda; \boldsymbol{x}^{-1}\right) = (-1)^{\dim K_j} F\left(K_j^0, \Lambda; \boldsymbol{x}\right) \\ &= F\left((-1)^{\dim K_j} 1_{K_j^0}, \Lambda; \boldsymbol{x}\right) = F\left((1_{K_j^0})^*, \Lambda; \boldsymbol{x}\right), \end{aligned}$$

we have the following proposition by the linearity of the dual operator *.

PROPOSITION 5.1. (Reciprocity Law for Lattice Cone Functions) *If f is a Λ-lattice cone function, then*

$$F(f,\Lambda;\boldsymbol{x}^{-1}) = F(f^*,\Lambda;\boldsymbol{x}).$$

In particular, if M is a Λ-lattice cone and a manifold with boundary ∂M, then

$$F(M,\Lambda;\boldsymbol{x}^{-1}) = (-1)^{\dim M} F(M-\partial M,\Lambda;\boldsymbol{x}).$$

Throughout this section we assume that $\mathbb{R}^N \subset \mathbb{R}^{N+1}$ by identifying $\boldsymbol{x} \in \mathbb{R}^N$ to $(0,\boldsymbol{x}) \in \mathbb{R}^{N+1}$. Let Λ_1 be the lattice of $\mathbb{R}^{N+1}$ generated by Λ and the vector $\boldsymbol{\xi}_0 = (1,0,\ldots,0) \in \mathbb{R}^{N+1}$. Let σ be a Λ-simplex of $\mathbb{R}^N$ with the vertex set $\{\boldsymbol{\alpha}_j\}_{j=0}^n$. We denote by σ_1 the Λ_1-simplex with the vertex set $\{\boldsymbol{\beta}_j = (1,\boldsymbol{\alpha}_j)\}_{j=0}^n$. Then $\#(m\sigma\cap\Lambda) = \#(m\sigma_1\cap\Lambda_1)$. Let

$$V = \bigoplus_{j=0}^n \mathbb{R}\boldsymbol{\beta}_j, \quad M = V\cap\Lambda_1, \quad C = \bigoplus_{j=0}^n \mathbb{R}_{\geq 0}\boldsymbol{\beta}_j. \tag{5.5}$$

Then the set $C\cap\Lambda_1$ of lattice points is a disjoint union of $m\sigma_1\cap\Lambda_1$, $m\geq 0$.

COROLLARY 5.2. (Reciprocity Law in Generating Functions) *If f is Λ-lattice polyhedral function, then*

$$\sum_{m=-\infty}^{\infty} L(f,\Lambda;m)t^m = 0.$$

PROOF. It follows from the linearity of the dual operator * and the fact that for any Λ-simplex σ of dimension n, if $\boldsymbol{x} = (t,1,\ldots,1)$ with $N+1$ entries, then

$$\sum_{m=-\infty}^{\infty} L(\sigma,\Lambda;m)t^m = F(C,\Lambda_1;\boldsymbol{x}) - (-1)^{n+1}F(C^0,\Lambda_1;\boldsymbol{x}^{-1}) = 0.$$

□

For the case of lattice cone manifolds with boundaries, Proposition 5.1 can be found in [**30**] and Corollary 5.2 in [**20**], using different methods.

LEMMA 5.3. *Let L_1 and M_1 be lattices of finite rank of a vector space V_1. Let L_2 and M_2 be lattices of the same rank of a vector space V_2 such that $L_2 \subset M_2$; $G = M_2/L_2$. Let $\langle\,,\,\rangle : V_1\times V_2 \to \mathbb{R}$ be a bilinear pairing satisfying the restriction $M_1\times M_2 \to \mathbb{Z}$.*

(i) *If $\langle\,,\,\rangle$ has the restriction $L_1\times L_2\to\mathbb{Z}$, then $\chi : L_1\times G\to\mathbb{C}^*$ by $\chi_g(\boldsymbol{\alpha}) = e^{-2\pi\imath\langle\boldsymbol{\alpha},g\rangle}$ is a well-defined group bi-homomorphism.*

(ii) *Let $\boldsymbol{\alpha}\in L_1 - M_1$. If there exist $g_{\boldsymbol{\alpha}}\in M_2$ such that $\langle\boldsymbol{\alpha},g_{\boldsymbol{\alpha}}\rangle\notin\mathbb{Z}$, then $\sum_{g\in G}\chi_g(\boldsymbol{\alpha}) = 0$.*

(iii) *Let $\{\boldsymbol{\gamma}_j\}_{j=0}^n$ be a lattice basis of L_1. Let $C = \bigoplus_{j=0}^n \mathbb{R}_{\geq 0}\boldsymbol{\gamma}_j$ and $C^0 = \bigoplus_{j=0}^n \mathbb{R}_{>0}\boldsymbol{\gamma}_j$. If $L_1\times L_2\to\mathbb{Z}$, $C\cap M_1\subset C\cap L_1$, and for each $\boldsymbol{\alpha}\in C\cap(L_1-M_1)$ there exist $g_{\boldsymbol{\alpha}}\in M_2$ such that $\langle\boldsymbol{\alpha},g_{\boldsymbol{\alpha}}\rangle\notin\mathbb{Z}$, then*

$$F(C,M_1;\boldsymbol{x}) = \sum_{\boldsymbol{\alpha}\in C\cap M_1}\boldsymbol{x}^{\boldsymbol{\alpha}} = \frac{1}{\#(G)}\sum_{g\in G}\prod_{j=0}^n\frac{1}{1-\chi_g(\boldsymbol{\gamma}_j)\boldsymbol{x}^{\boldsymbol{\gamma}_j}}; \tag{5.6}$$

$$F(C^0,M_1;\boldsymbol{x}) = \sum_{\boldsymbol{\alpha}\in C^0\cap M_1}\boldsymbol{x}^{\boldsymbol{\alpha}} = \frac{1}{\#(G)}\sum_{g\in G}\prod_{j=0}^n\frac{\chi_g(\boldsymbol{\gamma}_j)\boldsymbol{x}^{\boldsymbol{\gamma}_j}}{1-\chi_g(\boldsymbol{\gamma}_j)\boldsymbol{x}^{\boldsymbol{\gamma}_j}}. \tag{5.7}$$

PROOF. (i) Notice that when we write $\langle \boldsymbol{\alpha}, g\rangle$ for $\boldsymbol{\alpha} \in L_1$ and $g \in G$, it is assumed that g is taken to be a representative of its coset $g + L_2$. Now it becomes trivial that $\chi_g(\boldsymbol{\alpha})$ is a group bi-homomorphism.

(ii) Fix one $g_{\boldsymbol{\alpha}}$ such that $\langle \boldsymbol{\alpha}, g_{\boldsymbol{\alpha}}\rangle \notin \mathbb{Z}$. Let d denote the order of $g_{\boldsymbol{\alpha}}$ in G. Since $\chi_{g_{\boldsymbol{\alpha}}}(\boldsymbol{\alpha}) = e^{-2\pi\imath\langle\boldsymbol{\alpha}, g_{\boldsymbol{\alpha}}\rangle}$ is a root of unity, we have $\sum_{j=1}^{d} \chi_{jg_{\boldsymbol{\alpha}}}(\boldsymbol{\alpha}) = 0$. Let $\langle g_{\boldsymbol{\alpha}}\rangle$ denote the cyclic group generated by $g_{\boldsymbol{\alpha}}$. Write G as disjoint union $G = \bigcup_k (g_k + \langle g_{\boldsymbol{\alpha}}\rangle)$ of cosets for some representatives g_k; we then have

$$\sum_{g\in G} \chi_g(\boldsymbol{\alpha}) = \sum_k \chi_{g_k}(\boldsymbol{\alpha}) \sum_{g\in\langle g_{\boldsymbol{\alpha}}\rangle} \chi_g(\boldsymbol{\alpha}) = 0.$$

(iii) Let $\boldsymbol{\alpha} \in C \cap L_1$. If $\boldsymbol{\alpha} \in M_1$, then $\chi_g(\boldsymbol{\alpha}) = 1$ for all $g \in G$. It follows from (ii) that

$$\frac{1}{\#(G)} \sum_{g\in G} \chi_g(\boldsymbol{\alpha}) = \begin{cases} 1 & \text{for } \boldsymbol{\alpha} \in M_1 \\ 0 & \text{for } \boldsymbol{\alpha} \notin M_1. \end{cases}$$

Write $\boldsymbol{\alpha}$ uniquely as $\boldsymbol{\alpha} = \sum_{j=0}^{n} k_j \boldsymbol{\gamma}_j$ for some $k_j \geq 0$; we have

$$\begin{aligned}
F(C, M_1; \boldsymbol{x}) &= \sum_{\boldsymbol{\alpha}\in C\cap L_1} \boldsymbol{x}^{\boldsymbol{\alpha}} \left(\frac{1}{\#(G)} \sum_{g\in G} \chi_g(\boldsymbol{\alpha}) \right) \\
&= \frac{1}{\#(G)} \sum_{g\in G} \sum_{\boldsymbol{\alpha}\in C\cap L_1} \chi_g(\boldsymbol{\alpha}) \boldsymbol{x}^{\boldsymbol{\alpha}} \\
&= \frac{1}{\#(G)} \sum_{g\in G} \sum_{k_1=0}^{\infty} \cdots \sum_{k_n=0}^{\infty} \chi_g\left(\sum_{j=0}^{n} k_j \boldsymbol{\gamma}_j \right) \boldsymbol{x}^{\sum_{j=0}^{n} k_j \boldsymbol{\gamma}_j} \\
&= \frac{1}{\#(G)} \sum_{g\in G} \sum_{k_1=0}^{\infty} \cdots \sum_{k_n=0}^{\infty} \prod_{j=0}^{n} \chi_g^{k_j}(\boldsymbol{\gamma}_j) (\boldsymbol{x}^{\boldsymbol{\gamma}_j})^{k_j} \\
&= \frac{1}{\#(G)} \sum_{g\in G} \prod_{j=0}^{n} \frac{1}{1 - \chi_g(\boldsymbol{\gamma}_j) \boldsymbol{x}^{\boldsymbol{\gamma}_j}}.
\end{aligned}$$

The situation for $F(C^0, M_1; \boldsymbol{x})$ is similar, just pay attention to $k_0 \geq 1, k_1 \geq 1, \ldots, k_n \geq 1$ in the above summation. □

THEOREM 5.4. *Let* $V = \bigoplus_{j=0}^{n} \mathbb{R}\boldsymbol{\beta}_j$, $M = V \cap \Lambda_1$, $C = \bigoplus_{j=0}^{n} \mathbb{R}_{\geq 0}\boldsymbol{\beta}_j$ *be as defined in* (5.5); *and let* p_j *be positive integers such that* $C \cap M \subset C \cap L$, *where* $L = \bigoplus_{j=0}^{n} \mathbb{Z}\beta_j/p_j$. *Let* L' *and* M' *be lattices of the same rank of a vector space* V' *such that* $L' \subset M'$; $G = M'/L'$. *Let* $\langle\, ,\, \rangle : V \times V' \to \mathbb{R}$ *be a bilinear pairing satisfying the restrictions* $M \times M' \to \mathbb{Z}$ *and* $L \times L' \to \mathbb{Z}$. *If for any* $\boldsymbol{\alpha} \in C \cap (L - M)$ *there exist* $g_{\boldsymbol{\alpha}} \in M'$ *such that* $\langle \boldsymbol{\alpha}, g_{\boldsymbol{\alpha}}\rangle \notin \mathbb{Z}$, *then*

$$\text{(5.8)} \quad \sum_{m=0}^{\infty} L(\sigma, \Lambda; m) e^{-2\pi m s} = \frac{1}{2^{n+1}\#(G)} \sum_{g\in G} \prod_{j=0}^{n} \left(1 + \coth \frac{\pi}{p_j} \left(s + \imath\langle \boldsymbol{\beta}_j, g\rangle \right) \right);$$

$$\text{(5.9)} \quad \sum_{m=1}^{\infty} L(\sigma^0, \Lambda; m) e^{-2\pi m s} = \frac{(-1)^{n+1}}{2^{n+1}\#(G)} \sum_{g\in G} \prod_{j=0}^{n} \left(1 - \coth \frac{\pi}{p_j} \left(s + \imath\langle \boldsymbol{\beta}_j, g\rangle \right) \right).$$

PROOF. This is an immediate consequence of Lemma 5.3. Let $V_1 = V$, $M_1 = M$, $L_1 = L$, $V_2 = V'$, $M_2 = M'$, $L_2 = L'$, $\boldsymbol{x} = (e^{-2\pi s}, 1, \ldots, 1)$, $\boldsymbol{\gamma}_j = \boldsymbol{\beta}_j/p_j$. Notice that the first coordinate of $\boldsymbol{\gamma}_j$ is $1/p_j$. Then

$$\chi_g(\boldsymbol{\gamma}_j)\boldsymbol{x}^{\boldsymbol{\gamma}_j} = e^{-2\pi\imath\langle\boldsymbol{\beta}_j/p_j, g\rangle} e^{-2\pi s/p_j} = e^{-\frac{2\pi}{p_j}(s+\imath\langle\boldsymbol{\beta}_j, g\rangle)}.$$

Since $C = \bigoplus_{j=0}^{n} \mathbb{R}_{\geq 0}\boldsymbol{\gamma}_j$ and $C^0 = \bigoplus_{j=0}^{n} \mathbb{R}_{>0}\boldsymbol{\gamma}_j$, we have

$$\begin{aligned} F(C, M_1; \boldsymbol{x}) &= \sum_{m=0}^{\infty} \sum_{\boldsymbol{\alpha} \in m\sigma_1 \cap \Lambda_1} \boldsymbol{x}^{\boldsymbol{\alpha}} = \sum_{m=0}^{\infty} L(\sigma, \Lambda; m) e^{-2\pi m s}; \\ F(C^0, M_1; \boldsymbol{x}) &= \sum_{m=1}^{\infty} \sum_{\boldsymbol{\alpha} \in m\sigma_1^0 \cap \Lambda_1} \boldsymbol{x}^{\boldsymbol{\alpha}} = \sum_{m=1}^{\infty} L(\sigma^0, \Lambda; m) e^{-2\pi m s}. \end{aligned}$$

Notice that

$$\frac{1}{1-e^{-2z}} = \frac{1}{2}(1+\coth z) \quad \text{and} \quad \frac{e^{-2z}}{1-e^{-2z}} = -\frac{1}{2}(1-\coth z).$$

We further have

$$\begin{aligned} \frac{1}{1-\chi_g(\boldsymbol{\gamma}_j)\boldsymbol{x}^{\boldsymbol{\gamma}_j}} &= \frac{1}{2}\left(1+\coth\frac{\pi}{p_j}\left(s+\imath\langle\boldsymbol{\beta}_j, g\rangle\right)\right); \\ \frac{\chi_g(\boldsymbol{\gamma}_j)\boldsymbol{x}^{\boldsymbol{\gamma}_j}}{1-\chi_g(\boldsymbol{\gamma}_j)\boldsymbol{x}^{\boldsymbol{\gamma}_j}} &= -\frac{1}{2}\left(1-\coth\frac{\pi}{p_j}\left(s+\imath\langle\boldsymbol{\beta}_j, g\rangle\right)\right). \end{aligned}$$

Apply (iii) of Lemma 5.3; we obtain (5.8) and (5.9). □

A notable feature is that the group G in (5.8) and (5.9) can be selected as long as the lattices L, L', M, M' satisfy the required conditions. One can fix a pairing $V \times V' \to \mathbb{R}$ satisfying $M \times M' \to \mathbb{Z}$, then construct a lattice L such that $C \cap M \subset C \cap L$. Since L may contain some points not in M, we may not have $L \times M' \to \mathbb{Z}$. However, there always exists a sublattice $L' \subset M'$ of the same rank such that $L \times L' \to \mathbb{Z}$.

Let A be the $N \times n$ matrix whose columns are the vectors $\boldsymbol{\alpha}_1 - \boldsymbol{\alpha}_0, \ldots, \boldsymbol{\alpha}_n - \boldsymbol{\alpha}_0$. The matrix A can be viewed as an abelian group homomorphism from $\mathbb{Z}^n$ to $\mathbb{Z}^N$; the *torsion* of A is defined to be the order of the torsion subgroup of $\mathbb{Z}^N/\operatorname{Im} A$, denoted tor A. The matrix A can be transformed into an upper triangular form (5.10) by the *modified Gaussian elimination* (adding an integral multiple of one row to another row; switching two rows), having $a_{jj} > 0$.

$$\begin{pmatrix} a_{11} & a_{12} & \cdots & a_{1n} \\ 0 & a_{22} & \cdots & a_{2n} \\ \vdots & \vdots & & \vdots \\ 0 & 0 & \cdots & a_{nn} \\ 0 & 0 & \cdots & 0 \\ \vdots & \vdots & & \vdots \\ 0 & 0 & \cdots & 0 \end{pmatrix} \tag{5.10}$$

When $N = n$, this upper triangular matrix is known as the *Hermite normal form*; see [**25**].

LEMMA 5.5. *Let A be an $N \times n$ matrix in the upper triangular form (5.10). Let $\boldsymbol{u} = (u_1, \dots, u_n) \in [0,1)^n$. Then $A\boldsymbol{u} \in \mathbb{Z}^N$ if and only if there exists a unique $(j_1, \dots, j_n) \in \Pi_{i=1}^n [0, a_{ii}) \cap \mathbb{Z}$ such that*

$$(5.11) \quad \begin{cases} u_n = \frac{j_n}{a_{nn}}, \\ u_i = \frac{\lceil a_{i(i+1)}u_{i+1} + \cdots + a_{in}u_n \rceil}{a_{ii}} - \frac{a_{i(i+1)}u_{i+1} + \cdots + a_{in}u_n}{a_{ii}} + \frac{j_i}{a_{ii}}, \quad 1 \le i \le n-1. \end{cases}$$

In particular, $A\boldsymbol{u} \in \mathbb{Z}^N$ implies that $u_i = \frac{j}{a_{ii}\cdots a_{nn}}$ for some $j \in \mathbb{Z}$. Moreover, the map $\tilde{\boldsymbol{u}} : \prod_{i=1}^n [0, a_{ii}) \cap \mathbb{Z} \to D(\sigma, \boldsymbol{\alpha}_0)$ by $\tilde{\boldsymbol{u}}(j_1, \dots, j_n) = \sum_{i=1}^n u_i(\boldsymbol{\alpha}_i - \boldsymbol{\alpha}_0)$ is a bijection, where u_i are given by (5.11).

PROOF. We proceed by induction on the index of u_k from n to 1. Obviously, $u_n = \frac{j_n}{a_{nn}}$, where $j_n \in \mathbb{Z}$ and $0 \le j_n < a_{nn}$. Assume that $u_k = \frac{j_k'}{a_{kk}\cdots a_{nn}}$ and are determined by (5.11) for $k = i+1, \dots, n$. We are to determine u_i and $0 \le j_i < a_{ii}$. Let J_i be the smallest integer such that

$$a_{ii}u_i' + a_{i(i+1)}u_{i+1} + \cdots + a_{in}u_n = J_i$$

for some real number $u_i' \in [0,1)$. Then

$$\frac{J_i}{a_{ii}} - \frac{a_{i(i+1)}u_{i+1} + \cdots + a_{in}u_n}{a_{ii}} \ge 0$$

and

$$\frac{J_i - 1}{a_{ii}} - \frac{a_{i(i+1)}u_{i+1} + \cdots + a_{in}u_n}{a_{ii}} < 0.$$

This means that $J_i = \lceil a_{i(i+1)}u_{i+1} + \cdots + a_{in}u_n \rceil$. Hence

$$u_i' = \frac{\lceil a_{i(i+1)}u_{i+1} + \cdots + a_{in}u_n \rceil}{a_{ii}} - \frac{a_{i(i+1)}u_{i+1} + \cdots + a_{in}u_n}{a_{ii}}.$$

Since $u_i \ge u_i'$, we write $u_i = u_i' + u_i''$. Then $a_{ii}u_i'' \in \mathbb{Z}$. Thus $u_i'' = \frac{j_i}{a_{ii}}$ for some nonnegative integer j_i. Consequently,

$$u_i = \frac{\lceil a_{i(i+1)}u_{i+1} + \cdots + a_{in}u_n \rceil}{a_{ii}} - \frac{a_{i(i+1)}u_{i+1} + \cdots + a_{in}u_n}{a_{ii}} + \frac{j_i}{a_{ii}},$$

where $0 \le j_i \le a_{ii} - 1$, because when $j_i = a_{ii} - 1$,

$$\frac{\lceil a_{i(i+1)}u_{i+1} + \cdots + a_{in}u_n \rceil}{a_{ii}} - \frac{a_{i(i+1)}u_{i+1} + \cdots + a_{in}u_n}{a_{ii}} + \frac{a_{ii} - 1}{a_{ii}} < 1,$$

and when $j_i = a_{ii}$,

$$\frac{\lceil a_{i(i+1)}u_{i+1} + \cdots + a_{in}u_n \rceil}{a_{ii}} - \frac{a_{i(i+1)}u_{i+1} + \cdots + a_{in}u_n}{a_{ii}} + \frac{a_{ii}}{a_{ii}} \ge 1.$$

It is clear that $(j_1, \dots, j_n)$ is unique. By definition of $D(\sigma, \boldsymbol{\alpha}_0)$, the map $\tilde{\boldsymbol{u}}$ is a bijection. □

THEOREM 5.6 (Diaz and Robins). *Let σ be a lattice n-simplex of $\mathbb{R}^n$ with the vertex set $\{\boldsymbol{\alpha}_j\}_{j=0}^n \subset \mathbb{Z}^n$. Let A be the matrix whose columns are the vectors $\boldsymbol{\alpha}_j - \boldsymbol{\alpha}_0$, $1 \le j \le n$; and let B be the matrix whose columns are $\boldsymbol{\beta}_j = (1, \boldsymbol{\alpha}_j)$, $1 \le j \le n$. Let A be in upper triangular form (5.10); set $p_j = a_{jj} \cdots a_{nn}$, $0 \le j \le n$, where $a_{00} = 1$; $G = \prod_{j=0}^n \mathbb{Z}/p_j\mathbb{Z}$. Then we have (5.8) and (5.9).*

PROOF. Let $N = n$, $V = V' = \mathbb{R}^{n+1}$, $M = M' = \Lambda_1 = \mathbb{Z}^{n+1}$. Let $\langle\,,\,\rangle : V \times V' \to \mathbb{R}$ be the standard inner product of $\mathbb{R}^{n+1}$; that is, $\langle \boldsymbol{\alpha}, \boldsymbol{\beta} \rangle = \sum_{j=0}^{n} u_j v_j$ for $\boldsymbol{\alpha} = \sum_{j=0}^{n} u_j \boldsymbol{e}_j$ and $\boldsymbol{\beta} = \sum_{j=0}^{n} v_j \boldsymbol{e}_j$, where $\{\boldsymbol{e}_j\}_{j=0}^{n}$ is the standard lattice basis of $\mathbb{Z}^{n+1}$. Obviously, $\langle\,,\,\rangle$ satisfies the restriction $M \times M' \to \mathbb{Z}$.

Let $\boldsymbol{\gamma}_j = \boldsymbol{\beta}_j / p_j$, $\boldsymbol{\gamma}'_j = p_j \boldsymbol{e}_j$, $L = \sum_{j=0}^{n} \mathbb{Z}\boldsymbol{\gamma}_j$, $L' = \sum_{k=0}^{n} \mathbb{Z}\boldsymbol{\gamma}'_k$. By Lemma 5.5, we have $C \cap M \subset C \cap L$. Notice that

$$\langle \boldsymbol{\gamma}_j, \boldsymbol{\gamma}'_k \rangle = \frac{p_k}{p_j} \langle \boldsymbol{\beta}_j, \boldsymbol{e}_k \rangle = \begin{cases} \frac{a_{kj} a_{kk} \cdots a_{nn}}{a_{jj} \cdots a_{nn}} & \text{for } j \geq k \\ 0 & \text{for } j < k \end{cases}$$

are integers, the pairing satisfies the restriction $L \times L' \to \mathbb{Z}$.

For $\boldsymbol{\alpha} \in C \cap (L - M)$, we write $\boldsymbol{\alpha} = \sum_{j=0}^{n} u_j \boldsymbol{e}_j$. Since $\boldsymbol{\alpha} \notin M$, some u_j must be non-integers, say $u_{j'}$; we then have $e_{j'} \in M'$ and $\langle \boldsymbol{\alpha}, \boldsymbol{e}_{j'} \rangle = u_{j'} \notin \mathbb{Z}$. Finally,

$$G = M'/L' = \mathbb{Z}^{n+1} / \prod_{j=0}^{n} p_j \mathbb{Z} = \prod_{j=0}^{n} \mathbb{Z}/p_j\mathbb{Z}.$$

Now all the conditions of Theorem 5.4 are satisfied. We thus recover the main result of [**13**], which was first obtained by the Poisson summation and Fourier analysis. □

In the following we apply Theorem 5.4 to find a convenient group G to express the generating functions in terms of hyperbolic cotangent functions. Recall the lattice basis $\{\boldsymbol{\xi}_k\}_{k=0}^{N}$ for the lattice Λ_1 in $\mathbb{R}^{N+1}$; there is a canonical inner product on $\mathbb{R}^{N+1}$: for $\boldsymbol{\alpha} = \sum_{k=0}^{N} u_k \boldsymbol{\xi}_k$ and $\boldsymbol{\beta} = \sum_{k=0}^{N} v_k \boldsymbol{\xi}_k$, define

$$\langle \boldsymbol{\alpha}, \boldsymbol{\beta} \rangle := \sum_{k=0}^{N} u_k v_k. \tag{5.12}$$

COROLLARY 5.7. *Let A be the $N \times n$ matrix whose columns are the coordinate vectors of $\boldsymbol{\alpha}_1 - \boldsymbol{\alpha}_0, \ldots, \boldsymbol{\alpha}_n - \boldsymbol{\alpha}_0$ relative to the lattice basis $\{\boldsymbol{\xi}_k\}_{k=1}^{N}$. Let B be the $(N+1) \times (n+1)$ matrix whose columns are the coordinate vectors of $\boldsymbol{\beta}_0, \boldsymbol{\beta}_1, \ldots, \boldsymbol{\beta}_n$ relative to the lattice basis $\{\boldsymbol{\xi}_k\}_{k=0}^{N}$. Let $p = \operatorname{tor} A$. Then*

$$\sum_{m=0}^{\infty} L(\sigma, \Lambda, m) e^{-2\pi m s} = \tag{5.13}$$

$$\frac{1}{2^{n+1} p^{N+1}} \sum_{\substack{h_0, h_1, \ldots, h_N = 0 \\ \boldsymbol{h} = (h_0, h_1, \ldots, h_N)}}^{p-1} \prod_{j=0}^{n} \left(1 + \coth \frac{\pi}{p} \left(s + \imath \langle \boldsymbol{\beta}_j, \boldsymbol{h} \rangle\right)\right);$$

$$\sum_{m=1}^{\infty} L(\sigma^0, \Lambda, m) e^{-2\pi m s} = \tag{5.14}$$

$$\frac{(-1)^{(n+1)}}{2^{n+1} p^{N+1}} \sum_{\substack{h_0, h_1, \ldots, h_N = 0 \\ \boldsymbol{h} = (h_0, h_1, \ldots, h_N)}}^{p-1} \prod_{j=0}^{n} \left(1 - \coth \frac{\pi}{p} \left(s + \imath \langle \boldsymbol{\beta}_j, \boldsymbol{h} \rangle\right)\right).$$

PROOF. Let $V = \bigoplus_{j=0}^{n} \mathbb{R}\boldsymbol{\beta}_j$, $V' = \mathbb{R}^{N+1}$, $M = V \cap \Lambda_1$, $M' = \Lambda_1$, $C = \bigoplus_{j=0}^{n} \mathbb{R}_{\geq 0} \boldsymbol{\beta}_j$. The inner product (5.12) gives a pairing $V \times V' \to \mathbb{R}$ satisfying the restriction $M \times M' \to \mathbb{Z}$. Let $\boldsymbol{\gamma}_j = \boldsymbol{\beta}_j / p$, $\boldsymbol{\gamma}'_k = p\boldsymbol{\xi}_k$, $L = \bigoplus_{j=0}^{n} \mathbb{Z}\boldsymbol{\gamma}_j$, $L' = \bigoplus_{k=0}^{N} \mathbb{Z}\boldsymbol{\gamma}'_k$. By Theorem 5.5, $C \cap M \subset C \cap L$. Obviously, $\langle \boldsymbol{\gamma}_j, \boldsymbol{\gamma}'_k \rangle = \langle \boldsymbol{\beta}_j, \boldsymbol{\xi}_k \rangle \in \mathbb{Z}$; this means that the pairing also satisfies the restriction $L \times L' \to \mathbb{Z}$. For $\boldsymbol{\alpha} \in C \cap (L - M)$, we write $\boldsymbol{\alpha} = \sum_{k=0}^{N} u_k \boldsymbol{\xi}_k$. Since $\boldsymbol{\alpha} \notin M$, some u_k must be non-integers, say $u_{k'}$; we then have $\boldsymbol{\xi}_{k'} \in M'$ and $\langle \boldsymbol{\alpha}, \boldsymbol{\xi}_{k'} \rangle = u_{k'} \notin \mathbb{Z}$. Finally, $G = M'/L'$ can be identified to the set $[0, p)^{N+1} \cap \mathbb{Z}^{N+1}$. Now all conditions of Theorem 5.4 are satisfied. We thus obtain (5.14) and (5.15). □

6. Reciprocity law of Dedekind sums

In this section we demonstrate that the reciprocity law for Dedekind sums can be obtained by using the coefficient formulas (2.14) and (2.15) about Ehrhart polynomials. The reciprocity law for the higher dimensional Dedekind sums [**17, 31**] may be treated in a similar way. Recall the periodic function $(\!(t)\!)$, defined by

$$(\!(t)\!) := \begin{cases} t - \lceil t \rceil + \frac{1}{2} & \text{for } t \notin \mathbb{Z} \\ 0 & \text{for } t \in \mathbb{Z}; \end{cases}$$

and the *Dedekind sum* $s(q, p)$ for positive integers p and q, defined by

$$s(q,p) := \sum_{k=1}^{p-1} \left(\!\!\left(\frac{qk}{p}\right)\!\!\right) \left(\!\!\left(\frac{k}{p}\right)\!\!\right).$$

Since $(\!(t)\!)$ is a periodic odd function of period 1, it is routine to verify the following properties (see [**27**]):

$$\begin{aligned} \sum_{k=0}^{p-1} \left(\!\!\left(\frac{k+t}{p}\right)\!\!\right) &= (\!(t)\!); \\ \sum_{k=0}^{p-1} \left(\!\!\left(\frac{k}{p}\right)\!\!\right)^2 &= \frac{p}{12} + \frac{1}{6p} - \frac{1}{4}. \end{aligned}$$

Let $\Delta(a, b)$ be the lattice triangle of $\mathbb{R}^2$ with the vertices $(0, 0)$, $(a, 0)$, $(0, b)$, where a and b are positive integers. On the one hand, the constant coefficient c_0 of $L(\Delta(a, b), t)$ is 1; on the other hand, by using (2.14) the coefficient can be computed as

$$c_0 = \frac{1}{2} \sum_{i=0}^{a-1} \sum_{j=0}^{b-1} s_2 \left(\left\lceil \frac{i}{a} + \frac{j}{b} \right\rceil - 1, \left\lceil \frac{i}{a} + \frac{j}{b} \right\rceil - 2 \right).$$

Let $d = \gcd(a, b)$. If $\frac{i}{a} + \frac{j}{b} = k$ is an integer, that is, $(\frac{b}{d})i + (\frac{a}{d})j = (\frac{ab}{d})k$, then $i \equiv 0 \pmod{\frac{a}{d}}$ and $j \equiv 0 \pmod{\frac{b}{d}}$. Let $i = i'a/d$ and $j = j'b/d$, where $0 \le i' < d$ and $0 \le j' < d$. Then $i' + j' = kd$; and this implies that either $k = 0$ or $k = 1$. Thus, if $d \ge 2$ and

$$(i,j) \notin \left\{ (0,0), \left(\frac{a}{d}, \frac{b(d-1)}{d}\right), \left(\frac{2a}{d}, \frac{b(d-2)}{d}\right), \ldots, \left(\frac{a(d-1)}{d}, \frac{b}{d}\right) \right\},$$

we have

$$\left\lceil \frac{i}{a} + \frac{j}{b} \right\rceil = \left(\frac{i}{a} + \frac{j}{b} \right) - \left(\!\!\left(\frac{i}{a} + \frac{j}{b} \right)\!\!\right) + \frac{1}{2}.$$

Write $u = \frac{i}{a} + \frac{j}{b}$; we have

$$c_0 = \frac{1}{2} + \frac{d}{8} + \frac{1}{2} \sum \left(u^2 - 2u(\!(u)\!) + (\!(u)\!)^2 - 2u + 2(\!(u)\!) + \frac{3}{4} \right), \tag{6.1}$$

where the sum is extended over $0 \le i \le a-1$ and $0 \le j \le b-1$. A straightforward calculation shows that

$$\begin{aligned} \sum u &= ab - \frac{a+b}{2}, \\ \sum u^2 &= \frac{7ab}{6} - (a+b) + \frac{1}{6}\left(\frac{a}{b} + \frac{b}{a}\right) + \frac{1}{2}. \end{aligned}$$

Let $a' = b/d$, $b' = a/d$, and $c = \operatorname{lcm}(a,b) = ab/d$. If $a'i + b'j \equiv a'i' + b'j' \pmod{c}$, that is, $a'(i-i') \equiv b'(j'-j) \pmod{c}$, then $i \equiv i' \pmod{b'}$ and $j \equiv j' \pmod{a'}$ because $\gcd(a',b') = 1$; hence $\frac{i-i'}{b'} \equiv \frac{j'-j}{a'} \pmod{d}$. Fix an integer-pair $(i,j) \in [0,a) \times [0,b)$; there are exactly d integers $i' \in [0,a)$ such that $i - i' = kb'$. Now fix each such i'; set $(j'-j)/a' = k + ld$ to find possible integers $j' \in [0,b)$ by choosing integer l. Then $j' = j + ka' + lda' = j + ka' + lb$. Clearly, there is exactly one choice for l such that $j' \in [0,b)$. We thus conclude that each residue class of modulo c appears exactly d times in the residue classes of $a'i + b'j$, where $(i,j) \in [0,a) \times [0,b)$. Therefore

$$\begin{aligned} \sum (\!(u)\!) &= \sum_{i=0}^{a-1}\sum_{j=0}^{b-1} \left(\!\!\left(\frac{a'i+b'j}{c}\right)\!\!\right) = d\sum_{k=0}^{c-1}\left(\!\!\left(\frac{k}{c}\right)\!\!\right) = 0, \\ \sum (\!(u)\!)^2 &= d\sum_{k=0}^{c-1}\left(\!\!\left(\frac{k}{c}\right)\!\!\right)^2 = \frac{ab}{12} + \frac{d^2}{6ab} - \frac{d}{4}. \end{aligned}$$

However, $\sum u(\!(u)\!)$ can only be simplified to the following Dedekind sums:

$$\begin{aligned} \sum u(\!(u)\!) &= \sum_{i=0}^{a-1}\frac{i}{a}\sum_{j=0}^{b-1}\left(\!\!\left(\frac{\frac{bi}{a}+j}{b}\right)\!\!\right) + \sum_{j=0}^{b-1}\frac{j}{b}\sum_{i=0}^{a-1}\left(\!\!\left(\frac{\frac{aj}{b}+i}{a}\right)\!\!\right) \\ &= \sum_{i=0}^{a-1}\frac{i}{a}\left(\!\!\left(\frac{bi}{a}\right)\!\!\right) + \sum_{j=0}^{b-1}\frac{j}{b}\left(\!\!\left(\frac{aj}{b}\right)\!\!\right) \\ &= \sum_{i=0}^{a-1}\left[\left(\!\!\left(\frac{i}{a}\right)\!\!\right) + \frac{1}{2}\right]\left(\!\!\left(\frac{bi}{a}\right)\!\!\right) + \sum_{j=0}^{b-1}\left[\left(\!\!\left(\frac{j}{b}\right)\!\!\right) + \frac{1}{2}\right]\left(\!\!\left(\frac{aj}{b}\right)\!\!\right) \\ &= s(b,a) + s(a,b). \end{aligned}$$

Substitute all these data into the constant coefficient formula (6.1) and simplify it carefully; we obtain explicitly

$$c_0 = \frac{1}{12}\left(\frac{a}{b} + \frac{b}{a} + \frac{d^2}{ab}\right) + \frac{3}{4} - s(a,b) - s(b,a).$$

Using the fact that $c_0 = 1$, we obtain the *Reciprocity Law* for Dedekind sums: For any positive integers a and b,

$$s(a,b) + s(b,a) = \frac{1}{12}\left(\frac{a}{b} + \frac{b}{a} + \frac{\gcd(a,b)^2}{ab}\right) - \frac{1}{4}. \tag{6.2}$$

There are different methods to interpret (6.2), but all use the same idea of counting lattice points of a lattice triangle; see [**3, 5, 26**]. Mordell [**23**] calculated the number of lattice points inside a tetrahedron $\Delta(a,b,c)$ with integral vertices $(0,0,0)$, $(a,0,0)$, $(0,b,0)$, $(0,0,c)$, which is equivalent to evaluating the coefficient c_1 for the Ehrhart polynomial of $\Delta(a,b,c)$; see [**9, 5, 13, 26**]. In higher dimensions, the constant coefficient formula for the Ehrhart polynomial of $\Delta(a_1,\ldots,a_n)$ will give rise to a reciprocity law for higher dimensional Dedekind sums in even dimensions; while in odd dimensions all the terms will be cancelled out.

Acknowledgement: The author thanks Dr. Qing-Hu Hou for some helpful discussion of Section 3 during the writing of this paper.

References

[1] A. I. Barvinok, *Computing the Ehrhart polynomial of a convex lattice polytope*, Discrete Comput. Geom. **12** (1994), 35–48.

[2] A. I. Barvinok and J. Pommersheim, *An algorithmic theory of lattice points in polyhedra. New perspectives in algebraic combinatorics* (Berkeley, CA, 1996–97), 91–147, Math. Sci. Res. Inst. Publ., **38**, Cambridge Univ. Press, Cambridge, 1999.

[3] M. Beck, *The reciprocity law for Dedekind sums via the constant Ehrhart coefficient*, Amer. Math. Monthly **106** (1999), 459–462.

[4] M. Beck, *Multidimensional Ehrhart reciprocity*, J. Combin. Theory Ser. A **97** (2002), 187–194.

[5] M. Brion and M. Vergne, *Lattice points in simple polytopes*, J. Amer. Math. Soc. **10** (1997), 371–392.

[6] S. E. Cappell and J. L. Shaneson, *Genera of algebraic varieties and counting lattice points*, Bull. Amer. Math. Soc. **30** (1994), 62–69.

[7] B. Chen, *On the Euler characteristics of finite unions of convex sets*, Discrete Comput. Geom. **10** (1993), 79–93.

[8] B. Chen, Weight functions, *double reciprocity laws, and volume formulas for lattice polyhedra*, Proc. Nat. Acad. Sci. USA, **95** (1998), 9093–9098.

[9] B. Chen, *Minkowski algebra I: A convolution theory of closed convex sets and relative open convex sets*, Asian J. Math. **3** (1999), 609–634.

[10] B. Chen, *Lattice points, Dedekind sums, and Ehrhart polynomials of lattice polyhedra*, Discrete Comput. Geom. **28** (2002), 175–199.

[11] B. Chen and V. Tauraev, *Counting lattice points of rational polyhedra*, Advances in Math. **155** (2000), 84–97.

[12] B. Chen and M. Yan, *Eulerian stratification of polyhedra*, Advances in Appl. Math. **21** (1998), 22–57.

[13] R. Diaz and S. Robins, *The Ehrhart polynomial of a lattice polytope*, Ann. of Math. **135** (1997), 503–518.

[14] E. Ehrhart, *Sur un problème de geómétrie diophantienne linéaire I*, J. Reine Angrew. Math. **226** (1967), 1–29.

[15] E. Ehrhart, *Sur un problème de geómétrie diophantienne linéaire II*, J. Reine Angrew. Math. **227** (1967), 25–49.

[16] H. Groemer, *On the extension of additive functionals on classes of convex sets*, Pacific J. Math. **75** (1978), 397–410.

[17] F. Hirzebruch and D. Zagier, *The Atiyah-Singer Index Theorem and Elementary Number Theory*, Publish or Perish, Boston, MA, 1974.

[18] J. M. Kantor and A. G. Khovanskii, *Une application du Théoréme de Riemann-Roch combinatorie au polynôme d'Ehrhart des polytopes entiers de $\mathbb{R}^n$*, C.R. Acad. Sci. Paris, Sér. I **317** (1993), 501–507.

[19] D. A. Klain and G.-C. Rota, *Introduction to Geometric Probability*, Cambridge Univ. Press, 1997.

[20] J. Lawrence, *Minkowski ring*, preprint, 1985.

[21] I. G. Macdonald, *The volume of a lattice polyhedron*, Math. Proc. Cambridge Philos. Soc. **59** (1963), 719–726.

[22] I. G. Macdonald, *Polynomials associated with finite cell complexes*, J. London Math. Soc. (2) **4** (1971), 181–192.

[23] L. J. Mordell, *Lattice points in a tetrahedron and Dedekind sums*, J. Indian Math. **15** (1951), 41–46.

[24] R. Morelli, *Pick's theorem and the Todd class of toric variety*, Advances in Math. **100** (1993), 183–231.

[25] M. Newman, *Integral Matrices*, Pure & Applied Math., v.45, Academic Press, New York, 1972.

[26] J. Pommersheim, *Toric varieties, lattice points and Dedekind sums*, Math Ann. **295** (1993), 1–24.
[27] H. Rademacher and E. Grosswald, *Dedekind Sums*, The Carus Mathematical Monographs, vol. 16, The Mathematical Association of America, Washington, DC, 1972.
[28] R. P. Stanley, Combinatorial reciprocity theorems, Advances in Math. **14** (1974), 194-253.
[29] R. P. Stanley, Linear diophantine equations and local cohomology, Inventiones Math. **6** (1980), 333-342.
[30] R. P. Stanley, *Enumerative Combinatorics*, Vol. I, Cambridge Univ. Press, 1997.
[31] D. Zagier, *Higher dimensional Dedekind sums*, Math. Ann. **202** (1973), 149–172.

DEPARTMENT OF MATHEMATICS, HONG KONG UNIVERSITY OF SCIENCE AND TECHNOLOGY, CLEAR WATER BAY, KOWLOON, HONG KONG

E-mail address: mabfchen@ust.hk

Contemporary Mathematics
Volume **374**, 2005

Lattice points, contingency tables, and sampling

Yuguo Chen, Ian Dinwoodie, Adrian Dobra, and Mark Huber

ABSTRACT. Markov chains and sequential importance sampling (SIS) are described as two leading sampling methods for Monte Carlo computations in exact conditional inference on discrete data in contingency tables. Examples are explained from genotype data analysis, graphical models, and logistic regression. A new Markov chain and implementation of SIS are described for logistic regression.

1. Introduction

This paper is partly a survey of some recent theory on statistical problems of discrete data, and partly a description of new results for problems of sparse contingency tables where existing theory is not adequate.

Counts from statistical experiments are put in contingency tables that may be considered vectors of nonnegative integers. These are typically frequencies of events from an experiment where two or more outcomes are possible in a series of trials. Algebraic and geometric theory of lattice points and polytopes become useful when one wants to make inferences about the statistical model in place during the sampling and the data is multidimensional. One is led to computations over a collection of tables with certain constraints that are often linear and define a polytope S_0 whose elements correspond to constrained tables of integers where each cell in a table is a dimension in the space containing the polytope.

The statistical ideas of conditional inference that make polytopes an essential sample space were developed by Ronald Fisher to deal with two fundamental statistical issues: to determine if a family of probabilities (a model) could include the prevailing probability distribution when the family involves several unknown parameters; and to compute measures of distance from the observed data to the collection of tables consistent with the model without using asymptotic approximations. Conditional inference is described in Agresti (1990). The number of lattice points in the polytope S_0 representing tables of interest may be 10^{20} or larger, and

1991 *Mathematics Subject Classification.* Primary 62F03, 13P10; Secondary 05A99.
Key words and phrases. Contingency table, Markov chain, Gröbner basis, toric ideal.
This work was supported by SAMSI under grant DMS-0112069. The first author was supported under NSF grant DMS 0203762. The second author was supported under NSF grant DMS-0200888.

over this set we will want to compute expectations $E_\pi(f(\mathbf{n}))$ for certain functions $f : S_0 \to \mathbf{R}$, and distributions π that may be uniform or conditional on sufficient statistics, often the hypergeometric distribution. Sometimes there are formulas or *ad hoc* sampling methods for efficient Monte Carlo computation, but this only happens for certain distributions π and special polytopes S_0. In general it is not efficient or not possible to list all the elements of S_0 for further exact computation.

All expectations of interest can be computed as accurately as desired with a random sample from S_0 from a known probability distribution that is positive over all elements of S_0. Generating the random sample is the goal. To this end we discuss two particularly useful methods for sampling from polytopes: Markov chains and sequential importance sampling (SIS).

Markov chains are usually easy to program and memory efficient and have been used for Monte Carlo computations for decades. They require a "Markov basis" to make them irreducible–to run through all tables so time averages approximate space averages. The basis can be described as a generating set for a toric ideal, which is one of the fundamental results of Diaconis and Sturmfels (1998). With an irreducible and aperiodic Markov chain in S_0 one has the ergodic theorem that allows us to approximate expectations with sample averages:

$$E_\pi(f) = \lim_{m\to\infty} \frac{1}{m}\sum_{i=1}^{m} f(\mathbf{n}_i)$$

where $\mathbf{n}_i$ are random tables with stationary distribution π. The size of m for a good approximation is usually not clear. Theoretical results on the time to stationarity are hard to prove and hard to apply. Perfect sampling methods such as coupling-from-the-past that make unnecessary the analysis of convergence have not yet been found for most applications of conditional inference.

In some cases the Markov basis is very hard or impossible to compute completely. Odds-ratio models such as logistic regression are hard cases like this, and these are the examples we focus on here, beginning in §3. The constraints defining the polytope are Lawrence liftings, and the generators of the toric ideal are hard to enumerate and can have high total degree. We show that slightly larger sets of tables S_1, suggested by intuition, saturation, or primary decomposition, may be much easier to sample, and the set $S_0 \subset S_1$ can be studied by "conditioning." This idea is not new, but designing S_1 so it is not much larger than S_0 can use new technology from algebra. Our main new result is Theorem 3.1, which gives an efficient relaxation of the logistic regression problem that allows easy computations. The relaxation uses ideas of primary decomposition from work of Diaconis, Eisenbud, and Sturmfels (1998) and illustrates results of Hosten and Shapiro (2000).

In §4 we describe sequential importance sampling, with specific application to logistic regression. SIS has proven to be much more efficient than Markov chains for sampling from the uniform distribution from rectangular tables with fixed row and column sums. This has been shown in Chen, Diaconis *et al.* (2003) and in follow-up work. To run the Markov chain for this application, one increments the present table with a random increment of the form

$$\pm \begin{bmatrix} \cdots & + & \cdots & - & \cdots \\ \cdots & - & \cdots & + & \cdots \end{bmatrix}$$

and it can be shown that the resulting sequence of tables will visit all tables eventually. But the time to stationarity is long compared to the time required by SIS to go through the cells in sequence, sampling uniformly from an interval of possible values for each cell computed with up-to-date Fréchet bounds, and keeping track of weights to measure the variation from the uniform distribution. We make some informal connections between SIS and commutative algebra that could be further developed.

2. Notation and Examples

A family of positive probabilities $(\mu_\theta)_{\theta \in \mathbf{R}^p}$ on nonnegative integer vectors $\mathbf{n} \in Z_+^d$, whose entries sum to a known sample size n and may have other constraints, is given by

$$\mu_\theta(\mathbf{n}) = h(\mathbf{n}) \frac{e^{\theta' A_0 \mathbf{n}}}{z_\theta}$$

where A_0 is a $p \times d$ nonnegative integer matrix, z_θ is a normalizing constant, $\theta \in \mathbf{R}^p$ is a real parameter, and $h(\mathbf{n}) \geq 0$ may involve multinomial coefficients. The set of integers will be denoted Z, and the nonnegative integers in dimension d are denoted Z_+^d. The conditional distribution given statistics $A_0\mathbf{n} = \mathbf{t}$ is parameter free:

$$\mu_\theta(\mathbf{n} \mid A_0\mathbf{n} = \mathbf{t}) \propto h(\mathbf{n}),$$

defined on tables or lattice points $\mathbf{n} \in Z_+^d$ that satisfy $A_0\mathbf{n} = \mathbf{t}$ and that may also satisfy *a priori* constraints that combine into a single constraint matrix A, say

$$S_0 := \{\mathbf{n} \in Z_+^d : A\mathbf{n} = \begin{bmatrix} \mathbf{t} \\ \mathbf{s} \end{bmatrix}\}.$$

Our main goal is to sample from S_0 according to the conditional probability distribution proportional to $h(\mathbf{n})$, in order to compute expectations for tests of goodness-of-fit and parameter significance based on the theory of exact conditional inference.

A general method for constructing an irreducible chain was described in Diaconis and Sturmfels(1998). Suppose

$$G := \{\mathbf{x}^{\mathbf{a}_1} - \mathbf{x}^{\mathbf{b}_1}, \ldots, \mathbf{x}^{\mathbf{a}_g} - \mathbf{x}^{\mathbf{b}_g}\}$$

is a Gröbner basis of monomial differences for the toric ideal

$$I_A := \langle \mathbf{x}^{\mathbf{n}} - \mathbf{x}^{\mathbf{m}} : A\mathbf{n} = A\mathbf{m} \rangle$$

in $Q[\mathbf{x}]$. The vector increments represented by the differences of the exponents

$$M_G := \{\mathbf{a}_1 - \mathbf{b}_1, \ldots, \mathbf{a}_g - \mathbf{b}_g\}$$

chosen randomly with random signs will connect all points of the set S_0, eventually, so the process is an irreducible Markov chain. A generating set of binomials is sufficient for irreducibility (Diaconis and Sturmfels (1998), p. 375) , but anything less than a generating set could have two or more connected components within S_0, depending on the actual values of the constraints defining the polytope. A fundamental and useful result of Diaconis, Eisenbud, and Sturmfels (see Sturmfels (2002), p. 110) is that two tables $\mathbf{n}$ and $\mathbf{m}$ in the polytope S_0 will be connected by the Markov chain based on moves in some collection C whose corresponding ideal $I_C \subset I_A$ if the binomial $\mathbf{x}^{\mathbf{n}} - \mathbf{x}^{\mathbf{m}} \in I_C$. If the collection C is a Gröbner basis, the path between $\mathbf{n}$ and $\mathbf{m}$ can be constructed by long division. The theory of toric ideals, lattice bases, and the connection with Markov chains is explained in

Sturmfels (1996), and Diaconis and Sturmfels (1998). Some aspects of the theory are in Pistone, Riccomagno, and Wynn (2000). The algebra for Markov chains can be quite useful for understanding SIS– arguments in §4 will use notions of square-free lead terms and adjacent minors for establishing implementation details of SIS.

These Markov chains are similar to reflecting random walks, although tight corners of the polytope make a precise analogy difficult. Their convergence rates in simple cases can be estimated based on eigenvalue computations. The strongest results for random walks in lattice points are in Diaconis and Saloff-Coste (1998).

EXAMPLE 2.1 (Genotype Data). Genotype data is a table of counts of unordered pairs of alleles, like the following experimental data on Gaucher disease from a paper of Le Coutre *et al.* (1997). It is known that the genotype pair IVS2+1/IVS2+1 is lethal, and therefore constitutes a structural zero in the triangular table of genotypes (or possibly a missing entry, whose analysis is slightly different, but we will assume a structural zero). The table rows and columns normally are labelled with the allele names, but we omit these since they are not needed here.

0	5	2	1	0	0	10
	2	0	0	0	1	2
		0	0	0	0	0
			–	0	0	0
				0	0	1
					0	0
						1

The probability model for Hardy-Weinberg equilibrium postulates that the cell probabilities are the result of independent combination of alleles. For parameters $\mathbf{p} = (p_1, \ldots, p_7)$ that give the population proportions of each allele, the (unconditional) probability $\mu_{\mathbf{p}}$ on upper triangular tables $\mathbf{n} = (n_{ij})_{1 \le i \le j \le 7}$ with fixed total sum of n is given by the multinomial formula

$$\mu_{\mathbf{p}}(\mathbf{n}) = \binom{n}{(n_{ij})} p_1^{f_1} \cdot \ldots \cdot p_7^{f_7} \, 2^{\sum_{i<j} n_{ij}}$$

where f_i is the number of times allele i appears in the table. f_1 for example is $2 \times 0 + 5 + 2 + 1 + 0 + 0 + 10 = 18$, a sum over the genetic pairs that contain allele 1. These frequency counts are called sufficient statistics. The conditional probability distribution on tables with the same allele frequencies $\mathbf{f} = (f_1 = 18, f_2 = 12, \ldots, f_7 = 15)$, ignoring the structural zero, is given by

$$\mu_{\mathbf{f}}(\mathbf{n}) = \frac{\binom{n}{(n_{ij})} \, 2^{\sum_{i<j} n_{ij}}}{\binom{2n}{\mathbf{f}}},$$

where n is defined by $2n = f_1 + f_2 + \cdots + f_7$ and can be interpreted as the number of individuals in the sample, who contribute a total of $2n$ alleles of 7 types. This model cannot hold with the structural zero as in the data above, which complicates both the model and the analysis. We show how to modify the analysis to handle the structural zero.

The constraints that come from fixing the sufficient statistics for Hardy-Weinberg proportions are entry-wise dot products with 7 "row vectors" in A like

$$\begin{array}{ccccccc} 0 & 1 & 0 & 0 & 0 & 0 & 0 \\ & 2 & 1 & 1 & 1 & 1 & 1 \\ & & 0 & 0 & 0 & 0 & 0 \\ & & & - & 0 & 0 & 0 \\ & & & & 0 & 0 & 0 \\ & & & & & 0 & 0 \\ & & & & & & 0 \end{array}$$

which counts the number of alleles of type 2 in the table. We also force entry $(4,4)$ to be zero. With $\mathbf{f} = (f_1 = 18, f_2 = 12, \ldots, f_7 = 15)$ the allele counts, our polytope S_0 for the exact analysis with structural zero is the set of triangular tables that have these same allele counts and satisfy the additional linear constraint $n_{4,4} = 0$. The conditional probability distribution on S_0 is defined by $\mu_{\mathbf{f}}(\mathbf{n}) \propto \binom{n}{(n_{ij})} 2^{\sum_{i<j} n_{ij}}$. The Markov chain of Guo and Thompson (1992) for the original test of Hardy-Weinberg equilibrium can be described as the collection of moves that arise by folding the traditional $\begin{array}{cc} + & - \\ - & + \end{array}$ minors over the diagonal: $g(\mathbf{n})_{ij} = n_{ij} + n_{ji}$, $i < j$, $g(\mathbf{n})_{ii} = n_{ii}$:

$$x_{11}x_{34} - x_{14}x_{31} = \begin{array}{cccc} + & 0 & 0 & - \\ 0 & 0 & 0 & 0 \\ - & 0 & 0 & + \\ 0 & 0 & 0 & 0 \end{array} \quad \overset{g}{\mapsto} \quad \begin{array}{cccc} + & 0 & - & - \\ & 0 & 0 & 0 \\ & & 0 & + \\ & & & 0 \end{array} = x_{11}x_{34} - x_{13}x_{14}.$$

To handle the structural zero at entry $(4,4)$ we can use a Lawrence lifting to get a larger collection of Markov moves. The folded images under g of the well-known "circuit moves" that are the universal Gröbner basis for the independence model (fixed row and column sums) give an irreducible chain in the upper triangular tables with the given constraints, with arbitrarily placed zeros. These moves described differently appear in Takemura and Aoki (2002), but a proof of irreducibility can be made quite simple by using the algebraic description above.

Computationally, it is easiest for this example to just produce the binomials corresponding to the Markov basis from the constraint matrix by using a saturation algorithm, such as the one implemented in Cocoa, because there are fewer than 200 total moves. The calculation requires simply typing the constraint matrix A, with $49 = 7 \times 7$ columns and with 7 rows for the allele constraints, one row for forcing the lower triangle to be zeros, and one for forcing entry n_{44} to be zero:

```
Use R::=Q[x[1..7,1..7]];
Toric(A);
```

The work of De Loera, Haws *et al.* (2003) and the Latte software can be used to enumerate the elements of the polytope S_0. The value for statistics of enumeration is significant. It can be used to benchmark sequential importance sampling, which requires some fine tuning usually that can be done with an enumeration step. Enumeration can help understand convergence to stationarity of Markov chains, which depends partly on the number of points in the polytope as well as other geometric features such as diameter and shape. And enumeration can show the dependence of the size of the polytope on variations in the constraint matrix and constraint values. A typical problem of this type is feasibility, or whether the polytope S_0 is nonempty for a particular constraint vector $\mathbf{t}$. This is important for applications in SIS and disclosure limitation.

There is an efficient sequential method that fills in entries successively with conditional distributions, which can also handle one diagonal zero. This is described in Huber, Chen *et al.* (2003), and has the advantage that its complexity does not depend on the sizes of the table entries, but only the number of cells. Lazzeroni and Lange (1997) have extensions to multi-locus data and stopping times for exact sampling.

EXAMPLE 2.2 (Graphical Model). We illustrate with an example from the monograph of Sturmfels (2002) which describes many new results on graphical models. Consider the notation $X_1 - -X_2 - -X_3 - -X_4$, where X_i is a random variable that takes values in $\{0, 1\}$. The notation is meant to suggest that X_1 is independent of $\{X_3, X_4\}$ given X_2, and $\{X_1, X_2\}$ are independent of X_4 given X_3. The data would be a four-way table $\mathbf{n} = (n_{ijkl})$ with $2^4 = 16$ cells each containing a count. The notation $\mathbf{n}_{\{1,2\}}$ used by statisticians is the 2-way table that is the projection onto dimensions 1 and 2, with $\mathbf{n}_{\{1,2\}}(00) = \sum_{kl} \mathbf{n}_{00kl}$, called a margin.

The family of positive probabilities $\left(p_\theta(ijkl)\right)_{\theta \in \mathbf{R}^{12}} \subset \mathbf{R}^{16}$ on the space of 2^4 outcomes that satisfy these independence conditions is parameterized by an exponential map

$$p_\theta(ijkl) = \frac{e^{(\theta_{12}|\theta_{23}|\theta_{34})A_{(\,\cdot\,,ijkl)}}}{z_\theta}$$

where θ_{12} is a real vector of four components, z_θ is a normalizing constant, and A is the 12×16 matrix with columns indexed by $0000, 0001, 0010, \ldots, 1111$ and rows indexed by 2-way margin summaries for each of the 3 cliques $\{1,2\}, \{2,3\}, \{3,4\}$ written by statisticians as $\mathbf{n}_{\{1,2\}}, \mathbf{n}_{\{2,3\}}, \mathbf{n}_{\{3,4\}}$:

$$A = \begin{array}{r} \\ \mathbf{n}_{\{1,2\}}(00) \\ \mathbf{n}_{\{1,2\}}(01) \\ \mathbf{n}_{\{1,2\}}(10) \\ \mathbf{n}_{\{1,2\}}(11) \\ \mathbf{n}_{\{2,3\}}(00) \\ \mathbf{n}_{\{2,3\}}(01) \\ \mathbf{n}_{\{2,3\}}(10) \\ \mathbf{n}_{\{2,3\}}(11) \\ \mathbf{n}_{\{3,4\}}(00) \\ \mathbf{n}_{\{3,4\}}(01) \\ \mathbf{n}_{\{3,4\}}(10) \\ \mathbf{n}_{\{3,4\}}(11) \end{array} \begin{array}{cccccccccccccccc} 0000 & 0001 & 0010 & & & & & & & & & & & & & 1111 \\ 1 & 1 & 1 & 1 & 0 & 0 & 0 & 0 & 0 & 0 & 0 & 0 & 0 & 0 & 0 & 0 \\ 0 & 0 & 0 & 0 & 1 & 1 & 1 & 1 & 0 & 0 & 0 & 0 & 0 & 0 & 0 & 0 \\ 0 & 0 & 0 & 0 & 0 & 0 & 0 & 0 & 1 & 1 & 1 & 1 & 0 & 0 & 0 & 0 \\ 0 & 0 & 0 & 0 & 0 & 0 & 0 & 0 & 0 & 0 & 0 & 0 & 1 & 1 & 1 & 1 \\ 1 & 1 & 0 & 0 & 0 & 0 & 0 & 0 & 1 & 1 & 0 & 0 & 0 & 0 & 0 & 0 \\ 0 & 0 & 1 & 1 & 0 & 0 & 0 & 0 & 0 & 0 & 1 & 1 & 0 & 0 & 0 & 0 \\ 0 & 0 & 0 & 0 & 1 & 1 & 0 & 0 & 0 & 0 & 0 & 0 & 1 & 1 & 0 & 0 \\ 0 & 0 & 0 & 0 & 0 & 0 & 1 & 1 & 0 & 0 & 0 & 0 & 0 & 0 & 1 & 1 \\ 1 & 0 & 0 & 0 & 1 & 0 & 0 & 0 & 1 & 0 & 0 & 0 & 1 & 0 & 0 & 0 \\ 0 & 1 & 0 & 0 & 0 & 1 & 0 & 0 & 0 & 1 & 0 & 0 & 0 & 1 & 0 & 0 \\ 0 & 0 & 1 & 0 & 0 & 0 & 1 & 0 & 0 & 0 & 1 & 0 & 0 & 0 & 1 & 0 \\ 0 & 0 & 0 & 1 & 0 & 0 & 0 & 1 & 0 & 0 & 0 & 1 & 0 & 0 & 0 & 1 \end{array}$$

Each probability point in $\mathbf{R}^{16}$ in the parameterization lies in the toric variety of the toric ideal $I_A \subset Q[x_{0000}, \ldots, x_{1111}]$. The parameterization seems at first to have 12 parameters, but there are in fact only 7 free parameters which is the dimension of the projective variety. To sample from tables with fixed margins $\mathbf{n}_{\{1,2\}}, \mathbf{n}_{\{2,3\}}, \mathbf{n}_{\{3,4\}}$, one can run a uniform Markov chain with vector increments that are the differences of the exponents in a generating set of binomials for the toric ideal I_A. An important result of Sturmfels says one can obtain the toric ideal with 20 generators by saturating the smaller ideal of 12 "pairwise independence relations." These are obtained by fixing the values of variables $\{2, 4\}$ in 4 ways for

specifying conditional independence of variables $\{1,3\}$, then considering variables $\{1,4\}$ fixing $\{2,3\}$ and finally considering variables $\{2,4\}$ fixing $\{1,3\}$. Then $I_A = IP : p^\infty$ where $IP =$
$\langle x_{0000}\,x_{1010} - x_{0010}\,x_{1000}, x_{0001}\,x_{1011} - x_{0011}\,x_{1001}, x_{0100}\,x_{1110} - x_{0110}\,x_{1100},$
$x_{0101}\,x_{1111} - x_{1101}\,x_{0111}, x_{0000}\,x_{1001} - x_{0001}\,x_{1000}, x_{0010}\,x_{1011} - x_{1010}\,x_{0011},$
$x_{0100}\,x_{1101} - x_{0101}\,x_{1100}, x_{0110}\,x_{1111} - x_{0111}\,x_{1110}, x_{0000}\,x_{0101} - x_{0001}\,x_{0100},$
$x_{0010}\,x_{0111} - x_{0011}\,x_{0110}, x_{1000}\,x_{1101} - x_{1001}\,x_{1100}, x_{1010}\,x_{1111} - x_{1011}\,x_{1110}\rangle$
and $p = \prod_{ijkl} x_{ijkl}$.

An efficient way to build the Markov chain for a large class of graphical models is described in Dobra and Sullivant (2003). Applications of the Markov chains for graphical and more general log-linear models are in conditional statistical inference, and in the developing area of disclosure limitation (Duncan, Fienberg *et al.* (2001)). This area will provide some computational challenges since some applications involve tables of more than ten factors, resulting in at least 2^{10} indeterminates in the polynomial ring for simple two-level factors.

3. Logistic Regression

Hosmer and Lemeshow (1989, p. 3) present data that relates presence or absence of coronary heart disease of 100 patients to age. The age covariate extends from year 20 to year 69. The data can be summarized in a table that looks like

Age:	1	2	3	4	......	c
yes:	$n_{1,1}$	$n_{1,2}$	$n_{1,3}$	$n_{1,4}$	......	$n_{1,c}$
no:	$n_{2,1}$	$n_{2,2}$	$n_{2,3}$	$n_{2,4}$	......	$n_{2,c}$
	$n_{+,1}$	$n_{+,2}$	$n_{+,3}$	$n_{+,4}$	......	$n_{+,c}$

The data of Hosmer and Lemeshow has $c = 50$ columns, one for each age level, some of which have 0 counts in both the rows, meaning no one in the study had that age level. A simple statistical model for evaluating the effect of age on the presence of coronary heart disease is the logistic regression model, which specifies that $\mu_{\alpha,\beta}(\mathbf{n}) \propto e^{(\alpha,\beta)\cdot A_0\cdot(n_{1,\cdot})}$ where $A_0 = \begin{pmatrix} 1 & 1 & \dots & 1 \\ 1 & 2 & \dots & c \end{pmatrix}$ and $(n_{1,\cdot})$ is the top row of data as a column vector. The constraints for conditional inference are: the total number of successes $T_1(\mathbf{n}) := n_{1,+}$ (top row sum) must be fixed at integer $t_1 \geq 0$; and the weighted sum $T_2(\mathbf{n}) := (1,2,\dots,c)\cdot(n_{1,\cdot}) = \sum_{i=1}^{c} i\cdot n_{1,i}$ must be fixed at integer $t_2 \geq 0$; and the column sums $n_{+,i}$ must be fixed at integers $c_i \geq 0$, which comes either from a design constraint on the number of subjects at each age, or a conditioning constraint in an odds-ratio model. With the data ordered $\mathbf{n} = (n_{11},\dots,n_{1,c},n_{2,1},\dots,n_{2,c})$ these can be built into a single constraint matrix

$$(1) \qquad A = \begin{pmatrix} A_0 & \mathbf{0} \\ I_{c\times c} & I_{c\times c} \end{pmatrix}.$$

The Markov chains for computing in the set S_0 of nonnegative tables with constraints from A have been studied in Diaconis, Graham, and Sturmfels (1996). Their conclusion is that an irreducible chain in the collection of tables with arbitrary fixed column sums (possibly zero), and arbitrary fixed $n_{1,+}, \sum_{i=1}^{c} i\, n_{1,i}$ consists of vector increments that correspond to homogeneous primitive partition identities (hppi's), such as $2+2 = 1+3$. Computing these moves is difficult because their number grows in c faster than any polynomial, and their degree (the total number of +'s in the vector increment) grows linearly in c. The number of moves corresponding to hppi's as a function of $c = 3, 4, 5, \ldots$ is $1, 5, 16, 51, 127, 340, 798, \ldots$ and this number has only been computed up to $c = 20$ using software 4ti2 of Ralf and Raymond Hemmecke. This will not help with a data set of 50 columns. The network method of Mehta, Patel, and Senchaudhuri (2000) will handle this data set, but the memory requirements are large compared to the Markov chain described below, and it is likely that some larger, more complex data sets may not be possible with the network method.

Consider the following collection M of $\binom{c-1}{2}$ vector increments, which together with their negative images form moves for a Markov chain in tables (nonnegative integer lattice points) with the above constraints.

- Choose an ordered pair of columns $1 \leq i < j \leq c$.
- Put a $+$ in the top row in columns i, j and put a $-$ in adjacent columns $i+1$ and $j-1$.
- Put the opposite signs in the bottom row.

With $c = 6$, there are 10 such moves like

+	- -	+	0	0	0
-	+ +	-	0	0	0

,

0	+	-	0	-	+
0	-	+	0	+	-

,

etc. These moves can be seen as differences of adjacent minors, and also as a subset of the partition identities. The first table above corresponds to the hppi $1+3 = 2+2$, and the second corresponds to $2+6 = 3+5$. The collection M is generally a strict subset of hppi's of degree four, because the adjacent $+, -$ changes do not include three degree 4 hppi's ($1+6 = 3+4, 2+6 = 4+4, 1+5 = 3+3$). The collection M does not include thirty-eight other hppi's of higher degree.

Let I_M denote the ideal in $Q[x_1, \ldots, x_c, y_1, \ldots, y_c]$ generated by the monomial differences corresponding to the moves described above: $I_M = \langle x_1 x_3 y_2^2 - x_2^2 y_1 y_3, \ldots \rangle$.

These vectors M are a lattice basis for the kernel of the constraint matrix A, so the ideal I_M saturates to the toric ideal I_A. Also, note that if we leave off the bottom row corresponding to the y-variables and work in $Q[x_1, \ldots, x_c]$, then the collection of moves on the top row corresponds in fact to a Gröbner basis for lex order for the toric ideal corresponding to $A_0 = \begin{pmatrix} 1 & 1 & 1 & \ldots & 1 \\ 1 & 2 & 3 & \ldots & c \end{pmatrix}$ with square-free exponent on the lead indeterminate.

PROPOSITION 3.1. Let $S_0 = \{\mathbf{n} \in Z_+^d : A\mathbf{n} = (t_1, t_2, \mathbf{c})'\}$ be the set of nonnegative integer tables with fixed row sum $n_{1,+}$, fixed weighted sum $\sum i \, n_{1,i}$ and fixed column sums $n_{+,i}$. Let $S_2 = \{\mathbf{z} \in Z^d : A\mathbf{z} = (t_1, t_2, \mathbf{c})', z_{1,i} \geq 0, z_{2,i} \geq -1\}$. Then the Markov chain with moves M connects any two tables in S_0 through S_2.

The result is a corollary of Theorem 3.1 below, so a careful proof will not be included. Although it appears to be very similar to the Theorem, its computational value is less.

In the data set of Hosmer and Lemeshow with 50 columns the set S_2 is about 10^6 times as big as S_0, so calculations are possible by conditioning on S_0. The goodness-of-fit calculation required several hours of time on a 48-node Linux cluster.

Proposition 3.1 also follows from a saturation property of the ideal I_M. Since

$$I_M : (y_1 y_2 \cdots y_c) = I_A,$$

it follows that any two tables $\begin{array}{|c|}\hline \mathbf{m}_1 \\ \hline \mathbf{m}_2 \\ \hline \end{array}$ and $\begin{array}{|c|}\hline \mathbf{n}_1 \\ \hline \mathbf{n}_2 \\ \hline \end{array}$ in S_0 satisfy $(y_1 \cdot y_2 \cdot \ldots \cdot y_c)(\mathbf{x}^{\mathbf{n}_1}\mathbf{y}^{\mathbf{n}_2} - \mathbf{x}^{\mathbf{m}_1}\mathbf{y}^{\mathbf{m}_2}) \in I_M$ implying by results of Sturmfels (1996) that $\begin{array}{|c|}\hline \mathbf{m}_1 \\ \hline \mathbf{m}_2 + \mathbf{1} \\ \hline \end{array}$ (the two tables with 1 added to each cell in the bottom row) can be connected through nonnegative integer points to $\begin{array}{|c|}\hline \mathbf{n}_1 \\ \hline \mathbf{n}_2 + \mathbf{1} \\ \hline \end{array}$. By comparison, the ideal for the Markov moves (that are also a lattice basis) of Bigatti *et al.* (1999) saturate slowly (with 5 columns, saturation occurs at step six).

A more careful look at the Markov chain with moves M above shows that it is irreducible even when only the bottom row entries in columns whose sums are fixed at 0 are allowed to drop down to -1, which is a significant computational advantage. If S_1 is this polytope that contains S_0 for the data set with 50 columns, we computed numerically by comparing time averages that $|S_1| \approx 10^3 |S_0|$, which is a factor of 10^3 better than $|S_2|/|S_0|$.

THEOREM 3.1. Let

$$S_1 := \{\mathbf{z} \in Z^d : A\mathbf{z} = (t_1, t_2, \mathbf{c})', z_{1,i} \geq 0, z_{2,i} \geq -1 \texttt{ if } c_i = 0, z_{2,i} \geq 0 \texttt{ if } c_i \geq 1\}$$

be the set of integer tables with the desired constraints, but allowing bottom row entries to be -1 in columns i where $c_i = 0$. Then the Markov chain with moves M connects any two tables in S_0 through S_1. It is irreducible in S_0 if all column sums c_i are positive.

PROOF. Assume for the moment the second assertion of irreducibility in S_0 if the column sums are positive. If this were true, and we wanted to connect nonnegative integer vectors $\mathbf{n}$ and $\mathbf{m}$ in S_1, then we could connect $\mathbf{n} + \begin{array}{|c|}\hline \mathbf{0} \\ \hline I_{\{i:c_i=0\}} \\ \hline \end{array}$ to $\mathbf{m} + \begin{array}{|c|}\hline \mathbf{0} \\ \hline I_{\{i:c_i=0\}} \\ \hline \end{array}$ as points in S_0 with constraint values $t_1, t_2, \mathbf{c} + I_{\{i:c_i=0\}}$. Then the path from $\mathbf{n}$ to $\mathbf{m}$ is obtained by subtracting $\begin{array}{|c|}\hline \mathbf{0} \\ \hline I_{\{i:c_i=0\}} \\ \hline \end{array}$ from each intermediate table in the connecting path. Thus the first assertion about irreducibility in S_1 follows from the second.

The second assertion is proved by showing that the L^1 distance between two tables can always be reduced by using one of the moves. One can show that a path between two tables is possible with length at most $c^2 \cdot t_1$.

There is an algorithm for constructing the connecting path, assuming $c_i \geq 1$. The moves of adjacent minors $\{x_i y_{i+1} - x_{i+1} y_i\}$ connect all $2 \times c$ tables with positive column sums and the same row sums. Further, they are a Gröbner basis for their ideal in two term orders: lex order, reading left to right, and weighted term order with weight vector $w = (1, 2, \ldots, c, 0, 0, \ldots, 0)$, with lex for ties. Consider two tables coded as a binomial with $\mathbf{x}^{\mathbf{n}_1}\mathbf{y}^{\mathbf{n}_2} - \mathbf{x}^{\mathbf{m}_1}\mathbf{y}^{\mathbf{m}_2}$ The lead term of one of the adjacent minors divides its lead term, in lex order. Do the division and save the adjacent minor. Now, divide the lead term of the intermediate dividend in weighted term order, to lower the weight that just increased by 1 using also an adjacent minor with a "−" sign. The two divisions yield a pair of adjacent minors that leave the weights fixed. The division will terminate by connecting the two tables. As an example, consider connecting the two tables

1	2	0	0	2
0	1	3	1	0

,

0	1	3	1	0
1	1	0	0	2

.

The procedure is a sequence of pairs of divisions that proceeds as below, with the divisors on the right:

```
0)lead  12002   lex 11102  weight   ;  0-+00,  000+-
        00310    -> 01210    ->        0+-00   000-+

1)lead  11111   lex 02111   weight  ;  -+000,  00+-0
        01201    -> 10201     ->       +-000   00-+0

2)lead  02201   lex 01301   weight  ;  0-+00,  000+-
        10111    -> 11011     ->       0+-00   000-+

3)end   01310
        11002
```

□

The result above can be seen in the primary decomposition of I_M as presented (in edited form) by Singular (Greuel *et al.* (2001)) below for five columns. At this time there do not appear to be general theorems that would give a proof based on the primary decomposition, but for five columns it is clear. The primary decomposition has four components, the toric part with 16 generators corresponding to the hppi's, and three others. Recall the basic result of Diaconis, Eisenbud, and Sturmfels: if the binomial $\mathbf{x}^{\mathbf{n}_1}\mathbf{y}^{\mathbf{n}_2} - \mathbf{x}^{\mathbf{m}_1}\mathbf{y}^{\mathbf{m}_2}$ corresponding to the two tables $\boxed{\begin{matrix}\mathbf{n}_1\\ \mathbf{n}_2\end{matrix}}$, $\boxed{\begin{matrix}\mathbf{m}_1\\ \mathbf{m}_2\end{matrix}}$ belongs to the ideal I_M generated by the moves M, then the moves in M connect the two tables. If the column sum c_i is positive for all i, then either x_i or y_i is present in each monomial, which implies membership in each of the primary components, and hence membership of the binomial in the ideal I_M.

```
i;
i[1]=x(1)*x(4)*y(2)*y(3)-x(2)*x(3)*y(1)*y(4)
i[2]=x(1)*x(5)*y(2)*y(4)-x(2)*x(4)*y(1)*y(5)
i[3]=x(2)*x(5)*y(3)*y(4)-x(3)*x(4)*y(2)*y(5)
i[4]=x(1)*x(3)*y(2)^2-x(2)^2*y(1)*y(3)
i[5]=x(2)*x(4)*y(3)^2-x(3)^2*y(2)*y(4)
i[6]=x(3)*x(5)*y(4)^2-x(4)^2*y(3)*y(5)
> primarydecomposition;
[1]:
      _[1]=x(3)*x(5)*y(4)^2-x(4)^2*y(3)*y(5)
      _[2]=x(2)*x(5)*y(3)*y(4)-x(3)*x(4)*y(2)*y(5)
      _[3]=x(2)*x(5)^2*y(4)^3-x(4)^3*y(2)*y(5)^2
      _[4]=x(2)*x(4)*y(3)^2-x(3)^2*y(2)*y(4)
      _[5]=x(2)^2*x(5)*y(3)^3-x(3)^3*y(2)^2*y(5)
      _[6]=x(1)*x(5)*y(3)^2-x(3)^2*y(1)*y(5)
      _[7]=x(1)*x(5)*y(2)*y(4)-x(2)*x(4)*y(1)*y(5)
      _[8]=x(1)*x(5)^2*y(3)*y(4)^2-x(3)*x(4)^2*y(1)*y(5)^2
      _[9]=x(1)*x(5)^3*y(4)^4-x(4)^4*y(1)*y(5)^3
      _[10]=x(1)*x(4)*y(2)*y(3)-x(2)*x(3)*y(1)*y(4)
      _[11]=x(1)*x(4)^2*y(3)^3-x(3)^3*y(1)*y(4)^2
      _[12]=-x(2)^2*x(5)*y(1)*y(4)^2+x(1)*x(4)^2*y(2)^2*y(5)
      _[13]=x(1)*x(3)*y(2)^2-x(2)^2*y(1)*y(3)
      _[14]=x(1)^2*x(5)*y(2)^2*y(3)-x(2)^2*x(3)*y(1)^2*y(5)
      _[15]=x(1)^2*x(4)*y(2)^3-x(2)^3*y(1)^2*y(4)
      _[16]=x(1)^3*x(5)*y(2)^4-x(2)^4*y(1)^3*y(5)
[2]:
      _[1]=y(3)
      _[2]=x(3)
      _[3]=x(1)*x(5)*y(2)*y(4)-x(2)*x(4)*y(1)*y(5)
```

```
[3]:
      _[1]=y(4)
      _[2]=x(4)
      _[3]=x(1)*x(3)*y(2)^2-x(2)^2*y(1)*y(3)
[4]:
      _[1]=y(2)
      _[2]=x(3)*x(5)*y(4)^2-x(4)^2*y(3)*y(5)
      _[3]=x(2)
> quit;
```

Logistic regression is a special type of odds-ratio model, a challenging class of models for which the constraint matrices are the higher Lawrence liftings of Santos and Sturmfels (2002).

4. Sequential Importance Sampling

Consider the three 2×3 tables below with the same row and column sums. It is clear that the marginal distribution of the count in cell $(1,1)$ is not uniform when the tables are equally likely.

$$\begin{array}{|c|c|c|}\hline 0 & 0 & 1 \\ \hline 1 & 1 & 0 \\ \hline \end{array}, \quad \begin{array}{|c|c|c|}\hline 0 & 1 & 0 \\ \hline 1 & 0 & 1 \\ \hline \end{array}, \quad \begin{array}{|c|c|c|}\hline 1 & 0 & 0 \\ \hline 0 & 1 & 1 \\ \hline \end{array}.$$

One can generate tables randomly from a known distribution as follows. First, recall the well-known Fréchet bounds on entries: $\max\{0, r_i + c_j - n\} \leq n_{ij} \leq \min\{r_i, c_j\}$ where the row sums are r_i, the column sums are c_j and n is the total count over all cells. In particular $n_{11} \in \{0, 1\}$. Choose uniformly from this interval to get a value for n_{11}, say 0, to get a partial table $\begin{array}{|c|c|c|}\hline 0 & & \\ \hline 1 & & \\ \hline \end{array}$ chosen with probability $\frac{1}{2}$. Now the bounds applied to the remaining part of the table for entry $(1, 2)$ are $n_{2,1} \in [1 + 1 - 2, \min\{1, 1\}]$, so 1 can be chosen with probability $\frac{1}{2}$ to get a further partial table $\begin{array}{|c|c|c|}\hline 0 & 1 & \\ \hline 1 & 0 & \\ \hline \end{array}$. The final column is then determined. The procedure generates the three tables above with probabilities $q_1 = \frac{1}{4}, q_2 = \frac{1}{4}, q_3 = \frac{1}{2}$ respectively, and sample averages can be reweighted with the reciprocals $1/q_i$ for approximate expectations with respect to the uniform distribution. A description of SIS can be found in Liu (2001) with many applications.

Consider now the problem of sampling from the $2 \times c$ tables for logistic regression introduced in §3. Recall the set of tables

$$S_0 := \{\mathbf{n} = (n_{11}, n_{12}, \ldots, n_{1c}, n_{21}, \ldots, n_{2,c}) : A\mathbf{n} = (t_1, t_2, \mathbf{c})'\}$$

where A is the matrix of equation (1) of §3. SIS attempts to 1) choose n_{11} for the first cell in a range of values, say an interval $[l_1, u_1]$, that allows ultimate completion to a table in S_0, then 2) choose $n_{1,2}$ from a new updated interval $[l_2(n_{11}), u_2(n_{11})]$ that allows ultimate completion, etc. It is useful if these intervals can be accurately computed during the sampling, and it is better if the range of values are intervals without gaps. That is, at each stage it is useful if the projection onto the next dimension of the polytope, after having fixed the first coordinate values, is an unbroken interval.

PROPOSITION 4.1. *If the column sums c_i are constrained to be positive, then every integer in a subinterval of the Fréchet bounds for n_{11} can yield a valid table in S_0.*

PROOF. With positive column sums, all tables in S_0 are connected with the moves in M, which correspond to a lex Gröbner basis with square-free exponent on first indeterminate. Since entries $n_{1,1}$ for cell $(1,1)$ are incremented by 1 in a connecting path between tables, the set of feasible values for cell $(1,1)$ is an interval of integers with no gaps. □

By comparison, consider the two tables $\begin{array}{|c|c|c|c|}\hline 0 & 3 & 0 & 0 \\ \hline 2 & 0 & 0 & 1 \\ \hline \end{array}$, $\begin{array}{|c|c|c|c|}\hline 2 & 0 & 0 & 1 \\ \hline 0 & 3 & 0 & 0 \\ \hline \end{array}$. There are no tables with the same constraint values and the value 1 in cell (1,1), but these two tables are not in the same connected component of the Markov chain built with the moves in M. The simplest example of this connectivity issue has $A := \begin{pmatrix} 1 & 3 \end{pmatrix}$. The set $S_0 = \{(n_1, n_2) : A\mathbf{n} = 3\}$ includes points $(0,1), (3,0)$ so the first coordinate is not in a complete interval. Here the Gröbner basis for I_A is $\{x^3 - y\}$ and does not have the square-free property . When structural zeros are imposed on tables, the sequential-interval property may be lost, because the Markov chain for irreducibility may no longer have the square-free property, which occurs for example in the genotype data of §2.

For logistic regression there is a recurring structure that reduces the problem of determining sampling intervals $[l_i, u_i]$ to the case of $[l_1, u_1]$, the first column. Recall that we need to fix the row sum $T_1(\mathbf{n}) := n_{1,+} = t_1$, and the weighted row sum $T_2(\mathbf{n}) := \sum_i i \cdot n_{1,i} = t_2$, and the column sums $n_{+,i} = c_i$. The following diagram illustrates how the problem of interval values recurs with different constraints after the first column is filled:

$$\begin{array}{|c|c|c|c|c|c|}\hline [l_1, u_1] & n_{1,2} & n_{1,3} & n_{1,4} & \cdots & n_{1,c} \\ \hline c_1 - [l_1, u_1] & n_{2,2} & n_{2,3} & n_{2,4} & \cdots & n_{2,c} \\ \hline \end{array} \; : \; \begin{array}{l} T_1(\mathbf{n}) = t_1 \\ T_2(\mathbf{n}) = t_2 \end{array}$$

$$\begin{array}{|c|}\hline x \\ \hline c_1 - x \\ \hline \end{array} \; \begin{array}{|c|c|c|c|c|}\hline [l_2(x), u_2(x)] & n_{1,3} & n_{1,4} & \cdots & n_{1,c} \\ \hline c_2 - [l_2(x), u_2(x)] & n_{2,3} & n_{2,4} & \cdots & n_{2,c} \\ \hline \end{array} \; : \; \begin{array}{l} T_1(\mathbf{n}_{(\cdot,2:c)}) = t_1 - x \\ T_2(\mathbf{n}_{(\cdot,2:c)}) = t_2 - t_1 \end{array}.$$

Computing the interval $[l_1, u_1]$ is made more efficient with the following feasibility test.

PROPOSITION 4.2. Assume $c_i \geq 1$. Let $n = \sum_{i=1}^{c} c_i$ be the table sum. For each $x \in [0, c_1]$ define tables $\mathbf{n}_L(x), \mathbf{n}_U(x)$ by

$$\mathbf{n}_L(x) := \begin{array}{|c|c|c|c|}\hline x & \min\{c_2, r_1 - x\} & \min\{c_3, r_1 - x - n_{12}\} & \cdots \\ \hline c_1 - x & c_2 - n_{1,2} & c_3 - n_{1,3} & \cdots \\ \hline \end{array}$$

$$\mathbf{n}_U(x) := \begin{array}{|c|c|c|c|}\hline x & \max\{c_2 + r_1 - x - n, 0\} & \max\{c_3 + r_1 - x - n_{12} - n, 0\} & \cdots \\ \hline c_1 - x & c_2 - n_{1,2} & c_3 - n_{1,3} & \cdots \\ \hline \end{array}$$

Then a nonnegative integer $x \in [l_1, u_1]$ if and only if $T_2(\mathbf{n}_L(x)) \leq t_2 \leq T_2(\mathbf{n}_U(x))$.

PROOF. The tables $\mathbf{n}_L(x), \mathbf{n}_U(x)$ minimize and maximize the value of T_2 over the collection of $2 \times c$ tables with top row sum t_1 and column sums c_i. $\mathbf{n}_L(x)$ puts the largest values possible to the left in the upper row, consistent with the Fréchet bounds for rectangular tables. Similarly, $\mathbf{n}_U(x)$ puts the smallest values possible to the left, and hence the largest on the right, consistent with the Fréchet bounds. Then it is clear that both inequalities must hold if a value x is consistent with

$T_2 = t_2$. Conversely, if both inequalities hold, then there are two tables $\mathbf{m}$ and $\mathbf{n}$ with the right row and column sums, but $T_2(\mathbf{m}) \leq t_2, T_2(\mathbf{n}) \geq t_2$. With column sums positive the table $\mathbf{m}$ can be connected to $\mathbf{n}$ with a sequence of adjacent minor moves (Sturmfels (2002), p. 64) $\begin{matrix} \cdots & - & + & \cdots \\ \cdots & + & - & \cdots \end{matrix}$, each of which changes the value of T_2 by ± 1. Therefore the intermediate value theorem proves that some table will have the right value $T_2 = t_2$. □

EXAMPLE 4.1 (Cancer data from Sugiura and Otake (1974)). Binary data on death from leukemia is classified by dose at 6-levels.

5	4	6	1	3	6
5973	11811	2620	771	792	820

It took only a few seconds to obtain 10^6 samples with SIS, which gave an estimate of 3053 with standard error 27.8 for the total number of tables with the same constraints. Based on these samples, the exact p-value for goodness-of-fit can be estimated at 0.0875 with standard error 0.005.

When some column sums are zero, SIS can still work without the sequential-interval property. The results for the 50 column cancer data of Hosmer and Lemeshow agreed closely with the Markov chain analysis.

For decomposable graphical models, it is known that there is a Markov basis with square free initial terms (Dobra and Sullivant (2003)) , and there are sharp bounds analogous to the Fréchet bounds (Dobra and Fienberg (2000)) so some of the elements for SIS are in place and a general theory may be possible. The use of algebra in SIS is relatively undeveloped compared to its use in designing Markov chains.

Acknowledgements. We thank Chong Tu for computing the p-value for Example 4.1 with sequential importance sampling, Seth Sullivant and Raymond Hemmecke for comments and discussions on this paper, and referees for valuable comments and suggestions.

References

[1] Agresti, A. (1990). *Categorical Data Analysis.* Wiley, New York.

[2] Aoki, S. (2001). Network algorithm for the exact test of Hardy-Weinberg proportion for multiple alleles. METR Technical Report 01-06. http://www.stat.t.u-tokyo.ac.jp/~aoki/research.html.

[3] Bigatti, A. M., La Scala, R., and Robbiano, L. (1999). Computing toric ideals. *Journal of Symbolic Computation*, **27**, 351-365.

[4] Capani, A., Niesi, G., and Robbiano, L. (2002). *Cocoa: a system for doing computations in commutative algebra.* ftp://cocoa.dima.unige.it.

[5] Chen, Y., Diaconis, P., Holmes, S. P., and Liu, J. S. (2003). Sequential Monte Carlo methods for statistical analysis of tables. *Journal of the American Statistical Association*, in press.

[6] De Loera, J. A., Haws, D., Hemmecke, R., Huggins, P., Sturmfels, B., and Yoshida, R. (2003). Short rational functions for toric algebra and applications. Manuscript.

[7] Diaconis, P., Eisenbud, D., and Sturmfels, B. (1998). *Lattice walks and primary decomposition.* In *Mathematical Essays in Honor of Gian-Carlo Rota,* eds. B. Sagan and R. Stanley. Birkhauser, Boston, pp. 173-193.

[8] Diaconis, P., Graham, R. L., and Sturmfels, B. (1996). Primitive partition identities. *Combinatorics: Paul Erdös is Eighty, Vol. 2*, 173-192.

[9] Diaconis, P., and Saloff-Coste, L. (1998). Nash inequalities for finite Markov chains. *Journal of Theoretical Probability*, **9**, 459-510.

[10] Diaconis, P., and Sturmfels, B. (1998). Algebraic algorithms for sampling from conditional distributions. *Annals of Statistics*, **26**, 363-397.

[11] Dobra, A., and Fienberg, S.E. (2000). Bounds for cell entries in contingency tables given marginal totals and decomposable graphs. *Proceedings of the NAS*, **97**, 11885-11892.

[12] Dobra, A., and Sullivant, S. (2003). A divide-and-conquer algorithm for generating Markov bases of multi-way tables. *Computational Statistics*, to appear August 2004.

[13] Duncan, G. T., Fienberg, S. E., Krishnan, R., Padman, R., and Roehrig, S. F. (2001). Disclosure limitation methods and information loss for tabular data. In *Confidentiality, Disclosure and Data Access: Theory and Practical Applications for Statistical Agencies,* eds. P Doyle, J. Lane, J. Theeuwes, L. Zayatz. Elsevier, New York, pp. 135-166.

[14] Greuel, G.-M., Pfister, G., and Schönemann, H. (2001). SINGULAR 2.0. A Computer Algebra System for Polynomial Computations. Centre for Computer Algebra, University of Kaiserslautern. http://www.singular.uni-kl.de.

[15] Guo, S. W., and Thompson, E. (1992). Performing the exact test of Hardy-Weinberg proportion for multiple alleles. *Biometrics*, **48**, 361-372.

[16] Hemmecke, R., and Hemmecke, R. (2003). *4ti2 Version 1.1 – Computation of Hilbert bases, Graver bases, toric Gröbner bases, and more.* http://www.4ti2.de.

[17] Hosmer, D. W., and Lemeshow, S. (1989). *Applied Logistic Regression.* Wiley, New York.

[18] Hosten, S., and Shapiro, J. (2000). Primary decomposition of lattice basis ideals. *Journal of Symbolic Computation*, **29**, 625-639.

[19] Huber, M., Chen, Y., Dobra, A., Dinwoodie, I. H., and Nicholas, M. (2003). Monte Carlo algorithms for Hardy-Weinberg proportions. Discussion Paper 03-09, Institute of Statistics and Decision Sciences, Duke University.

[20] Lazzeroni, L. C., and Lange, K. (1997). Markov chains for Monte Carlo tests of genetic equilibrium in multidimensional contingency tables. *Annals of Statistics*, **25**, 138-168.

[21] Le Coutre, P., Demina, A., Beutler, E., Beck, M., and Petrides, P. E. (1997). Molecular analysis of Gaucher disease: distribution of eight mutations and the complete gene deletion in 27 patients from Germany. *Human Genetics*, **99**, 816-821.

[22] Liu, J. S. (2001). *Monte Carlo Strategies in Scientific Computing.* Springer, New York.

[23] Mehta, C., Patel, N. R., and Senchaudhuri, P. (2000). Efficient Monte Carlo methods for conditional logistic legression. *Journal of the American Statistical Association,* **95**, 99-108.

[24] Pistone, G., Riccomagno, and Wynn, H. P. (2000). *Algebraic Statistics: computational commutative algebra in statistics.* Chapman and Hall, New York.

[25] Santos, F., and Sturmfels, B. (2002). Higher Lawrence liftings. Manuscript.

[26] Sturmfels, B. (1996). *Gröbner Bases and Convex Polytopes.* AMS, Providence Rhode Island.

[27] Sturmfels, B. (2002). *Solving Systems of Polynomial Equations.* AMS, Providence Rhode Island.

[28] Sugiura, N., and Otake, M. (1974). An extension of the Mantel-Haenszel procedure to K $2 \times c$ contingency tables and the relation to the logit model. *Communications in Statistics*, **A 3**, 829-842.

[29] Takemura, A., and Aoki, S. (2002). Some characterizations of minimal Markov basis for sampling from discrete conditional distributions. METR Technical Report 02-04, http://www.stat.t.u-tokyo.ac.jp/~aoki/research.html.

ISDS, 214 OLD CHEMISTRY BUILDING, BOX 90251, DUKE UNIVERSITY, DURHAM, NC 27708

Contemporary Mathematics
Volume **374**, 2005

Kostka Numbers and Littlewood-Richardson Coefficients

Charles Cochet

ABSTRACT. We apply some recent developments of Baldoni-Vergne [**BV01**] and Baldoni-DeLoera-Vergne [**BdLV03**] on vector partition function, to Kostant's and Steinberg's formulas, in the case of A_r. We therefore get a fast MAPLE program that computes for A_r: the multiplicity c_λ^μ of the weight μ in the representation $V(\lambda)$ of highest weight λ; the multiplicity $c_{\lambda\ \mu}^{\ \nu}$ of the representation $V(\nu)$ in $V(\lambda)\otimes V(\mu)$. The computation also gives the piecewise polynomial functions c_λ^μ and $c_{\lambda\ \mu}^{\ \nu}$.

1. Introduction

In this note, we are interested in the two following computational problems in the case of A_r:

- Kostka numbers, that is the multiplicity c_λ^μ of the weight μ in the representation $V(\lambda)$ of highest weight λ.
- Littlewood-Richardson coefficients, that is the multiplicity $c_{\lambda\ \mu}^{\ \nu}$ of the representation $V(\nu)$ in the tensor product of representations of highest weights λ and μ.

Recall that A_r is the Lie algebra $\mathfrak{sl}_{r+1}(\mathbb{C})$ of traceless matrices. Its positive root system is the set A_r^+ of vectors $(e_i - e_j)$ $(1 \leq i < j \leq r+1)$, where $(e_i)_i$ is the canonical basis of $\mathbb{R}^{r+1}$.

Softwares LiE (from van Leeuwen *et al.*, see [**vL94**]) and GAP (see [**GAP**]), and MAPLE packages `coxeter/weyl` (from Stembridge, see [**Ste95**]) use Freudenthal's and Klymik's formulas, and work for any semi-simple Lie algebra $\mathfrak{g}$ (and not only $\mathfrak{sl}_{r+1}(\mathbb{C})$). Unfortunately, these formulas are really sensitive to the size of coefficients of weights. Moreover these formulas cannot handle for A_r the computation of associated *Ehrhart polynomials* $t \mapsto c_{t\lambda}^{t\mu}$ and $t \mapsto c_{t\lambda\ t\mu}^{\ t\nu}$.

Here, the approach to these two problems is through vector partition function, namely number of integral points in lattice polytopes. The vector partition function arises in many areas of mathematics: representation theory, flows in networks, magic squares, statistics, crystal bases of quantum groups. Its complexity

2000 *Mathematics Subject Classification.* Primary 22E46, 52B55.

Key words and phrases. Representation theory, Kostka numbers, Littlewood-Richardson coefficients, convex polytopes.

is polynomial in the size of input when the dimension of the polytope is fixed, and NP-hard if it can vary ([**B94**], [**B97**] and [**BP99**]).

There are several approaches to this problem, such as Barvinok's decomposition algorithm ([**B94**]), recently implemented by the LattE team ([**DHTY03**]). In this note, we use recent results of Baldoni-Vergne ([**BV01**]) and Baldoni-DeLoera-Vergne ([**BdLV03**]), *via* inverse Laplace formula and involving iterated residues of rational functions, to get a fast algorithm for vector partition function.

We combine the resulting procedure with Kostant's and Steinberg's formulas (2.1) and (2.2). We then obtain a practical MAPLE program efficient up to A_8 for Kostka numbers and up to A_6 for Littlewood-Richardson coefficients. A remarkable fact is that one can compute associated Ehrhart polynomials $t \mapsto c_{t\lambda}^{t\mu}$ and $t \mapsto c_{t\lambda}{}^{t\nu}{}_{t\mu}$ with only a reasonable increase of time: we still can go up to A_8 and A_6 respectively. This program, named `multiplicities.mws` ([**C03**]), is also designed specially for large parameters in A_r: there is virtually no bound on the weights. However, other softwares and packages are not limited by the rank of the algebra like our program (for example computation of non-trivial examples in A_{11} is possible in LiE). These facts make our program complementary to traditional softwares.

This paper is organised as follows. The second section is devoted to explaining the problem of computing Kostka numbers and Littlewood-Richardson coefficients, and linking it with the vector partition function. The third section deals with the problem of counting integral points in rational convex polytopes, and describes the Baldoni-DeLoera-Vergne formula implemented in our program. The fourth section explains how to use the program and shows its internal mechanisms. The fifth and last section features several tests of the program.

Acknowledgements: The author would like to thank referees for their careful work and judicious remarks.

2. Representation theory and convex polytopes

Let us fix notations once and for all. Let G be a complex semi-simple Lie group of rank r and Lie algebra $\mathfrak{g}$. Choose a Cartan subgroup H of G, of Lie algebra $\mathfrak{h}$. The Weyl group of G for H is $W = W(G, H) = N_G(H)/H$.

Denote by $\mathbb{Z}_G^* \subset \mathfrak{h}^*$ the weight lattice. Let Δ^+ be a positive root system. The root lattice is defined as $\mathbb{Z}[\Delta^+]$. Let $C(\Delta^+)$ be the cone spanned by linear combinations with non-negative coefficients of positive roots.

The character $\mathrm{ch}(V)$ of a representation V of $\mathfrak{g}$ is by definition the trace of the representation. It expresses as the sum $\chi(V) = \sum_\mu \dim(V_\mu)e^\mu$, where e^μ is the element of the group ring over $\mathbb{Z}$ of the weight lattice such that $e^\mu(X) = \exp(\langle \mu, X \rangle)$ for any $X \in \mathfrak{g}$. Recall that the irreducible finite-dimensional representation of $\mathfrak{g}$ of highest weight λ is denoted by $V(\lambda)$.

Multiplicities $c_\lambda^\mu = \dim(V(\lambda)_\mu)$ are called *Kostka numbers* when $\mathfrak{g} = \mathfrak{sl}_{r+1}(\mathbb{C})$. On the other hand, multiplicities of representations $V(\nu)$ in the tensor product $V(\lambda) \otimes V(\mu)$ are called *Littlewood-Richardson coefficients* (or *Clebsch-Gordan coefficients*).

Evaluating Kostka numbers and Littlewood-Richardson coefficients in the case of A_r is a difficult task. For A_1, computing Kostka numbers is immediate and Clebsch-Gordan's formula gives Littlewood-Richardson coefficients. For A_2, one can still compute small examples by hand. But for A_r ($r \geq 3$) or for weights which components are big, direct computation is usually intractable.

There exist many formulas from representation theory for c_λ^μ and $c_{\lambda\ \mu}^{\ \nu}$. The first one, valid in any complex semi-simple Lie algebra $\mathfrak{g}$, is Weyl's character formula:

$$\mathrm{ch}(V(\lambda)) = \frac{F_{\lambda+\rho}}{F_\rho}, \quad \text{where } F_\mu = \sum_{w\in W} (-1)^{\varepsilon(w)} e^{w(\mu)},$$

where ρ is half of the sum of positive roots. Littlewood-Richardson coefficients are obtained from this formula, since the character of the tensor product $V(\lambda) \otimes V(\mu)$ is

$$\mathrm{ch}(V(\lambda) \otimes V(\mu)) = \mathrm{ch}(V(\lambda)) \cdot \mathrm{ch}(V(\mu)) = \textstyle\sum_\nu c_{\lambda\ \mu}^{\ \nu}\, \mathrm{ch}(V(\nu)).$$

But these two formulas do not lead to efficient computations when the rank of $\mathfrak{g}$ or the size of coefficients of weights grow. Moreover, computing the whole character is untractable: for $\mathfrak{g} = \mathfrak{sl}_4(\mathbb{C})$ and $\lambda = (3, 2, 1, -6)$, the character $\mathrm{ch}(V(\lambda))$ has 26 monomials and the character $\mathrm{ch}(V(10\lambda))$ has 11601 monomials.

In order to compute non-trivial examples of Kostka numbers and Littlewood-Richardson coefficients in the case of A_r, we hence need more efficient formulas computing only the needed coefficients.

Let us describe Kostant's and Steinberg's formulas in the case of any complex semi-simple Lie algebra $\mathfrak{g}$. Denote by $k_{\mathfrak{g}}(\beta)$ the number of ways one can write the element β of the root lattice, as a linear combination with non-negative coefficients of positive roots. This number satisfies the equation

$$\frac{1}{\prod_{\alpha\in\Delta^+}(1-e^\alpha)} = \sum_{\beta\in\mathbb{Z}_G^*} k_{\mathfrak{g}}(\beta)\, e^\beta.$$

Let λ and μ be respectively a dominant weight and a weight. A Weyl group element $w \in W$ is *valid* for λ and μ if the root lattice element $w(\lambda+\rho) - (\mu+\rho)$ is in the cone $C(\Delta^+)$. The set of such w's is denoted by $\mathcal{W}(\lambda, \mu)$. Then Kostant's formula asserts that the Kostka number c_λ^μ equals

$$c_\lambda^\mu = \sum_{w\in\mathcal{W}(\lambda,\mu)} (-1)^{\varepsilon(w)}\, k_{\mathfrak{g}}(w(\lambda+\rho) - (\mu+\rho)). \tag{2.1}$$

Similarly, let λ, μ, ν, be three dominant weights. The couple $(w, w') \in W \times W$ is *valid* for λ, μ, ν, if the root lattice element $w(\lambda+\rho) + w'(\mu+\rho) - (\nu+2\rho)$ is in $C(\Delta^+)$. The set of such couples is denoted by $\mathcal{W}(\lambda, \mu, \nu)$. Then Steinberg's formula asserts that the Littlewood-Richardson coefficient equals

$$c_{\lambda\ \mu}^{\ \nu} = \sum_{(w,w')\in\mathcal{W}(\lambda,\mu,\nu)} (-1)^{\varepsilon(w)+\varepsilon(w')} k_{\mathfrak{g}}(w(\lambda+\rho) + w'(\mu+\rho) - (\nu+2\rho)). \tag{2.2}$$

In the case of $\mathfrak{g} = \mathfrak{sl}_{r+1}(\mathbb{C})$, we speak of *valid permutations* and *valid couples of permutations*. Sets of valid permutations and valid couples of permutations turn out to be relatively small, when compared to Σ_{r+1} and $\Sigma_{r+1} \times \Sigma_{r+1}$.

From now on, the Lie algebra considered will always be A_r, that is $\mathfrak{g} = \mathfrak{sl}_{r+1}(\mathbb{C})$.

The Kostka number c_λ^μ is piecewise polynomial in coordinates of weights λ and μ ([**Stu95**]). Kirillov and Reshetikin proved that the *Ehrhart function* $\mathbb{N} \ni t \mapsto c_{t\lambda}^{t\mu}$ is polynomial ([**KR86**]). This result was proved again by DeLoera and MacAllister ([**DM03**]). Billey-Guillemin-Rassart ([**BGR03**]), using Gelfand-Tsetlin patterns, proved that, on particular polyhedral chambers, the function $(\lambda, \mu) \mapsto c_\lambda^\mu$ is a polynomial of degree $r(r-1)/2$ in coordinates of μ, which coefficients are polynomials of degree $r(r-1)/2$ in coordinates of λ.

Similarly, the Littlewood-Richardson coefficient $c_{\lambda\ \mu}^{\ \nu}$ is piecewise polynomial in coordinates of weights λ, μ and ν. Rassart, following [**BGR03**] and using the work of Knutson and Tao on honeycombs ([**KT99**]), proved that the Ehrhart function $t \mapsto c_{t\lambda\ t\mu}^{\ t\nu}$ is polynomial and that on particular polyhedral chambers the function $(\lambda, \mu, \nu) \mapsto c_{\lambda\ \mu}^{\ \nu}$ is a polynomial ([**R03**]).

The common point to Kostant's and Steinberg's formulas is the function counting the number of decompositions of a root lattice element as a linear combination with non-negative integral coefficients of positive roots of the Lie algebra. The next section deals with an efficient method to compute it.

3. Counting integral points in rational convex polytopes

3.1. Vector partition function. Let Φ be a $r \times N$ integral matrix, with columns $\Phi_1, \ldots, \Phi_N$. Choose an integral r-dimensional vector a. We define the *vector partition function* as the formal power series development

$$\prod_{i=1}^{N} \frac{1}{(1 - X^{\Phi_i})} = \sum_{a \in \mathbb{N}^r} k(\Phi, a)\ X^a,$$

using the standard convention that $X^a = X_1^{a_1} \cdots X_r^{a_r}$ if $a = (a_1, \ldots, a_r)$. In fact $k(\Phi, a)$ equals the number of integral points in the rational convex polyhedron

$$P(\Phi, a) \quad = \quad \left\{ x \in \mathbb{R}^N ;\ \sum_{i=1}^{N} x_i \Phi_i = a,\ x_i \geq 0 \right\}.$$

Remark that every convex polyhedron can be realized in the form $P(\Phi, a)$. We assume that a is in the cone $C(\Phi)$ spanned by non-negative linear combinations of vectors Φ_i, so that $P(\Phi, a)$ is non-empty. We also assume that $\ker(\Phi) \cap \mathbb{R}_+^N = \{0\}$, so that the cone $C(\Phi)$ is strongly convex and $P(\Phi, a)$ is a polytope (*i.e.* bounded). Finally, we assume that the matrix Φ has rank r.

REMARK 3.1. If $\Phi = \Phi(\mathfrak{g})$ is the matrix which columns are positive roots for a Lie algebra $\mathfrak{g}$, then have the equality $k_{\mathfrak{g}}(\beta) = k(\Phi(\mathfrak{g}), \beta)$ for any root lattice element β. Consequently an efficient way of computing the vector partition function opens the path to Kostka numbers and Littlewood-Richardson *via* Kostant's and Steinberg's formulas.

If Φ is the matrix of positive roots of A_r, then the associated function $a \mapsto k(A_r, a)$ is the *Kostant partition function.* Note that Φ is the $(r+1) \times (r(r+1)/2)$ matrix with columns $e_i - e_j$ $(1 \leq i < j \leq r+1)$, where $(e_i)_i$ is the canonical basis of $\mathbb{R}^{r+1}$. Note that the matrix $\Phi(A_r)$ has rank r (and not $r+1$), since the sum over a line is zero. For example

$$\Phi(A_3) = \begin{pmatrix} 1 & 1 & 1 & 0 & 0 & 0 \\ -1 & 0 & 0 & 1 & 1 & 0 \\ 0 & -1 & 0 & -1 & 0 & 1 \\ 0 & 0 & -1 & 0 & -1 & -1 \end{pmatrix}.$$

3.2. Baldoni-DeLoera-Vergne formula for A_r. Let us describe the formula, computing the number of integral points in rational convex polytopes, that was implemented in our program.

Consider a $(r+1)$-dimensional real vector space. Let E_r be the vector space spanned by positive roots $(e_i - e_j)$, that is the r-dimensional vector space of $(r+1)$-dimensional vectors $\mathbf{a} = \sum_{i=1}^{r+1} a_i e_i$ such that $\sum_{i=1}^{r+1} a_i = 0$. The map defined by

$$a = (a_1, a_2, \ldots, a_r) \quad \longmapsto \quad \mathbf{a} = a_1 e_1 + \cdots + a_r e_r - (a_1 + \cdots + a_r) e_{r+1}$$

explicitly provides an isomorphism of the Euclidean space $\mathbb{R}^r$ with E_r. The hyperplane arrangement (setting $z_{r+1} = 0$) generated by the set A_r^+ of positive roots of A_r is given by the following set of hyperplanes in $\mathbb{C}^r$:

$$\{z_i \,;\, 1 \le i \le r\} \cup \{(z_i - z_j) \,;\, 1 \le i < j \le r\}.$$

Let R_{A_r} be the set of rational functions $f(z_1, z_2, \ldots, z_r)$ on $\mathbb{C}^r$ that may have poles only on hyperplanes $z_i = z_j$ or $z_i = 0$. For a permutation $w \in \Sigma_r$, define a linear form $\mathrm{IRes}_{z=0}^{w}$ on R_{A_r} by

$$\begin{aligned} \mathrm{IRes}_{z=0}^{w} f &= \mathrm{Res}_{z_{w(1)}=0} \mathrm{Res}_{z_{w(2)}=0} \cdots \mathrm{Res}_{z_{w(r)}=0} f(z_1, z_2, \ldots, z_r) \\ &= \mathrm{Res}_{z_1=0} \mathrm{Res}_{z_2=0} \cdots \mathrm{Res}_{z_r=0} f(z_{w^{-1}(1)}, z_{w^{-1}(2)}, \ldots, z_{w^{-1}(r)}), \end{aligned}$$

where $\mathrm{Res}_{z_k=0}$ denotes the usual residue operator. In particular for $w = \mathrm{id}$ the linear form $f \mapsto \mathrm{IRes}_{z=0} f$ defined by

$$\mathrm{IRes}_{z=0} f = \mathrm{Res}_{z_1=0} \mathrm{Res}_{z_2=0} \cdots \mathrm{Res}_{z_r=0} f(z_1, z_2, \ldots, z_r)$$

is called the *iterated residue*. Remark that Σ_r is smaller than the Weyl group of A_r, which is Σ_{r+1}.

Another ingredient of Baldoni-Deloera-Vergne formula is the following. Let $\mathrm{Sp}(\mathbf{a})$ be the set of permutations $w \in \Sigma_r$ such that for all i:

(3.1) if $a_{w(1)} + \cdots + a_{w(i)} \ge 0$ then $w(i) < w(i+1)$, else $w(i) > w(i+1)$.

An element of $\mathrm{Sp}(\mathbf{a})$ is called a *special permutation* for a. Remark that if $a_i \ge 0$ for all $i \le r$, then $\mathrm{Sp}(\mathbf{a}) = \{\mathrm{id}\}$.

We need one last tool before describing the formula computing Kostant partition function $k(A_r, \mathbf{a})$. Let $C(A_r^+) \subset E_r$ be the cone generated by positive roots. A *polyhedral wall* is a hyperplane in E_r spanned by $(r-1)$ vectors of A_r^+. A vector $v \in C(A_r^+)$ is *regular* if it is on no polyhedral wall. This means that for every strict subset $I \subset \{1, \ldots, r+1\}$ we have $\sum_{i \in I} v_i \neq 0$. Remark that notions of polyhedral wall and regular vectors in this article refer to polyhedra theory and not representation theory.

Given $\mathbf{a} \in C(A_r^+) \cap \mathbb{Z}^{r+1}$, define the *deformed vector* of $\mathbf{a}$ as

$$\mathrm{def}(\mathbf{a}) = \mathbf{a} + \varepsilon\left(\sum_{i=1}^{r} e_i - r e_{r+1}\right)$$

with $\varepsilon = \dfrac{1}{2r}$. The purpose of this small deformation is to make regular any element $\mathbf{a}$ of $C(A_r^+) \cap \mathbb{Z}^{r+1}$.

Now we can state the formula that was implemented:

THEOREM 3.2 (Baldoni-DeLoera-Vergne [**BdLV03**]). *Let* $\mathbf{a} = a_1e_1 + \cdots + a_re_r - (\sum_{i=1}^{r} a_i)e_{r+1}$ *be an integral vector lying in the cone generated by positive roots of the Lie algebra* A_r. *Then the Kostant partition function* $k(A_r, \mathbf{a})$ *is given by:*

$$\sum_{w \in Sp(\mathbf{a}')} (-1)^{n(w)}\,\mathrm{IRes}_{z=0}^{w}\left(\frac{\prod_{i=1}^{r}(1+z_i)^{a_i+r-i}}{(\prod_{i=1}^{r} z_i)\prod_{1\le i<j\le r}(z_i - z_j)}\right) \tag{3.2}$$

where

$$\mathbf{a}' = \begin{cases} \mathbf{a} & \textit{if } \mathbf{a} \textit{ is regular,} \\ \mathrm{def}(\mathbf{a}) & \textit{otherwise.} \end{cases}$$

Here $n(w)$ is the number of consecutive inversions of the permutation $w \in \Sigma_r$, that is the number of couples (i, j) such that $i < j$ and $w(i) > w(j)$.

REMARK 3.3. The result is unchanged if we take $\mathbf{a}' = \mathrm{def}(\mathbf{a})$ even when $\mathbf{a}$ is regular. In order to simplify the implementation, we will always make this choice of $\mathbf{a}'$.

4. *Deus ex machina*

This section features a brief description of the algorithms that were implemented in the language of the computer algebra system MAPLE.

4.1. How to use the program. The initial data are only vectors: two for computing Kostka numbers c_λ^μ, three for computing Littlewood-Richardson coefficients $c_{\lambda\ \mu}^{\ \nu}$.

Our program works with weights represented in the canonical basis of $\mathbb{R}^{r+1}$, and not in the fundamental weights basis of A_r. More precisely a weight λ is a list of $r+1$ rationals such that $\lambda_i - \lambda_{i+1} \in \mathbb{Z}$. It is dominant if moreover $\lambda_i - \lambda_{i+1} \in \mathbb{N}$. The translation between these two approaches is performed *via* procedures FromFundaToCano and FromCanoToFunda. For example let $\lambda = \sum_{i=1}^{r} b_i\omega_i$ be a weight expressed in the fundamental weights basis, with $\omega_i = e^1 + \cdots + e^i$ ($i = 1, \ldots, r$). It rewrites as $b = (b_1 + \cdots + b_r)e^1 + \cdots + b_re^r$. Then FromFundaToCano(lambda) is the projection $(b_1 + \cdots + b_r, \ldots, b_r, 0) - (S, \ldots, S) \in \mathbb{R}^{r+1}$ of λ on the hyperplane of vectors with zero sum. Here $S = (b_1 + 2b_2 + \cdots + rb_r)/(r+1)$.

Therefore computing the Kostka number c_λ^μ and the Ehrhart polynomial $t \mapsto c_{t\lambda}^{t\mu}$ is done by typing in

```
multiplicity(lambda,mu);
polynomial_multiplicity(lambda,mu);
```

where λ is a dominant weight and μ a weight such that $\sum_{i=1}^{r+1} \lambda_i = \sum_{i=1}^{r+1} \mu_i$.

The syntax for computing the Littlewood-Richardson coefficient $c_{\lambda}{}^{\nu}{}_{\mu}$ and the Ehrhart polynomial $t \mapsto c_{t\lambda}{}^{t\nu}{}_{t\mu}$ is

```
tensor_product(lambda,mu,nu);
polynomial_tensor_product(lambda,mu,nu);
```

where λ, μ and ν are dominant weights such that $\sum_{i=1}^{r+1}(\lambda_i + \mu_i) = \sum_{i=1}^{r+1} \nu_i$.

4.2. Implementation. In our program, we use lists instead of vectors and we re-wrote several MAPLE procedures to save time. More generally, a great care about efficiency was taken during the implementation.

There are three important points in the implementation of formulas (2.1) and (2.2) (and underlying formula (3.2)): the deformed vector def($\mathbf{a}$), permutation sets and iterated residues.

Because of Remark 3.3, we may use def($\mathbf{a}$) instead of $\mathbf{a}$ and we do this to simplify procedures. We compute the vector def($\mathbf{a}$) in a straightforward procedure. This takes care of the first part.

Formulas implemented in our program contain sums over three sets, namely valid permutations (in (2.1)), valid couples of permutations (in (2.2)) and special permutations (in (3.1)). For example, a permutation w is valid for λ and μ if for all i we have

$$\sum_{j=1}^{i}(\lambda_{w(j)} + \rho_{w(j)}) \geq \sum_{j=1}^{i}(\mu_j + \rho_j).$$

Valid permutations are built step-by-step, by determining all possible $w(1)$'s, then all possible $(w(1), w(2))$'s, etc. After r steps, we get the whole subset of valid permutations $\mathcal{W}(\lambda, \mu)$ of Σ_r (see figure 1). Valid couples of permutations and special permutations are determined in a similar way. Computation time for these three sets is negligible, because this method is efficient: we do not run over the whole permutation set (thus saving speed) and we only store permutations that fit (thus saving memory). We stress that the MAPLE function `combinat[permute]` is impractical and does not go very far because of memory limitations.

The last ingredient of our program is iterated residue of a specific rational fraction (see formula (3.2)). This part is much more difficult to implement, due to multiple variables. We compute these residues iteratively. The function F which residue we need to compute is a product of a certain number of functions. This allows to take residues by introducing little by little the part of the function F containing the needed variable. We optimized the algorithm described in [**BdLV03**]. More precisely, the algorithm to compute

$$R(w, \mathbf{a}) = \mathrm{IRes}_{z=0}^{w}\left(\frac{\prod_i (1+z_i)^{a_i+r-i}}{(\prod_i z_i)\prod_{i<j}(z_i - z_j)}\right)$$

is described in figure 2.

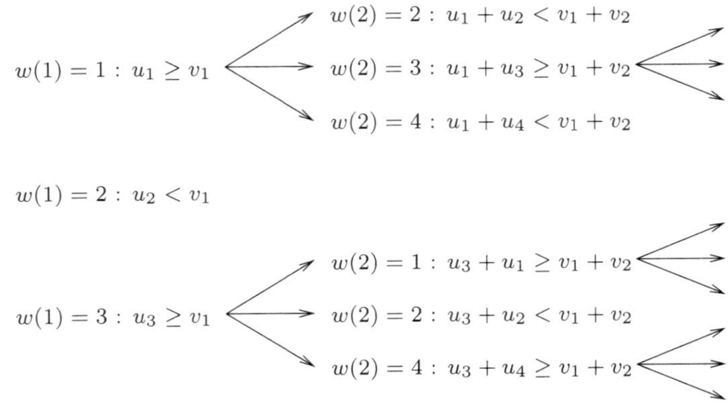

FIGURE 1. Step-by-step building of valid permutations (here $u = \lambda + \rho$ and $v = \mu + \rho$)

let $C = 1$
for i from r downto 1 by -1 do
    take the part P of the rational fraction containing $z_{w(i)}$
    compute the Taylor expansion T of $P \times C$
    extract the coefficient C of T in $\dfrac{1}{z_{w(i)}}$
end do
return C

FIGURE 2. Outline of the computation of iterated residues

Note that this algorithm relies on Taylor expansion, polynomial multiplication and coefficient extraction. Last two parts need roughly speaking two thirds of the computing time of the whole program.

Speed has also been increased thanks to the following empirical sorting

```
C := normal(expand(coeff(T,z[w[i]],-1)));
```

of the extracted coefficient.

REMARK 4.1. This algorithm computing iterated residues allows the parameter $\mathbf{a}$ to be formal. Hence we can compute polynomials $t \mapsto k(A_r, u+tv)$, for root lattice elements u, v. This is the key for the calculation of Ehrhart polynomials.

Computing Kostant function consumes much time. Even if $w = 1$, its evaluation is untractable when the root lattice element $\mathbf{a}$ lies in a high rank algebra. Moreover, we call $|\operatorname{Sp}(\mathbf{a})|$ times this procedure to compute Kostka numbers and Littlewood-Richardson coefficients; even if this number is really small when compared to the total number of permutations, it remains rather big.

5. Test of the program

Let θ be the $(r + 1)$-dimensional vector

$$\theta \quad = \quad re_1 + (r-1)e_2 + \cdots + 1e_{r-1} - \frac{r(r+1)}{2} e_{r+1}.$$

Its decomposition in the fundamental weights basis is the r-dimensional vector $(1, \ldots, 1, 1 + r(r+1)/2)$. With this vector, we checked that the multiplicity of the zero weight in the representation of A_r of highest weight $N\theta$ equals the dimension of the representation of A_{r-1} of highest weight $N\rho$, that is $(N+1)^{r(r-1)/2}$. This computation has also been done with the formal parameter N, leading as predicted to $(N+1)^{r(r-1)/2}$.

The tests were performed on a bi-processor Pentium III 1,13GHz computer at the "Institut de Mathématiques de Jussieu (UMR 7586)". The notation "−" in the table 3 means that we did not try the computation, and not that it failed.

The table 3 emphasizes the fact that counting integral points in a lattice polytope is polynomial in the size of input if dimension is fixed, and NP-hard if dimension is not fixed.

Algebra	Time	Multiplicity c_θ^0	Time	Polynomial $N \mapsto c_{N\theta}^0$
A_2	< 0.1 s	$2 = 2^1$	< 0.1 s	$(N+1)^1$
A_3	< 0.1 s	$8 = 2^3$	< 0.1 s	$(N+1)^3$
A_4	< 0.1 s	$64 = 2^6$	< 0.1 s	$(N+1)^6$
A_5	0.4 s	$1024 = 2^{10}$	1.4 s	$(N+1)^{10}$
A_6	7.6 s	$32768 = 2^{15}$	36.2 s	$(N+1)^{15}$
A_7	169.3 s	$2097152 = 2^{21}$	2091 s	$(N+1)^{21}$
A_8	9401 s	$268435456 = 2^{28}$	−	−

FIGURE 3. Time of computation for Kostka numbers, when the rank increases

Weight dilatation coefficient	Time for $c_{N\theta}^0$ in A_7 $\lvert\mathcal{W}(N\theta, 0)\rvert = 5040$ $\lvert\Sigma_8\rvert = 40320$	Time for $c_{N\theta,N\theta}^{2[9N/10]\theta}$ in A_6 $\lvert\mathcal{W}(N\theta, N\theta, 2[9N/10]\theta)\rvert = 1070$ $\lvert\Sigma_7 \times \Sigma_7\rvert = 25401600$
$N = 10^1$	204.9 s	40.5 s
$N = 10^2$	221.0 s	47.0 s
$N = 10^3$	235.6 s	50.3 s
$N = 10^4$	251.5 s	52.6 s
$N = 10^5$	259.1 s	53.8 s
$N = 10^6$	261.5 s	57.0 s
$N = 10^7$	283.8 s	58.4 s
$N = 10^8$	297.0 s	59.9 s
$N = 10^9$	297.2 s	62.3 s

FIGURE 4. Time of computation for Kostka numbers and Littlewood-Richardson coefficients, when the size of input grows

As announced in the introduction, our program is really efficient for weights with huge coefficients (see table 4).

Recall that our algorithms depend on sums over three permutation sets. For Kostka numbers, when the weight μ is close to λ, the set $\mathcal{W}(\lambda,\mu)$ of the valid permutations (defined in section 2) occuring in Kostant's formula (2.1) is small. This set grows as μ is approaching to the zero weight. The same phenomenon occurs when studying Littlewood-Richardson coefficients. The table 5 represents in the case of A_6 the number of valid permutations for weights $\lambda = 10\,\theta$ and $\mu = N\theta$, for N ranging from 0 to 10. Remark that the total permutation group is Σ_7, of cardinal 5040.

N	$\lvert\mathcal{W}(10\,\theta, N\,\theta)\rvert$	$c_{10\,\theta}^{N\,\theta}$	Time for $c_{10\,\theta}^{N\,\theta}$ in A_6
0	720	4177248169415651	10.8 s
1	720	4177248169415651	11.0 s
2	600	4177092544868045	11.1 s
3	576	4153282006799937	11.2 s
4	456	3899635283050694	11.2 s
5	402	2931789395526035	10.8 s
6	274	1297765345789869	9.8 s
7	132	210740227073233	6.3 s
8	72	5616154382787	3.3 s
9	10	3549690125	0.5 s
10	1	1	< 0.1 s

FIGURE 5. Number of valid permutations and computation time for Kostka numbers, when weights vary

The translation of `multiplicities.mws` in the programming language OBJECTIVE CAML is in progress. This will considerably increase the speed of computations.

There are several forthcoming developments of this MAPLE program. First, computation of the volume of a rational convex polytope with a formula of Baldoni-Vergne similar to (3.2). Next, computation of coefficients of multiple tensor product, that is multiplicity of $V(\mu)$ in $V(\lambda_1) \otimes \cdots \otimes V(\lambda_n)$. A harder task would be to find formulas similar to Baldoni-DeLoera-Vergne's (3.2) for other semi-simple Lie algebras.

References

[BdLV03] Baldoni-Silva W., De Loera J.A., and Vergne M., *Counting Integer flows in Networks*, available at math.ArXiv, CO/0303228, to be published in *Foundations of Computational Mathematics*.

[BV01] Baldoni-Silva W. and Vergne M., *Residues formulae for volumes and Ehrhart polynomials of convex polytopes*, manuscript 81 pages, 2001, available at math.ArXiv, CO/0103097.

[BP99] Barvinok A. and Pommersheim J., *An algorithmic theory of lattice points in polyhedra*, in: *New Perspectives in Algebraic Combinatorics* (Berkeley, CA, 1996-1997), 91-147, Math. Sci. Res. Inst. Publ. 38, Cambridge Univ. Press, Cambridge, 1999.

[B94] Barvinok A. I., *A polynomial time algorithm for counting integral points in polyhedra when the dimension is fixed*, Math. Oper. Res. 19 (1994), no. 4, p. 769–779.

[B97] Barvinok A., *Lattice points and lattice polytopes*, in *Handbook of discrete and computational geometry*, CRC Press Ser. Discrete Math. Appl., pages 133–152. CRC, Boca Raton, FL, 1997.

[BGR03] Billey S., Guillemin V. and Rassart É., *A vector partition function for the multiplicities of* $\mathfrak{sl}_k(\mathbb{C})$, (2003), math. CO/0307227, to appear in Journal of Alg. Combin.

[DHTY03] DeLoera J., Hemmecke R., Tauzer J. and Yoshida R., *Effective Lattice Point Counting in Rational Convex Polytopes*, (2003), to appear in Journal of Symbolic Computation.

[DM03] DeLoera J. and MacAllister T., *Vertices of Gelfand-Tsetlin Polytopes*, (2003), unpublished.

[DLS01] De Loera J.A. and Sturmfels M., *Algebraic Unimodular Counting*, math.CO/0104286

[GAP] `http://www.math.rwth-aachen.de/~GAP`.

[C03] `http://www.math.jussieu.fr/~cochet`.

[KR86] Kirillov A. N. and Reshetikhin N. Y., *The Bethe ansatz and the combinatorics of Young tableaux*, Zap. Nauchn. Sem. Leningrad. Otdel. Mat. Inst. Steklov. (LOMI) 155 (1986), no. Differentsialnaya Geometriya, Gruppy Li i Mekh. VIII, p. 65–115, 194, translation in Journal of Soviet Math. 41 (1988), 925-955.

[KT99] Knutson A. and Tao T., *The honeycomb model of $GL_n(\mathbb{C})$ tensor products. I. Proof of the saturation conjecture*, J. Amer. Math. Soc. 12 (1999), no. 4, p. 1055–1090.

[vL94] van Leeuwen M. A. A., *LiE, a software package for Lie group computations*, Euromath Bull., 1(2):83–94, 1994.

[R03] Rassart É, *A polynomiality property for Littlewood-Richardson coefficients*, (2003), manuscript, 14 pages, math.CO/0308101.

[Ste95] Stembridge, J., *A Maple package for symmetric functions*, (1995), J. Symbolic Computation, 20 (1995), 755–768.

[Stu95] Sturmfels B., *On vector partition functions*, J. Combin. Theory Ser. A 72 (1995), no. 2, p. 302–309.

Institut Mathématique de Jussieu, Université Paris 7 – Denis Diderot, 75251 Paris cedex 05, France

E-mail address: `cochet@math.jussieu.fr`

Contemporary Mathematics
Volume **374**, 2005

Polar decomposition and Brion's theorem

Christian Haase

Abstract. In this note we point out the relation between Brion's formula for the lattice point generating function of a convex polytope in terms of the vertex cones [**Bri88**] on the one hand, and the polar decomposition à la Lawrence/Varchenko [**Law91a, Var87**] on the other. We then go on to prove a version of polar decomposition for non-simple polytopes.

1. Introduction

The stars of this note are two formulas that express convex polytopes in terms of cones. On the one hand, Brion's theorem (equation (2) below) expresses lattice points in a polytope in terms of tangent cones at vertices [**Bri88**]. On the other hand, given a simple polytope and a generic objective function, one can write the indicator function of a simple polytope as a combination of polarized tangent cones at the vertices (equation (4)). We refer to this formula as polar decomposition [**Law91a, Var87**].

After we review these formulas in this introduction, we point out in section 2 some simple relations between the two formulas which, to our knowledge, have not appeared in the literature before. In particular, Brion's theorem (for simple polytopes) follows from polar decomposition. The last section is devoted to the formulation and proof of polar decomposition for non-simple polytopes. (Put together, sections 2 and 3 yield another proof of Brion's theorem.)

This is by no means an attempt to survey what is scattered throughout the literature[1] about decompositions of polytopes into cones. In fact, Matthias Beck made me aware of such a survey in the making [**BRSW04**]. We will concentrate on what we think is new. We will not state the most general form of the results, but leave generalizations (e.g., weighted versions) as exercises for the ambitious reader.

Acknowledgments. This paper has its origins in José Agapito's inspiring talk about his weighted version of polar decomposition for simple polytopes [**Aga03**]. After the talk, I asked the question whether one can obtain an analogous formula for non-simple polytopes by considering simple deformations. I got intrigued by

2000 *Mathematics Subject Classification.* 52B45 (Primary); 11L03, 11P21 (Secondary).
Work supported by NSF grant DMS–0200740.

[1]See [**Bar02**, p. 346] and the references therein for Brion's theorem, [**Law91a, Var87**] for polar decomposition, and [**Bri37, Gra74, She67**] for the Brianchon-Gram formula below.

this question, and, after computing several examples, was convinced that there is such a theorem out there.

I was lucky to get hold of a preliminary draft of the above mentioned survey [**BRSW04**]. Not only did it take away the pressure to try and write a comprehensive article about conic decompositions of polytopes, but it also made me aware of references and view points that I did not know about before. Much of the presentation in the present paper is influenced by (if not stolen from) this draft. Finally, I want to thank the referee who helped to clarify the exposition.

1.1. Basic definitions and the Brianchon-Gram formula. We will consider rational convex polytopes P and polyhedral cones C in $\mathbb{Q}^d$, where $\mathbb{Q}^d$ is endowed with the fixed lattice $\mathbb{Z}^d$. We will assume that polytopes and cones are full-dimensional. For standard polytope definitions and notation we refer to [**Zie95**]. To a subset $S \subseteq \mathbb{Q}^d$ we assign two objects:

- The indicator function $\mathbb{1}_S \colon \mathbb{Q}^d \to \mathbb{Z}$ vanishes outside S, and is one along S.
- The generating function (of the lattice points) is the formal Laurent power series $G_S = \sum_{m \in S \cap \mathbb{Z}^d} z^m \in \mathbb{C}[[z_1^{\pm 1}, \ldots, z_d^{\pm 1}]]$, where we write $z^m = z_1^{m_1} \cdot \ldots \cdot z_d^{m_d}$. If $S = P$ is a polytope, then G_P is a Laurent polynomial; when $S = C$ is a cone, then G_C is a rational function[2].

The essential features can be illustrated on the line. For $a < b \in \mathbb{Z}$, we have $G_{[a,b]}(x) = x^a + \ldots + x^b = \frac{x^a - x^{b+1}}{1-x}$, and $G_{[a,\infty)}(x) = x^a + \ldots = \frac{x^a}{1-x}$.

The mother of all conic decomposition theorems is the Brianchon-Gram formula. According to Shephard [**She67**], Brianchon [**Bri37**] and Gram [**Gra74**] independently proved the $d = 3$ case in 1837(!) and 1874 respectively. In 1927 Sommerville [**Som27**] published a proof for general d, which was corrected in the 1960's by Grünbaum [**Grü03**, §14.1].

Let $P \subset \mathbb{Q}^d$ be a convex polytope. For a face F of P, we define the tangent cone (of P at F) by

$$\mathcal{T}_F P = \{ f + x \in \mathbb{Q}^d \ : \ f \in F \text{ and } f + \varepsilon x \in P \text{ for some } \varepsilon > 0 \}$$

(See Figure 1.) Then $\mathbb{1}_P$ is the alternating sum of all tangent cones:

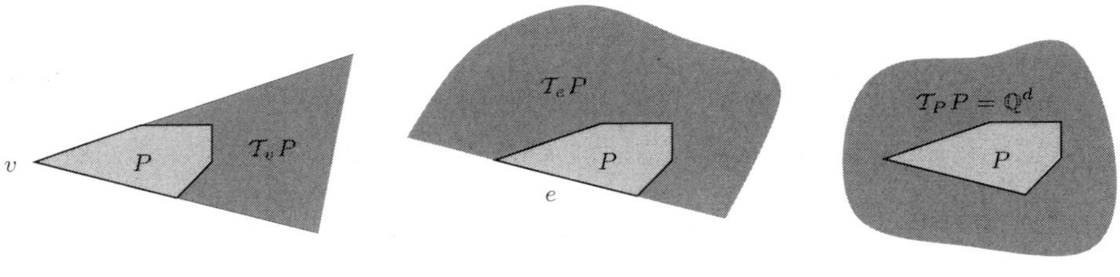

FIGURE 1. Tangent cones at vertex v, edge e, and at P.

$$\mathbb{1}_P = \sum_{F \preceq P} (-1)^{\dim F} \ \mathbb{1}_{\mathcal{T}_F P}. \tag{1}$$

[2]We are glossing over significant technicalities here and later on when we pass from formal power series to rational functions. For a first reading, we prefer to skip this step, and refer to [**Bar02, BV97**] for a careful treatment. The experts should be familiar with the standard arguments, anyway.

At every point $x \in P$, the right hand side computes the Euler characteristic of P (and P is contractible). While outside P, we have to subtract the Euler characteristic of the subcomplex that is visible from x (which again, is contractible, actually shellable). (Compare [**She67**].)

1.2. Brion's formula. Brion [**Bri88**] proved his decomposition using a Riemann-Roch type formula. In the mean time, more elementary proofs have been given [**Bar93, BV97, Law91b, PK92**]. The easiest way to formulate Brion's formula is in terms of generating functions. We can express the generating function for P as the sum of the generating functions of the tangent cones at vertices.

$$G_P = \sum_{v \text{ vertex of } P} G_{\mathcal{T}_v P} \tag{2}$$

In the example, $\mathcal{T}_a[a,b] = [a,\infty)$, and $\mathcal{T}_b[a,b] = (-\infty,b]$ with generating functions $G_{[a,\infty)} = \frac{x^a}{1-x}$ and $G_{(-\infty,b]} = \frac{(1/x)^{-b}}{1-1/x} = -\frac{x^{b+1}}{1-x}$. As predicted, they sum to $G_{[a,b]}$.

This version of Brion's formula is implied by the following two results. (Compare [**Bar02**].)

THEOREM 1. *The function $\mathbb{1}_P - \sum_{v \text{ vertex of } P} \mathbb{1}_{\mathcal{T}_v P}$ is a linear combination of indicator functions of cones that contain affine lines.*

This follows immediately from (1). It is also implied by the considerations in Sections 2 and 3. In our example, $\mathbb{1}_{[a,b]} = \mathbb{1}_{[a,\infty)} + \mathbb{1}_{(\infty,b]} - \mathbb{1}_{\mathbb{Q}}$.

LEMMA 2. *Let $C \subseteq \mathbb{Q}^d$ be a cone that contains an affine line. Then $G_C = 0$.*

The idea of the proof is to decompose the monoid $C \cap \mathbb{Z}^d$ into a direct sum $L \oplus R$ of a one dimensional lattice L and a complement R. Then G_C can be written as $G_R G_L$, and $G_L = 0$[3].

1.3. (Simple) polar decomposition. By flipping the edge vectors emanating from each vertex of P in a systematic way, we can get the so-called polar decomposition, which expresses the characteristic function of a convex simple polytope in terms of the characteristic functions of the cones supported at the vertices of P (no cones with straight lines needed this time). This was first obtained by Varchenko [**Var87**] and Lawrence [**Law91a**] independently. Lawrence used Brianchon-Gram's theorem together with the principle of inclusion-exclusion to derive the polar decomposition theorem. Then Karshon, Sternberg and Weitsman obtained a weighted version of this decomposition by assigning, in a consistent way, particular weights to the lattice points in the polytope and in the cones [**KSW03**]. Their work was also motivated by methods in differential and algebraic geometry. Finally, using the same source of motivation, Agapito [**Aga03**] gave a more general weighted version of the polar decomposition theorem for simple polytopes that includes Lawrence/Varchenko and Karshon-Sternberg-Weitsman versions as particular cases.

Polar decomposition is like Morse theory for simple polytopes. We sweep over the polytope and build it up from local contributions, critical point (vertex) by critical point. As input we use a generic[4] linear functional $\xi \in (\mathbb{Q}^d)^*$. Using ξ, we define the polarized tangent cones as follows.

---

[3]Glossing over alert: look under the rug [**Bar02**]!

[4]Generic means here that ξ is not constant on any edge of P. In Section 3 we ask a little bit more.

Because P is simple, at every vertex v, the tangent cone $\mathcal{T}_v P$ is generated by d linearly independent directions $t_1, \dots, t_d$ (remember P is full-dimensional).

$$\mathcal{T}_v P = v + \sum_{i=1}^{d} \mathbb{Q}_{\geq 0} t_i$$

Now the polarized tangent cone (with respect to ξ) is

$$\mathcal{T}_v^\xi P = v + \sum_{\xi(t_i)>0} \mathbb{Q}_{\geq 0} t_i + \sum_{\xi(t_i)<0} \mathbb{Q}_{<0} t_i.$$

This is a locally closed cone all whose points are ξ-higher than v. (Compare

FIGURE 2. Polarized tangent cones.

Figure 2.) Dually, if $n_1, \dots, n_d \in (\mathbb{Q}^d)^*$ are the inner facet normals to P at v, and $\xi = \sum \alpha_i n_i$, then $\mathcal{T}_v^\xi P$ can be defined by the inequalities

$$n_i(x) \geq n_i(v) \text{ for } \alpha_i > 0, \text{ and } n_i(x) < n_i(v) \text{ for } \alpha_i < 0. \tag{3}$$

Define the index $\operatorname{ind}_\xi(v)$ of the vertex v as the number of ξ-negative edge directions t_i, or equivalently, as the number of negative coefficients α_i. Then we can write P as the signed sum of polarized tangent cones.

$$\mathbb{1}_P = \sum_{v \text{ vertex of } P} (-1)^{\operatorname{ind}_\xi(v)} \mathbb{1}_{\mathcal{T}_v^\xi P} \tag{4}$$

Again, we can formulate a generating function version of this result: just replace all $\mathbb{1}$'s by G's[5].

Equation (4) follows from (1) if we group all faces according to where they achieve their ξ-maximum. Then it remains to check that

$$(-1)^{\operatorname{ind}_\xi(v)} \mathbb{1}_{\mathcal{T}_v^\xi P} = \sum_{\substack{v \preceq F \preceq P \\ \xi(F) \leq \xi(v)}} (-1)^{\dim F} \mathbb{1}_{\mathcal{T}_F P}. \tag{5}$$

There is a weighted version of equation (4) [**Aga03**]. The weighted indicator function of a polytope P is the function $\mathbb{1}_P^w \colon \mathbb{Q}^d \to \mathbb{Z}[z]$ which vanishes outside P, and takes the value z^k along the relative interior of codimension k faces.[6] One could even introduce a different variable for every facet of P, and assign their product to the corresponding intersection. Now the same formula holds if we modify the indicator functions of the polarized tangent cones as follows: A face of (the closure of) $\mathcal{T}_v^\xi P$ is defined as the set of points that satisfy equality for some of the inequalities (3). The value of $\mathbb{1}^w_{\mathcal{T}_v^\xi P}$ along a face defined by k_+ equalities $n_i(x) = n_i(v)$ for $\alpha_i > 0$ and k_- equalities for $\alpha_i < 0$ is the polynomial $z^{k_+}(1-z)^{k_-} \in \mathbb{Z}[z]$. One recovers the unweighted version for $z = 1$, and for $z = 0$ one obtains a decomposition of the interior of P.

[5]More glossing over.

[6]The formula in [**Aga03**] is stated using the substitution $z = \frac{1}{1+y}$.

2. This follows from that

In this section we show the relation between the two formulas. Essentially, for the indicator functions, Brion's formula (in the simple case) is polar decomposition modulo cones that contain lines.

LEMMA 3. *If P is a simple polytope and v is a vertex, then the different polarized tangent cones for various ξ partition $\mathbb{Q}^d$. Moreover, for $\xi_1, \xi_2 \in (\mathbb{Q}^d)^*$, $(-1)^{\mathrm{ind}_{\xi_1}(v)} \mathbb{1}_{\mathcal{T}_v^{\xi_1} P}$ and $(-1)^{\mathrm{ind}_{\xi_2}(v)} \mathbb{1}_{\mathcal{T}_v^{\xi_2} P}$ are equivalent modulo cones that contain lines.*

PROOF. If $n_i(x) \geq n_i(v)$ for $i = 1, \ldots, d$ are the inequalities of the facets incident to v, the hyperplanes $n_i(x) = n_i(v)$ subdivide $\mathbb{Q}^d$ into orthant cones. Every point $x \in \mathbb{Q}^d$ belongs to exactly one polarized tangent cone according to the signs of the $n_i(x)$.

We can get from every polarized tangent cone to every other by successively flipping inequalities. Now the sum of two adjacent polarized tangent cones is defined by $d-1$ (strict and non-strict) inequalities. Thus it is a cone that contains a line. □

For a simple polytope, the indicator function version of Brion's formula (Theorem 1) can be derived from equation (4) (modulo cones that contain lines). As Theorem 1 does not specify these cones, there is no converse. On the other hand, generating functions do not see cones that contain lines. So for generating functions, the two formulas actually coincide.

The same considerations carry through for the weighted indicator functions. One thus obtains a weighted Brion's formula. We leave this as an exercise for the reader.

3. Non-simple polar decomposition

We want to generalize polar decomposition to non-simple polytopes. We will compute the local contribution from a simple deformation of the vertex in question, and prove that the result does not depend on the chosen deformation.

As there are not too many non-simple polytopes in dimension ≤ 2, the new running example will be the pyramid with the five vertices $(0,0,0)$, and $(\pm 1, \pm 1, 1)$.

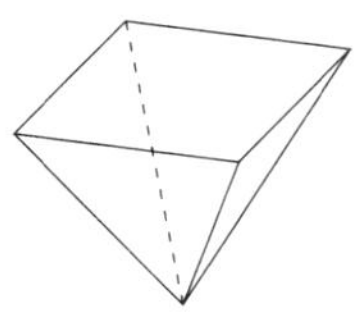

FIGURE 3. The simplest non-simple polytope.

Let $n_1, \ldots, n_N$ be the inner facet normals to P at v. They generate the rays of the normal cone $\mathcal{N}_v$. A virtual deformation of the vertex v is a regular triangulation[7] Δ of $\mathcal{N}_v$. That is a face-to-face subdivision of $\mathcal{N}_v$ into simplicial cones $\sigma_1, \ldots, \sigma_M$, so that there is a convex piecewise linear function on $\mathcal{N}_v$ with domains of linearity the σ_i.

[7]Compare [**Lee97**, § 14.3]. Moving the facets according to the values of the convex function at the n_i would yield an actual simple deformation of the vertex v.

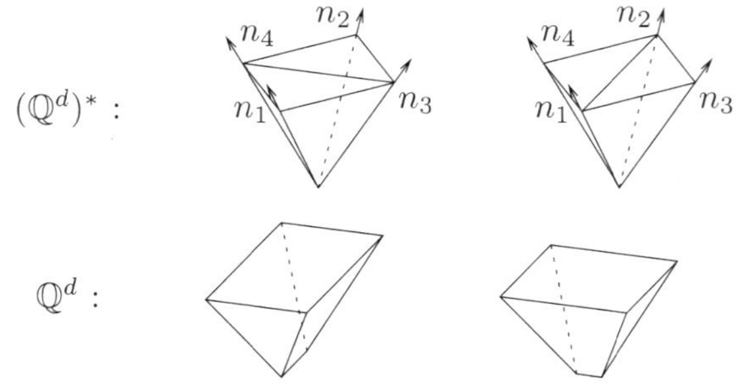

FIGURE 4. Triangulations and the corresponding deformations.

In the example, $v = (0,0,0)$ is the only non-simple vertex. The normal cone is generated by $n_1 = (1,0,1)$, $n_2 = (-1,0,1)$, $n_3 = (0,1,1)$, and $n_4 = (0,-1,1)$. It has two triangulations, both of which are regular: Δ_1 with maximal simplices spanned by n_1, n_3, n_4 and n_2, n_3, n_4, and Δ_2 with maximal simplices spanned by n_1, n_2, n_3 and n_1, n_2, n_4. The corresponding deformations are sketched in Figure 4.

Now the tangent cone $\mathcal{T}_v P$ can be written as the intersection of simple cones defined by the σ_i:

$$\mathcal{T}_v P = \bigcap_i \mathcal{T}_{\sigma_i} \text{ for } \mathcal{T}_{\sigma_i} = \{x \in \mathbb{Q}^d \, : \, n_j(x) \geq n_j(v) \text{ for all } n_j \in \sigma_i\}. \tag{6}$$

For generic[8] $\xi \in (\mathbb{Q}^d)^*$, we write ξ in terms of the $n_j \in \sigma_i$, and flip those inequalities in (6) where ξ has negative coefficients. Then we compute the index $\mathrm{ind}_\xi(\sigma_i)$ as the number of flipped inequalities, and define the local contribution at v as:

$$\mathbb{1}^\xi_{\Delta,v} = \sum_i (-1)^{\mathrm{ind}_\xi(\sigma_i)} \mathbb{1}_{\mathcal{T}^\xi_v \mathcal{T}_{\sigma_i}}$$

The cool thing is that $\mathbb{1}^\xi_{\Delta,v} = \mathbb{1}^\xi_v$ does not depend on the triangulation. So it really is a *local* contribution.

THEOREM 4. *Let $P \subset \mathbb{Q}^d$ be any polytope, and let $\xi \in (\mathbb{Q}^d)^*$ be generic. Then $\mathbb{1}^\xi_{\Delta,v} = \mathbb{1}^\xi_v$ does not depend on the choice of a regular triangulation Δ of the inner normal cone $\mathcal{N}_v$ at vertex v. Moreover,*

$$\mathbb{1}_P(x) = \sum_{v \text{ vertex of } P} \mathbb{1}^\xi_v(x). \tag{7}$$

The Δ-invariance can be shown, e.g., using the fact that all regular triangulations are connected by flips [**Lee97**, § 14.6]. In this note, however, we follow a different strategy. First, we will show the decomposition formula (7) for compatible choices of triangulations (Lemma 5). Then, we observe that under relatively weak conditions, all decompositions are the same (Lemma 6).

But let us first see how the Δ-invariance works out in our example. If we choose $\xi = (4,2,0)$, then we compute for triangulation Δ_1 as follows: $\xi = 4n_1 - n_3 - 3n_4 =$

[8]Now, generic means that ξ is not constant on any ray of any of the $\mathcal{T}_{\sigma_i}$'s, i.e., ξ does not lie on any hyperplane used in the triangulation.

$-4n_2 + 3n_3 + n_4$. Thus the polarized cones are

$$\mathcal{T}_v^\xi \mathcal{T}_{\sigma_1} = \{x \ : \ n_1(x) \geq 0, n_3(x) < 0, n_4(x) < 0\} \text{ with index 2 and}$$
$$\mathcal{T}_v^\xi \mathcal{T}_{\sigma_2} = \{x \ : \ n_2(x) < 0, n_3(x) \geq 0, n_4(x) \geq 0\} \text{ with index 1.}$$

The trace of $\mathbb{1}_{\mathcal{T}_v^\xi \mathcal{T}_{\sigma_1}} - \mathbb{1}_{\mathcal{T}_v^\xi \mathcal{T}_{\sigma_2}}$ in the $\xi = 12$ plane is sketched in Figure 5 on the left. Similarly, for Δ_2 we get

FIGURE 5. The local contribution from Δ_1 and from Δ_2.

$$\mathcal{T}_v^\xi \mathcal{T}_{\sigma_3} = \{x \ : \ n_1(x) \geq 0, n_2(x) < 0, n_3(x) \geq 0\} \text{ with index 1 and}$$
$$\mathcal{T}_v^\xi \mathcal{T}_{\sigma_4} = \{x \ : \ n_1(x) \geq 0, n_2(x) < 0, n_4(x) < 0\} \text{ with index 2.}$$

The trace of $\mathbb{1}_{\mathcal{T}_v^\xi \mathcal{T}_{\sigma_4}} - \mathbb{1}_{\mathcal{T}_v^\xi \mathcal{T}_{\sigma_3}}$ in the $\xi = 12$ plane is sketched in Figure 5 on the right. Observe that $\mathcal{T}_v^\xi \mathcal{T}_{\sigma_3}$ and $\mathcal{T}_v^\xi \mathcal{T}_{\sigma_4}$ cancel out on the overlap to yield the same contribution as $\mathcal{T}_v^\xi \mathcal{T}_{\sigma_1}$ and $\mathcal{T}_v^\xi \mathcal{T}_{\sigma_2}$.

LEMMA 5. *Let $P \subset \mathbb{Q}^d$ be any polytope, and let $\xi \in (\mathbb{Q}^d)^*$ be generic. Suppose Δ is a regular triangulation of the polar dual polytope P^*. It restricts to regular triangulations Δ_v of the normal cones $\mathcal{N}_v$. Then*

$$\mathbb{1}_P(x) = \sum_{v \text{ vertex of } P} \mathbb{1}^\xi_{\Delta_v, v}(x) \tag{8}$$

One can mimic the proof for the simple version (4) in order to show that the sum of the $\mathbb{1}^\xi_{\Delta_v,v}$ over all vertices of P does not depend on ξ, and that for every $x \in \mathbb{Q}^d$ there is a suitable ξ such that equation (8) is satisfied.

Now we actually use (8) together with the following lemma in order to show that $1^\xi_{\Delta,v} = 1^\xi_v$ is independent of the triangulation.

LEMMA 6. *Let $P \subset \mathbb{Q}^d$ be any polytope, and let ξ be a generic element of $(\mathbb{Q}^d)^*$. Suppose that there are two decompositions*

$$\mathbb{1}_P = \sum_{v \text{ vertex of } P} f_v = \sum_{v \text{ vertex of } P} g_v$$

that are both conic: *for every direction $t \in \mathbb{Q}^d \setminus 0$, the $f_v(v + \lambda t)$ and $g_v(v + \lambda t)$ are constant in $\lambda > 0$, and* positive: *$f_v(v + t) = g_v(v + t) = 0$ if $\xi(t) < 0$. Then $f_v = g_v$.*

PROOF. This proof illustrates what was meant by Morse theory earlier on. We sweep over the polytope bottom to top, and only at the vertices does something happen. That is, we compare the restrictions of the functions to the hyperplanes

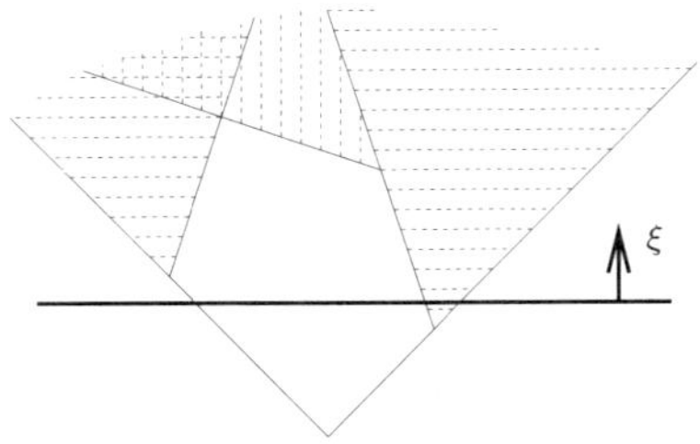

FIGURE 6. Positive conic decomposition of $\mathbb{1}_P$ is unique.

$\xi = c$. Order the vertices $v_1, \ldots, v_k$ in ξ-increasing order. For $c \in [\xi(v_i), \xi(v_{i+1})[$, we have

$$g_{v_i} \overset{(\mathbf{p})}{=} \sum_{j=i}^{k} g_{v_j} = \mathbb{1}_P - \sum_{j=1}^{i-1} g_{v_j} \overset{(\mathbf{c})}{=} \mathbb{1}_P - \sum_{j=1}^{i-1} f_{v_j} = \sum_{j=i}^{k} f_{v_j} \overset{(\mathbf{p})}{=} f_{v_i}$$

along $\xi = c$. For (**p**) we use positivity of the f_v's and g_v's, and for (**c**) we use induction, and the fact that the f_v's and g_v's are conic. □

Another incarnation of *the* positive conic decomposition is to group the summands in the Brianchon-Gram formula (1) as we did in equation (5), though it is less obvious that this is positive.

3.1. Homework. *or* **Why this is no good.** This is the wild speculation section. At the risk of exposing the full extent of my ignorance, I ask a bunch of questions that I stumbled over while compiling these notes.

If a good proof is one that makes us wiser [**Man77**], then this is a bad proof. It would be nicer to have a deformation independent definition of the local contribution in terms of the facet hyperplane arrangement at each vertex. How much geometry is needed? Can we compute the local contribution in purely combinatorial terms?

In some sense, the actual deformation of a vertex into several simple vertices converges to the non-simple vertex, and the local contribution is continuous. While this asks for a topology on the polytope algebra, I believe that this is a more combinatorial question. There must be a combinatorial deformation theory for (Eulerian?) posets.

Where are Gröbner bases? Often when regular triangulations show up, the corresponding convex function determines a term order. Does the result tell us something about commutative algebra?

References

[Aga03] José Agapito. A weighted version of quantization commutes with reduction principle for a toric manifold, July 2003. Preprint math.SG/0307318.

[Bar93] Alexander I. Barvinok. Computing the volume, counting integral points, and exponential sums. *Discrete Comput. Geom.*, 10(2):123–141, 1993.

[Bar02] Alexander Barvinok. *A course in convexity*, volume 54 of *Graduate Studies in Mathematics*. American Mathematical Society, Providence, RI, 2002.

[Bri37] C. J. Brianchon. Théorème nouveau sur les polyèdres. *J. Ecole Polytechnique*, 15:317–319, 1837.

[Bri88] Michel Brion. Points entiers dans les polyèdres convexes. *Ann. Sci. École Norm. Sup. (4)*, 21(4):653–663, 1988.

[BRSW04] Matthias Beck, Sinai Robins, Frank Sottile, and Jonathan Weitsman. Conic decompositions of polytopes, 2004. In preparation.

[BV97] Michel Brion and Michèle Vergne. Residue formulae, vector partition functions and lattice points in rational polytopes. *J. Amer. Math. Soc.*, 10(4):797–833, 1997.

[Gra74] J. P. Gram. Om rumvinklerne i et polyeder. *Tidsskrift for Math. (Copenhagen)*, 4(3):161–163, 1874.

[Grü03] Branko Grünbaum. *Convex polytopes*, volume 221 of *Graduate Texts in Mathematics*. Springer-Verlag, New York, second edition, 2003. Prepared and with a preface by Volker Kaibel, Victor Klee and Günter M. Ziegler.

[KSW03] Yael Karshon, Shlomo Sternberg, and Jonathan Weitsman. The Euler-Maclaurin formula for simple integral polytopes. *Proc. Natl. Acad. Sci. USA*, 100(2):426–433 (electronic), 2003.

[Law91a] Jim Lawrence. Polytope volume computation. *Math. Comp.*, 57(195):259–271, 1991.

[Law91b] Jim Lawrence. Rational-function-valued valuations on polyhedra. In *Discrete and computational geometry (New Brunswick, NJ, 1989/1990)*, volume 6 of *DIMACS Ser. Discrete Math. Theoret. Comput. Sci.*, pages 199–208. Amer. Math. Soc., Providence, RI, 1991.

[Lee97] Carl W. Lee. Subdivisions and triangulations of polytopes. In Jacob E. Goodman and Joseph O'Rourke, editors, *Handbook of Discrete and Computational Geometry*, pages 271–290. CRC–Press, New York, 1997.

[Man77] Yu. I. Manin. *A course in mathematical logic*. Springer-Verlag, New York, 1977. Translated from the Russian by Neal Koblitz, Graduate Texts in Mathematics, Vol. 53.

[PK92] A. V. Pukhlikov and A. G. Khovanskiĭ. The Riemann-Roch theorem for integrals and sums of quasipolynomials on virtual polytopes. *Algebra i Analiz*, 4(4):188–216, 1992.

[She67] G. C. Shephard. An elementary proof of Gram's theorem for convex polytopes. *Canad. J. Math.*, 19:1214–1217, 1967.

[Som27] D. M. Y. Sommerville. The relations connecting the angle-sums and volume of a polytope in space of n dimensions. *Proceedings Royal Soc. London (A)*, 115:103–119, 1927.

[Var87] A. N. Varchenko. Combinatorics and topology of the arrangement of affine hyperplanes in the real space. *Funktsional. Anal. i Prilozhen.*, 21(1):11–22, 1987.

[Zie95] Günter M. Ziegler. *Lectures on Polytopes*, volume 152 of *GTM*. Springer–Verlag, 1995.

E-mail address: haase@math.duke.edu

Department of Mathematics, Duke University, Durham, NC 27708–0320, USA

Contemporary Mathematics
Volume **374**, 2005

Gröbner basis degree bounds on $\mathrm{Tor}_{\bullet}^{k[\Lambda]}(k,k)_{\bullet}$ and Discrete Morse Theory for Posets

Patricia Hersh and Volkmar Welker

Abstract. The purpose of this paper is twofold.
▷ We give combinatorial bounds on the ranks of $\mathrm{Tor}_{\bullet}^{R}(k,k)_{\bullet}$ in the case where $R = k[\Lambda]$ is an affine semi-group ring, and in the process provide combinatorial proofs for bounds by Eisenbud, Reeves and Totaro on which Tor groups vanish. In addition, we show that if the bounds hold for a field k then they hold for $\mathbb{K}[\Lambda]$ and any field $\mathbb{K}$. Moreover, we provide a combinatorial construction for a free resolution of $\mathbb{K}$ over $\mathbb{K}[\Lambda]$ which achieves these bounds.
▷ We extend the lexicographic discrete Morse function construction of Babson and Hersh for the determination of the homotopy type and homology of order complexes of posets to a larger class of facet orderings that includes orders induced by monomial term orders.

Since it is known that the order complexes of finite intervals in the poset of monomials in $k[\Lambda]$ ordered by divisibility in $k[\Lambda]$ govern the Tor-groups, the newly developed tools are applicable and serve as the main ingredients for the proof of the bounds and the construction of the resolution.

1. Introduction.

Let Λ be a submonoid of $\mathbb{N}^e$ which is finitely generated by $\alpha_1, \ldots, \alpha_n$, and denote by $k[\Lambda]$ the affine semi-group ring of Λ generated over the field k by monomials $\underline{\mathbf{x}}^{\alpha_i} = x_1^{\alpha_{i1}} \cdots x_e^{\alpha_{ie}}$, $1 \leq i \leq n$. Thus, $k[\Lambda] \cong k[z_1, \ldots, z_n]/I_\Lambda$ is the coordinate ring of an affine, not necessarily normal, toric variety. The isomorphism results from sending z_i to $\underline{\mathbf{x}}^{\alpha_i}$, $1 \leq i \leq n$, and letting the toric ideal I_Λ record the syzygies among the generators. The monoid Λ is endowed with a partial order given by $\mu \leq \lambda$ if and only if $\lambda - \mu \in \Lambda$. Denote by $\Delta(\mu, \lambda)$ the simplicial complex of linearly ordered subsets $\mu < \mu_0 < \cdots < \mu_i < \lambda$ of the interval $[\mu, \lambda] := \{\gamma \in \Lambda \mid \mu \leq \gamma \leq \lambda\}$.

Based on work by Laudal and Sletjøe [**LS**] and Peeva, Reiner, Sturmfels [**PRS**], several recent papers (see e.g. [**HRW**], [**BW**]) have used this partial order on Λ and the simplicial homology of order complexes $\Delta(\mu, \lambda)$ for $\mu \leq \lambda$ in Λ as a tool for understanding minimal free resolutions of the field k as a $k[\Lambda]$-module. In general,

---

2000 *Mathematics Subject Classification.* 05E99,13A02.

Key words and phrases. Affine semigroup ring, free resolution, Koszul property, discrete Morse theory.

The first author was supported by an NSF Math Sciences Postdoctoral Research Fellowship. The second author was supported by EU Research Training Network "Algebraic Combinatorics in Europe," grant HPRN-CT-2001-00272.

the minimal free resolution of the field k over a k-algebra R is still a mysterious object (see [**Av**]) and even results known to hold by algebraic arguments pose hard and interesting combinatorial questions when $R = k[\Lambda]$ (see [**BjWe**]). Notably the Koszul property has attracted a lot of interest (for background motivating the study of the Koszul property see [**Fr**]). This property is equivalent by work of Peeva, Reiner, Sturmfels [**PRS**] to the property of all intervals in the poset Λ being Cohen-Macaulay over k. In general, it is known that a standard graded k-algebra is Koszul whenever its defining ideal has a quadratic Gröbner basis. In Peeva, Reiner, Sturmfels [**PRS**] and subsequent work [**HRW**] a combinatorial understanding of this implication is developed. In [**BW**] it is shown that if each interval in Λ is shellable then it is actually possible to construct a minimal free resolution for k as a $k[\Lambda]$-module.

In Section 5 we give an alternative combinatorial approach based on a discrete Morse function that also explains all these phenomena related to the Koszulness of $k[\Lambda]$ without requiring a shelling. The main idea behind our combinatorial approach is quite natural, and is explained in Remark 2.8 and the discussion that follows, after suitable notation is introduced. Section 6 uses the discrete Morse function of Section 5 to provide a minimal free resolution for k as a $k[\Lambda]$-module when I_Λ has a quadratic Gröbner basis, whether or not each interval in Λ is shellable.

In Section 7 we give the proof of our main result, a discrete Morse function on the order complex of Λ which provides combinatorial upper bounds on all the multigraded Tor groups for $k[\Lambda]$, most notably yielding the following:

THEOREM 1.1. *Let $\Lambda \subseteq \mathbb{N}^e$ be an affine semi-group generated by n elements of $\mathbb{N}^e$. Assume there is a field k such that for $k[\Lambda] \cong k[z_1, \ldots, z_n]/I_\Lambda$ the ideal I_Λ has a Gröbner basis of degree d, then*

(i) $\tilde{H}_i(\Delta(\hat{0}, \lambda); \mathbb{K}) = 0$ *for* $i < -1 + \frac{\deg(\lambda)-1}{d-1}$ *and any field* $\mathbb{K}$.

(ii) $\mathrm{Tor}_i^{k[\Lambda]}(\mathbb{K}, \mathbb{K})_\lambda = 0$ *for* $i < 1 + \frac{\deg(\lambda)-1}{d-1}$ *and any field* $\mathbb{K}$.

Moreover, the vanishing of Tor*-groups is achieved by a free cellular resolution resulting from a discrete Morse function on Λ.*

Here we denote by $\deg(\lambda)$ the length, i.e. cardinality minus one, of a saturated chain in the poset interval $[\hat{0}, \lambda]$. Note that if all generators of Λ lie on an affine hyperplane then this grading actually makes Λ a graded poset and $k[\Lambda]$ a standard graded k-algebra. In general, $\deg(\lambda)$ is the degree of the image of $\underline{\mathbf{x}}^\lambda$ in the associated graded ring of R.

By results of [**LS**], Theorem 1.1 (i) will immediately imply Theorem 1.1 (ii). The vanishing of Tor-groups in the case $\mathbb{K} = k$ also follows for general standard graded k-algebras, by a flat degeneration argument, from a result of Eisenbud, Reeves and Totaro (cf. [**ERT**]) about monomial ideals. Our arguments are completely combinatorial.

The main tool for the proof of Theorem 1.1 is discrete Morse theory, which was developed in the mid 90's by Forman [**Fo**]. Discrete Morse theory is a tool for determining the homology and homotopy type of a simplicial complex, or more general a regular CW-complex. In [**BH**] the authors develop tools that facilitate the use of discrete Morse theory in the case when the simplicial complex is the order complex of a poset whose edges are labeled. The edges $\mu < \lambda$ of the Hasse diagram of Λ are naturally labeled by $\lambda - \mu$ which by the definition of the order relation is one of the generators $\lambda_1, \ldots, \lambda_n$ of Λ. Thus, to any saturated chain

$\mu_0 < \cdots < \mu_i$ in λ there is associated the monomial $z_{j_1} \cdots z_{j_n}$ in $k[z_1, \ldots, z_n]$, where $\mu_i - \mu_{i-1} = \lambda_{j_i}$. In particular, any term order on $k[z_1, \ldots, z_n]$ induces a partial order on the finite saturated chains in Λ. In the case of Gröbner bases with properties analogous to requirements made in [**PRS**] (called 'supporting a poset"), we can apply the results from [**BH**] on lexicographic discrete Morse function which then easily give a degree d analogue of a lexicographic shelling, and imply the desired connectivity bound. In this Morse function, collections of at most d labels, given by descents and Gröbner basis leading terms, will play the role traditionally filled by the descents in a lexicographic shelling. In order to able to handle arbitrary Gröbner bases, whether or not they support a poset, we extend (Sections 3 and 4) the applicability of the methods from [**BH**] by using critical cell cancellation via gradient path reversal. Most notably, we introduce the notion of a content-lex facet ordering, which has also recently proven useful in work of [**HHS**]. The discrete Morse function on $\Delta(\Lambda) := \bigcup_{\lambda \in \Lambda} \Delta(\hat{0}, \lambda)$ also yields a free resolution of $\mathbb{K}$ over $\mathbb{K}[\Lambda]$ whose multigraded Betti-numbers can be read off from the number of critical cells of given dimension and given multidegree, i.e. from the Morse numbers, and from the gradient paths governing incidence among critical cells. Sections 6 and 8 will describe the critical cells in the Morse function as follows:

THEOREM 1.2. *The critical cells of the discrete Morse function on $\Delta(\Lambda)$ in bijection with the words of a language accepted by a finite state automaton, i.e. the words of a regular language. Thus, the generating function for Morse numbers is a rational function whose coefficients give upper bounds on all of the Betti numbers.*

The preceding theorem provides a construction of a free resolution of k over $k[\Lambda]$ for which the free modules in homological degree i are indexed by the critical cells of dimension i of the Morse-matching and whose differential can be calculated from the gradient paths.

In the quadratic Gröbner basis case, the resolution will be minimal and thus the generating function is exactly the Poincaré-Betti series, yielding yet another proof of its rationality in this case. There are known rational function bounds on the Poincaré-Betti series (see Proposition 3.3.2 in [**Av**]). But in contrast to these bounds our rational function comes close to the actual Betti numbers, in the sense that it exhibits the vanishing of Betti numbers as in Theorem 1.1.

We conclude the paper in Section 9 with remarks and open questions.

2. Background

2.1. Posets and Order Complexes. Let P be a poset with unique minimal element $\hat{0}$ and unique maximal element $\hat{1}$. We denote by $\Delta(P)$ the simplicial complex whose i-simplices are the chains $\hat{0} < p_0 < \cdots < p_i < \hat{1}$ in P. The maximal chains – with respect to inclusion – in P are sometimes called *saturated chains.* Notice that the saturated chains in P give rise to the facets (maximal faces) in $\Delta(P)$, and sometimes we will speak of saturated chains of P and facets of $\Delta(P)$ interchangeably. For $x \leq y$ in P, let $[x, y]$ be the closed interval $\{z \mid x \leq z \leq y\}$ and $(x, y) := [x, y] - \{x, y\}$ the open interval. We write $\Delta(x, y)$ for $\Delta([x, y])$.

For an arbitrary simplicial complex we denote by $H_i(\Delta; R)$ and $\widetilde{H}_i(\Delta; R)$ the non-reduced and reduced simplicial homology of Δ with coefficients in the ring R. The i-th Betti number b_i of Δ (with respect to $\mathbf{Z}$) is the rank of the free part of the i-th non-reduced homology group of Δ with coefficients in $\mathbf{Z}$. In order to calculate the non-reduced homology $H_i(\Delta; R)$ or the reduced homology $\widetilde{H}_i(\Delta; R)$ for the relevant simplicial complexes Δ we will use two basic facts from algebraic topology. First, if Δ is homotopy equivalent to a topological space X then the simplicial homology of Δ and the cellular/singular homology of X coincide. In our situation X will always be a CW-complex. The second important fact is, that if a CW-complex X has m_i cells of dimension i then the i-th Betti number b_i satisfies $b_i \leq m_i$. A poset P if called *homotopically Cohen-Macaulay* if the order complexes $\Delta(x, y)$ of all intervals $[x, y]$ in P are homotopy equivalent to a wedge of spheres of dimension $\dim(\Delta(x, y))$. In particular this holds if for each $\Delta(x, y)$ either $\dim(\Delta(x, y)) \leq 0$ or $\dim(\Delta(x, y)) \geq 1$ and $\Delta(x, y)$ is homotopy equivalent to a CW-complex with no cell in dimension $0 < i < \dim(\Delta(x, y))$ and a single cell in dimension 0.

Our main tool for the construction of a CW-complex X homotopy equivalent to a given simplicial complex Δ is discrete Morse theory.

2.2. Discrete Morse Theory: General Theory. This section reviews discrete Morse theory results we will need from [**Fo**], [**Ch**], [**Jo**], [**BH**], [**He2**] and [**BW**], along with some other requisite background. Forman [**Fo**] defines a function f which assigns real values to the cells in a regular CW-complex X to be a *discrete Morse function* if for each cell $\sigma \in X^{(*)}$ the sets

$$\left\{\tau \subseteq \overline{\sigma} \;\middle|\; \tau \in X^{(*)}, \dim(\tau) = \dim(\sigma) - 1, f(\tau) \geq f(\sigma)\right\}$$

and

$$\left\{\overline{\tau} \supseteq \sigma \;\middle|\; \tau \in X^{(*)}, \dim(\tau) = \dim(\sigma) + 1, f(\tau) \leq f(\sigma)\right\}$$

each have cardinality at most one. Here $X^{(*)}$ denotes the collection of open cells in X and $\overline{\sigma}$ denotes the closure of σ in X for $\sigma \in X^{(*)}$. The condition implies that for each σ, at most one of two sets is non-empty. When both are empty, then σ is called a *critical cell.* The main result on discrete Morse functions is the following:

THEOREM 2.1 ([**Fo**]). *If f is a discrete Morse function on the regular CW-complex X then X is homotopy equivalent to a (not necessarily regular) CW-complex X^M, such that for any given i the number of cells of dimension i in X^M equals the number of critical cells of dimensions i of the Morse function of f. Moreover, incidences among cells in X^M are governed by a collapsing procedure that leads from X to X^M while preserving homotopy type at each step.*

In [**Ch**], Chari reformulated discrete Morse functions for regular CW-complexes in terms of certain types of face poset matchings. Recall that the *face poset* of a CW-complex X is the partial order $F(X)$ on the cells in $X^{(*)}$ defined by $\tau \leq \sigma$ whenever τ is contained in the closure $\overline{\sigma}$ of σ. If X is the geometric realization of an abstract simplicial complex Δ then this order is just the inclusion relation between the simplices of Δ. The *Hasse diagram* of a poset is the graph whose vertices are the poset elements and whose edges are the covering relations $x \prec y$, i.e. pairs $x < y$ such that $x \leq z \leq y$ implies $z = x$ or $z = y$.

DEFINITION 2.2 ([**Ch**]). A matching on the Hasse diagram of the face poset $F(X)$ of a regular CW-complex X is called *acyclic* if the directed graph obtained by directing matching edges upward and all other poset edges downward has no directed cycles.

Notice that the non-critical cells of a discrete Morse function f come in pairs that prevent each other from being critical. Hence, this pairing gives a matching on the face poset of the CW-complex. Furthermore, this matching is acyclic, because Chari's edge orientation will orient all edges in the direction in which f weakly decreases. Conversely, many different (but in some sense equivalent) discrete Morse functions may be constructed from any face poset acyclic matching. For instance, one may obtain f by choosing a monotone function on any total order extension of the partial order given by the acyclic directed graph. The face poset elements that are left unmatched by an acyclic matching are exactly the critical cells in any corresponding discrete Morse function. We will work exclusively in terms of acyclic matchings rather than discrete Morse functions, but at times it is helpful to have both points of view in mind.

Denote by m_i the number of critical cells of dimension i in a discrete Morse function on a regular CW-complex X. As usual b_i is the i-th Betti number of X. By virtue of Theorem 2.1 X is homotopy equivalent to a complex X^M constructed from m_i cells of dimension i. The first of the following two results is an immediate corollary from Theorem 2.1, the second was first proved in [**Fo**]. Both results exhibit a strong analogy with traditional Morse theory:

$$m_j \geq b_j \text{ for } 0 \leq j \leq \dim(X) \tag{2.1}$$

$$\sum_{i=0}^{\dim(X)} (-1)^i m_{\dim(X)-i} = \sum_{i=0}^{\dim(X)} (-1)^i b_{\dim(X)-i} = \chi(X) \tag{2.2}$$

The inequality (2.1) will be used in later sections in oder to obtain bounds on Betti numbers. For our applications also the following special situation which we already mentioned in Section 2.1 will be of importance.

Corollary 2.3. *Let X be a regular CW-complex and f a discrete Morse functions on X with m_i critical cells of dimension i. If $m_i = 0$ for $0 < i < j$ and $m_0 = 1$ then X is $(j-1)$-connected. In particular, if for all order-complexes $\Delta(x, y)$ of intervals $[x, y]$ in a poset P such that $\dim(\Delta(x, y)) > 0$ there is a Morse function f_{xy} with no critical cell in dimension $0 < i < \dim(\Delta(x, y))$ and a single critical cell in dimension 0, then each interval is homotopy equivalent to a wedge of spheres of dimension $\dim((\Delta(x, y))$ and P is homotopically Cohen-Macaulay.*

Definition 2.4. Let X be a regular CW-complex and f a discrete Morse function on X. A *gradient path* from a critical cell τ to another critical cell σ of dimension $\dim(\tau) - 1$ is a directed path upon which the Morse function weakly decreases.

It is a simple consequence of the definition that a gradient path between τ and σ will alternate between cells of dimension $\dim(\tau)$ and $\dim(\tau) - 1$.

It will turn out that for our purposes we need to find a discrete Morse function on $\Delta(x, y)$ for intervals $x \leq y$ in an affine semigroup Λ such that inequality (2.1) becomes an equality. We will not always be able to achieve this goal and indeed it is open whether this is even possible. In our approach we construct a discrete Morse function on $\Delta(x, y)$ and then try to optimize the function in order to make (2.1) tight. For the latter we will employ the following observation.

OBSERVATION 2.5 ([**Fo**]). If there is a unique gradient path from τ to σ, then reversing the orientation of each edge in this path yields a new acyclic matching for which τ and σ are no longer critical.

We refer to the procedure described in Observation 2.5 as "cancelling critical cells."

The following lemma from [**Jo**] will be useful for combining several acyclic matchings to a single acyclic matching.

LEMMA 2.6 (Cluster Lemma). *Let X be a regular CW-complex which decomposes into collections X_p of cells indexed by the elements p in a partial order P with unique minimal element $\hat{0}$ as follows:*

(1) *X decomposes into the disjoint union $\cup_{p\in P} X_p$, that is, each cell belongs to exactly one X_p*
(2) *For each $p \in P$, $\cup_{q\leq p} X_p$ is a subcomplex of X*

For each $p \in P$, let M_p be an acyclic matching on the subposet $F(X) \cap X_p$ of $F(X)$ consisting of the cells in X_p. Then $\cup_{p\in P} M_p$ is an acyclic matching on $F(X)$.

2.3. Discrete Morse Theory: The Case of Poset Order Complexes. Let P be a poset with unique minimal element $\hat{0}$ and unique maximal element $\hat{1}$. Assume that the edges of the Hasse diagram of P are labeled by a labelling $\mathfrak{L}$ which takes values in a linearly ordered set. Via this labelling we can assign to each saturated chain $p_0 < p_1 < \cdots < p_{i-1}$ in P of cardinality i an $(i-1)$-tuple $(\mathfrak{L}(p_0,p_1), \ldots, \mathfrak{L}(p_{i-1},p_i))$. Thus any linear extension of the lexicographic ordering on the tuples induces a linear order on the saturated chains in P.

In our situation where $P = [x,y]$ for $x \leq y$ in an affine semigroup Λ generated by $\alpha_1, \ldots, \alpha_n$ the labelling $\mathfrak{L}$ is given by sending a cover relation $\lambda \prec \mu$ to $\alpha_i = \mu - \lambda$ or if we consider $k[\Lambda]$ as a quotient of $k[z_1, \ldots, z_n]$ we can equivalently label $\lambda \prec \mu$ with z_i. Note, that in the latter case the product over the labels of a saturated chain is a monomial in $[z_1, \ldots, z_n]$. In either case we can choose an arbitrary linear order on the sets $\{\alpha_1, \ldots, \alpha_n\}$ or $\{z_1, \ldots, z_n\}$. Since the two labellings are equivalent we will not distinguish between z_1 and λ_i in the rest of the paper.

It is shown in [**BH**] how to construct a discrete Morse function on $\Delta(P)$, respectively an acyclic matching on the face poset $F(\Delta(P))$ of $\Delta(P)$ for a labeled poset P, from the lexicographic order on the saturated chains of P. If this lexicographic ordering happens to be a shelling order (see [**BjW**]) then it is possible to infer directly from the constructed Morse function that the poset is homotopically Cohen-Macaulay – just as it can be deduced from the lexicographic shelling itself. We will refer to a Morse function resulting from a lexicographic ordering on the saturated chains as a *lexicographic discrete Morse function.* We will show in Section 3 that the construction of [**BH**] applies to a larger class of orders on saturated chains. This will allow us to construct a discrete Morse function on $\Delta(\Lambda)$ for an affine semigroup Λ generated by $\alpha_1, \ldots, \alpha_n$ from a facet order based on an arbitrary monomial term order on $k[z_1, \ldots, z_n]$.

Let $\mathfrak{L}$ be a labelling of the poset covering relations such that $\mathfrak{L}(u,v) \neq \mathfrak{L}(u,w)$ for $v \neq w$. Then the ordering of the label sequences on the saturated chains by the lexicographic order already gives a total order on saturated chains, i.e. an ordering $F_1, \ldots, F_r$ on facets in $\Delta(P)$. To describe the corresponding discrete Morse function from [**BH**], we will need to speak of the ranks of elements of a saturated chain, whether or not a poset is graded. We will do this by assigning to an element of a saturated chain the rank of the element within the chain and speak of the rank with respect to the chain. Indeed, we do not require consistency of the notion of rank between different saturated chains. Also we will identify a set of chains in P, resp. faces of $\Delta(P)$, with the simplicial complex generated by the chains.

Each maximal face in $F_j \cap (\cup_{i<j} F_i)$ has rank set (with respect to F_j) of the form $1, \dots, i, j, \dots, n$, i.e. it consists of all ranks in F_j except for a single interval $i+1, \dots, j-1$ of consecutive ranks that are omitted; this follows from the use of a lexicographic order on facets. Call each such list $i+1, \dots, j-1$ of ranks a *minimal skipped interval* of F_j, and say the interval has *height* $j-i-1$. Following [**BiH**] and [**BH**], call the collection of minimal skipped intervals for F_j the *interval system* or *set of I-intervals* of F_j.

For each facet F_j, [**BH**] constructs an acyclic matching on the set of faces in $F_j \setminus \cup_{i<j} F_i$ in terms of the interval system for F_j. This is done in such a way that the union (over all F_j) of these matchings is acyclic, and each $F_j \setminus (\cup_{i<j} F_i)$ includes at most one critical cell. We say F_j *contributes a critical cell* if $F_j \setminus (\cup_{i<j} F_i)$ contains a critical cell. F_j will contribute a critical cell if and only if the homotopy type changes with the attachment of F_j.

Description of critical cells:

(Case 1) If the I-intervals of F_j do not collectively have support covering all the ranks in F_j, then F_j does not contribute a critical cell.

(Case 2) If the I-intervals of F_j cover all ranks and have disjoint support, then the critical cell consists of the lowest rank from each of the I-intervals.

(Case 3) If there is some overlap in the minimal skipped intervals of F_j, then iterate the following procedure to obtain the critical cell, ordering I-intervals so that their minimal ranks are increasing:

1. Include the lowest rank from I_1 in the critical cell.
2. Truncate all the remaining minimal. skipped intervals by chopping off any ranks that they share with I_1.
3. Discard I_1 and any skipped intervals that are no longer minimal.
4. Re-index the remaining truncated minimal skipped intervals to begin with a new I_1.
5. Repeat until there are no more minimal skipped intervals.

The non-overlapping intervals obtained by the above truncation procedure are called the *J-intervals* of F_j.

REMARK 2.7. F_j contributes a critical cell if and only if its I-intervals cover all ranks of F_j. In this case, the dimension of the critical cell is one less than the number of J-intervals.

In order to cancel critical cells by reversing gradient paths, we will also need some information about the matching itself.

Description of the acyclic matching on $F_j \setminus \cup_{i<j} F_i$:

- If F_j has no critical cells, then there is at least one cone point in $F_j \cap (\cup_{i<j} F_i)$. In this situation we match by including/excluding cone point of lowest rank.

- If F_j contributes a critical cell, then match any non-critical cell based on the lowest I-interval of F_j where the cell differs from the critical cell. Specifically, match by including/excluding the lowest element of the I-interval, since the cell must include at least one element of the I-interval other than this lowest possible element, in order to cover the I-interval but differ from the critical cell.

REMARK 2.8. If there is some d such that every I-interval in a lexicographic discrete Morse function has height at most $d-1$, then the above construction immediately implies that each poset interval (x, y) is at least $(-1+\frac{\text{remark(y)}-\text{remark(x)}-1}{d-1})$-connected. In fact, it suffices for the average interval height to be at most $d-1$.

In our setting, where $P = \Delta(x, y)$ for $x \leq y$ in an affine semigroup Λ Remark 2.8 will turn out to be applicable for d the degree of a Gröbner basis of the ideal I_Λ (i.e. the maximal total degree of a polynomial in the Gröbner basis). We will use the fact that every I-interval results from a label sequence descent or from a syzygy leading term, by virtue of our use of a monomial term order to order the saturated chains. Furthermore, all leading terms will be divisible by Gröbner basis leading terms of degree at most d, which will allow us to show that the average interval height is at least $d-1$, yielding the desired lower bound on connectivity. In a sense, this gives a new combinatorial explanation for the connection between Gröbner basis degree and complexity, in which a shelling is the special case with $d = 2$. However, just as in the shelling in [PRS], only certain types of monomial term orders and Gröbner bases will immediately yield the degree d analogue of a shelling.

EXAMPLE 2.9. Consider $k[z_1, z_2, z_3, z_3]/(z_1z_4 - z_2^2)$ with a term order $\preceq$ such that $\text{in}_\preceq(z_1z_4 - z_2^2) = z_1z_4$. If z_1z_4 is a Gröbner basis leading term, but z_1z_3 and z_3z_4 are not Gröbner basis leading terms, Then $z_3(z_1z_4 - z_2^2) = 0$ precludes the shelling from [**PRS**], since z_1z_3 and z_3z_4 are not leading terms,

In order to be able to allow completely general monomial term orders, i.e. to deal with situations such as in Example 2.9, we need to extend the tools from [**BH**] described in this section by performing critical cell cancellation. This extension will be done in Section 3. The the method of critical cell cancellation is explained in the next section.

2.4. Discrete Morse Theory: Optimizing Discrete Morse Functions. This section reviews tools from [**He2**] for eliminating critical cells by cancelling pairs by a gradient path reversal. Later we will construct a lexicographic discrete Morse function for monoid posets, and then use these Morse function optimization tools to eliminate all of the low-dimensional critical cells.

Let Δ be a simplicial complex and f a discrete Morse function on Δ. Define the *multi-graph face poset*, denoted $F(\Delta)^M$, for the complex Δ^M of critical cells as follows:

(1) The vertices in $F(\Delta)^M$ are the cells in Δ^M, or equivalently the critical cells in the discrete Morse function on Δ.
(2) There is one edge between a pair of cells σ, τ of consecutive dimension $\dim(\tau) = \dim(\sigma) + 1$ for each gradient path from τ to σ.

THEOREM 2.10. *Any acyclic matching on* $F(\Delta)^M$ *specifies a collection of gradient paths in* $F(\Delta)$ *that may simultaneously be reversed to obtain a discrete Morse function* M' *whose critical cells are the unmatched cells in the matching on* $F(\Delta)^M$.

To cancel cells, we will need to know that a gradient path from a critical cell τ to a critical cell σ is the only gradient path from τ to σ.

DEFINITION 2.11. Let $u \prec v \prec w$ be covering relations in a poset P labeled by $\mathfrak{L}$. Assume we are given a linear order on the saturated chains in $[u, w]$. The labels $\mathfrak{L}(u, v)$ and $\mathfrak{L}(v, w)$ on covering relations $u \prec v \prec w$ are said to *commute* if the least saturated chain in $[u, w]$ is labeled by $\mathfrak{L}(u, v), \mathfrak{L}(v, w)$ arranged in ascending order.

Let $\mathfrak{L}(u, v)$ denote the sequence of edge labels on the least saturated chain from u to v.

DEFINITION 2.12. Let P be a poset labelled by $\mathfrak{L}$ and assume that within each interval the saturated chains are linearly ordered by an order depending only on the label sequence.

(i) The weakly increasing rearrangement of the label sequence of a saturated chain is called the *content* of the chain.
(ii) The labelling $\mathfrak{L}$ is called *least-increasing* if every interval has a (weakly) increasing chain as its least saturated chain.
(iii) The labelling $\mathfrak{L}$ is is called *least-content-increasing* if it is east-increasing and in addition the label sequence of the least chain equals or precedes the content of the label sequence of every other saturated chain in the interval.

Note, that since the definition assumes the linear order on the saturated chains only depends on the label-sequences, Definition 2.12(iii) we can consider this order as an order on label sequences. If the linear order on the saturated chains is given by the lexicographic order then the condition least-increasing is weaker than being an EL-labelling (see [**BjW**]), in that intervals may have several increasing chains.

REMARK 2.13. If two critical cells τ, σ in a least-content-increasing labelling are contributed by saturated chains of equal content, then every downward step in any gradient path from τ to σ must preserve content, and in fact must sort labels on the interval where the chain element was deleted.

Combining results (see Theorem 6.6) in [**He2**] yields the following.

THEOREM 2.14. *Let* P *be a poset labelled by a least-content-increasing labelling* $\mathfrak{L}$. *Let* M *be the lexicographic discrete Morse induced by* $\mathfrak{L}$ *and let* τ, σ, $\dim(\tau) = \dim(\sigma) + 1$, *be critical cells resulting from saturated chains whose label sequences* $\mathfrak{L}(\tau), \mathfrak{L}(\sigma)$ *have equal content. Suppose further that the permutation transforming* $\mathfrak{L}(\tau)$ *to* $\mathfrak{L}(\sigma)$ *is 321-avoiding. If there is a gradient path* γ *from* τ *to* σ *such that each downward step swaps a pair of consecutive labels by deleting an element* v *from a chain which also includes elements covering and covered by* v, *then* γ *is the unique gradient path from* τ *to* σ.

REMARK 2.15. Our upcoming discrete Morse function will use a content-lex facet order, as introduced in Section 3. Theorem 2.14 also applies in that setting without requiring any modification.

Theorem 4.2 will generalize the above to deal with non-saturated chain segments, as needed for cancelling critical cells in our upcoming Morse function.

2.5. Discrete Morse Theory: Application to Cellular Resolutions. Let M be a module over a commutative ring R. A *free resolution* of M over R is a complex of free R-modules F_i and R-module homomorphisms ∂_i

$$\mathcal{F} : \cdots \stackrel{\partial_i+1}{\to} F_i \stackrel{\partial_i}{\to} F_{i-1} \stackrel{\partial_{i-1}}{\to} \cdots \stackrel{\partial_1}{\to} F_0.$$

which is exact in all degrees $\neq 0$ (i.e., $\mathrm{Im}(\partial_i) = \mathrm{Ker}(\partial_{i-1})$ for $i \geq 2$) and $\mathrm{Coker}(\partial_1) \cong M$. In our case $R = k[\Lambda]$ is a k-algebra and carries an additional multigraded structure, Recall that a k-algebra $R = k[z_1, \ldots, z_n]/I$ is called $\mathbb{N}^d$-*multigraded* if $R = \bigoplus_{\alpha \in \mathbb{N}^d} R_\alpha$ as k-vector spaces and $R_\alpha R_\beta \subseteq R_{\alpha+\beta}$. If $d = 1$ and R is generated in degree 1 over $k = R_0$ then R is called *standard graded.* Analogously defined are $\mathbb{N}^d$-graded R-modules. In this situation we consider multigraded free resolutions $\mathcal{F}$. In addition to being a resolution one demands that the F_i are free multigraded R-modules and that the ∂_i are $\mathbb{N}^d$-homogeneous. A free multigraded module F is a direct sum $\oplus_{\alpha \in \mathbb{N}^d} R(-\alpha)^{\beta_\alpha}$ of free R-modules $R(-\alpha)$ of rank one whose grading is defined by assigning α as the degree of the unit element 1.

A $\mathbb{N}^d$-graded free resolution

$$\mathcal{F} : \cdots \stackrel{\partial_i+1}{\to} \bigoplus_{\alpha \in \mathbb{N}} R(-\alpha)^{\beta_{i,\alpha}} \stackrel{\partial_i}{\to} \bigoplus_{\alpha \in \mathbb{N}} R(-\alpha)^{\beta_{i-1,\alpha}} \stackrel{\partial_{i-1}}{\to} \cdots \stackrel{\partial_1}{\to} \bigoplus_{\alpha \in \mathbb{N}} R(-\alpha)^{\beta_{0,\alpha}}.$$

is called cellular, if there is a CW-complex X and a map $\mathrm{gr} : X^{(*)} \to \mathbb{N}^d$ from the set $X^{(*)}$ of its cells to $\mathbb{N}^d$ such that:

- There is a basis e_c of $F_i = \bigoplus_{\alpha \in \mathbb{N}} R(-\alpha)^{\beta_{i,\alpha}}$ indexed by the i-cells of X in such a way that if e_c belongs to $R(-\alpha)$ then $\mathrm{gr}(c) = \alpha$.
- For the i-cell c of X and its cellular differential $\delta_i(c) = \sum_{c'} [c : c']c'$, we have $\partial_i(e_c) = \sum_{c'} [c : c']\underline{\mathbf{x}}^{\mathrm{gr}(c') - \mathrm{gr}(c)} e_{c'}$.

In this situation we say that X *supports* the resolution $\mathcal{F}$.

Consider the face poset $P = F(X)$ of a CW-complex X supporting a resolution. [**BW**, Proposition 2.2] shows that an acyclic matching A on P leads to a chain homotopy between the original resolution and a smaller cellular resolution, given by the smaller CW complex X^M of critical cells in a discrete Morse function given by A, if A matches only cells that have the same value under gr.

It is well known that for any $\mathbb{N}^d$-graded module M there exists a *minimal* multigraded free resolution (i.e., a resolution that uses the least number of free modules in each degree). Now the results from [**BW**], as described above, allow one to construct smaller resolutions from a given resolution. It is also clear (see [**BW**]) that this process will not always allow one to produce the minimal free resolution.

Let us consider a 'big' cellular resolution in the situation treated in this paper (i.e, $R = k[\Lambda]$ and $M = k$). It is well known that the simplicial complex $\Delta(\Lambda)$ of all finite chains $\lambda_0 < \cdots < \lambda_r$ in Λ together with the grading $\mathrm{gr}(\lambda_0 < \cdots < \lambda_r) = \lambda_r$ gives a multigraded free cellular resolution of the maximal ideal $\mathbf{m} = \mathrm{Ker}(k[\Lambda] \to k)$ – the normalized Bar resolution. Since a free minimal resolution of k over $k[\Lambda]$ starts with $k[\Lambda]$, minimizing the normalized Bar resolution is equivalent to minimizing a resolution of k. A well known criterion for a resolution to be minimal is that no unit elements of R occur in the matrices representing the differentials. If in addition R is standard graded and all matrix entries are either 0 or elements of degree 1, then the resolution is called *linear*. If k has a linear resolution then R is called *Koszul*.

In our case, where $R = k[\Lambda]$, we know that R is standard graded if and only if the generators of Λ lie on an affine hyperplane. In this situation R carries two gradings, the standard grading and a multigrading given by Λ. The Tor-groups $\mathrm{Tor}_i^R(k,k) \cong k^{\beta_i}$, where $\beta_i = \sum_\alpha \beta_{i,\alpha}$, also carry a multigraded structure $\mathrm{Tor}_i^R(k,k)_\alpha = k^{\beta_{i,\alpha}}$. Thus Koszulness can be read off from the Tor-groups. Namely, R is Koszul if and only if $\mathrm{Tor}_i^R(k,k)_\alpha = 0$ for $\underline{\mathbf{x}}^\alpha$ with standard grading not equal to i.

A well known sufficient condition for a k-algebra $R = k[z_1,\ldots,z_n]/I$ which is standard graded to be Koszul is that I has a quadratic Gröbner basis. Recall that for a monomial order $\preceq$ on $k[z_1,\ldots,z_n]$ a *Gröbner basis* of I is a generating set $\mathcal{G}$ of polynomials in I such that the *initial ideal* $\mathrm{in}_\preceq(I) := \langle \mathrm{in}_\preceq(f) | f \in I\rangle$ is equal to the ideal generated by $\{\mathrm{in}_\preceq(f) | f \in \mathcal{G}\}$. Recall that the *leading monomial*, denoted $\mathrm{in}_\preceq(f)$, for a polynomial f is the largest monomial with respect to $\preceq$ occurring in f; we write $\mathrm{in}(I)$ and $\mathrm{in}(f)$ if the monomial order is clear from the context. Finally, a *monomial term order* on $k[z_1,\ldots,z_n]$ is a linear order $\preceq$ on the monomials in the ring such that (1) $1 \preceq m$ for all monomials m, and (2) $m \preceq m'$ implies $mn \preceq m'n$ for all monomials n, m, m'.

2.6. Basic Facts on Finite State Automata, Regular Languages and Rational Generating Functions. A central question in the theory of infinite resolutions is 'Which conditions on a module M imply that Poincaré-Betti series of its minimal free resolution is rational ?' (see [**Av**]). When $R = k[\Lambda]$ is a standard graded and multigraded k-algebra and $M = k$, the (graded and multigraded) *Poincaré-Betti series* is given by $\sum_{i,\alpha} \beta_{i,\alpha} t^i \underline{\mathbf{x}}^\alpha z^{\deg(\alpha)}$, where $\deg(\alpha)$ is the degree of $\underline{\mathbf{x}}^\alpha$ in the standard grading on R.

We will not be able to give a new criterion for the rationality of the Poincaré-Betti series. But we will be able to give in Section 8 a rational series which bounds the Poincaré-Betti series from above, i.e. all coefficients are greater or equal to the ones in the Poincaré-Betti series. In order to prove rationality of our series we will resort to the theory of regular languages. It is well known (see [**BR**]) that the generating series of a regular language is rational. A language L over a finite *alphabet* Σ is called *regular* if there is a finite state automaton which accepts exactly the words in L. The generating series of L is given by $\sum_{w\in L} t^{|w|}$, where $|w|$ is the number of letters in w. See for instance [**BR**] for additional information.

3. Monomial term orders and discrete Morse functions resulting from (not-necessarily-lexicographic) facet orders

This section will show how the lexicographic discrete Morse function construction of [**BH**] generalizes easily to a larger class of facet orders for poset order complexes; this will include facet orders for monoid posets based on arbitrary monomial term orders. First observe that the [**BH**] construction applies without modification to any facet order $F_1,\ldots,F_r$ which yields an interval system structure on each $F_j \setminus (\cup_{i<j} F_i)$. Equivalently, the construction will work for facet orderings satisfying the crossing condition, as introduced in [**He1**] and defined below.

Crossing condition. *Let $\le$ be a linear order on the saturated chains in a partially ordered set P of rank n and rank function* rk. *Let F be a saturated chain, $G \le F$ and $\sigma = F \cap G$. Suppose that $[n] - \{\mathrm{rk}(p) \mid p \in \sigma\}$ is not an interval of natural numbers. Then there is some facet $G' \le F$ such that $F \cap G \subsetneq F \cap G'$.*

The crossing condition implies that for a saturated chain F, maximal faces in $F \cap (\cup_{G<F} G)$ are supported on a set of ranks whose complement is a single interval of consecutive ranks.

THEOREM 3.1. *If a facet ordering on an order complex satisfies the crossing condition, then the acyclic matching construction of* [**BH**] *applies to this facet ordering.*

PROOF. The effect of the crossing condition for a particular facet order $F_1, ..., F_k$ is to ensure that each maximal face of $F_j \cap (\cup_{i<j} F_i)$ for $1 < j \leq k$ skips a single interval of consecutive ranks. Thus, the faces in $F_j \setminus (\cup_{i<j} F_i)$ are the ones that "hit" each of these intervals, implying that the matching construction from [**BH**] still applies. □

DEFINITION 3.2. A facet order on a poset order complex which satisfies the crossing condition is called *lex-like facet order.* The discrete Morse function obtained by applying the construction from [**BH**] to such a facet order is called a *lex-like discrete Morse function.*

Let $[x, v]$ be an interval in a labelled poset P. For a saturated chain in $[x, y]$, the set of all saturated chains in $[x, y]$ having the same content is called the fibre of the content. Now again turn to intervals in affine semigroups Λ. Let $\mathfrak{L}$ be the usual labelling of a covering relation $\lambda \prec \mu$ by the generator $\mu - \lambda$ of the semigroup. We assume that we are given a monomial term order on $k[z_1, \ldots, z_n]$ and again as usual identify the generators of Λ with the variables z_i. This identification allows us to order the saturated chains by the given monomial term order. Recall, that we identify a label sequence with the monomial which is the product over the labels, which in turn can be seen as the content of the saturated chain with the given label sequence. Lemma 3.3 considers linear orderings of saturated chains in intervals of Λ obtained by combining the (commutative) monomial term ordering with a lexicographic order on each fibre; that is, we extend the order given by the monomial term order on the content by the lexicographic ordering on chains that have the same content. The lexicographic order uses the monomial term order on degree 1 monomials to order the labels.

LEMMA 3.3. *Let $P = [x, y]$ be an interval in an affine semigroup Λ and assume that the saturated chains in P are ordered by a monomial term order refined by the lexicographic order. Then this ordering satisfies the crossing condition. In particular, we can construct a discrete Morse function just as in the case of a lexicographic order.*

PROOF. Let $\equiv$ be the equivalence relation on the set of saturated chains such that $m \equiv n$ if and only if m and n are two saturated chains in the same closed interval in P. We abuse notation and say $m = n$ if the two chains are labeled by the same commutative monomial, i.e. the labels on one saturated chain are a permutation of the labels on the other. Suppose that for two saturated chains $m_1 m_2$ and $n_1 n_2$ we have $n_1 \neq m_1$, $n_2 \neq m_2$ but $n_1 \equiv m_1$ and $n_2 \equiv m_2$. Assume further that $n_1 m_1 <_{monom} n_2 m_2$ in the given monomial order $<_{monom}$. This is one situation where a saturated chain has an overlap face with earlier saturated chains such that the complement of the ranks in the overlap face is disconnected.

We check that either $m_1 n_2 <_{monom} n_1 n_2$ or $n_1 m_2 <_{monom} n_1 n_2$, as follows. Suppose $m_1 n_2 >_{monom} n_1 n_2$, which implies $m_1 >_{monom} n_1$. Suppose

$n_1m_2 >_{monom} n_1n_2$ also holds, implying $m_2 >_{monom} n_2$. Combining these inequalities yields

$$m_1m_2 >_{monom} n_1m_2 >_{monom} n_1n_2,$$

a contradiction. Hence, at least one of the monomials m_1n_2 or n_1m_2 precedes n_1n_2 in our term order. This ensures that the maximal face shared by m_1m_2 and n_1n_2 is not a maximal face in the simplicial complex of faces shared by n_1n_2 and facets that precede it in lexicographic order, just as needed.

Now suppose there is a saturated chain F_j not dealt with above that has an overlap face with an earlier saturated chain F_i such that the complement of the ranks in the overlap face is disconnected. Then F_i, F_j are labeled $m_1m_2m_3, \pi(m_1)n_2\sigma(m_3)$, respectively, where π, σ are permutations on the labels in m_1, m_3 and $m_2 \equiv n_2$ but $m_2 \neq n_2$. Consider F_i' which is labeled $\pi(m_1)m_2\sigma(m_3)$. Since $F_i \cap F_j \subsetneq F_i' \cap F_j$, we are done. □

REMARK 3.4. Examples suggest that the following procedure may be convenient for posets with no particularly well-behaved global labelling.

(1) Label edges in a poset Hasse diagram in a natural way (or more generally, give a chain-labelling on saturated chains).
(2) Partition the set of saturated chains into groups called *content classes* according to the content of their label sequences.
(3) Put an ordering on these content classes.
(4) Within each content class, order saturated chains lexicographically.
(5) Prove that the resulting facet order satisfies the crossing condition.
(6) Cancel pairs of critical cells which have the same content, i.e. pairs in the same content class.

This approach can make critical cell cancellation manageable for posets with no particularly nice global labelling, because gradient paths that begin and end in the same content class must never leave that content class. This will be essential to our analysis of monoid posets and to arguments in [**HHS**].

DEFINITION 3.5. A *content-lex facet order* is a lex-like facet order such that:

- The ordering is constructed from a labelling by refining a linear order on fibres by a lexicographic order.
- The least saturated chain in each interval has weakly increasing labels.

One way content-lex facet orders arise is when each content class individually has an EL-labelling. This is the situation for our upcoming facet order on monoid posets as well as a $GL_n(q)$-analogue of the partition lattice examined in [**HHS**].

REMARK 3.6. Content-lex facet orders behave as least-content-increasing labellings for purpose of applying Theorem 2.14 to verify gradient path uniqueness.

4. Uniqueness of 321-avoiding gradient paths in discrete Morse functions from content-lex facet orders

Now we generalize Theorem 2.14 to allow non-saturated chain segments in gradient paths between critical cells in the same content class in a content-lex facet order, under certain additional assumptions. Remark 9.2 suggests that the 321-avoiding assumption is probably necessary for any general result about gradient path uniqueness for lex-like discrete Morse functions.

Upcoming sections will construct a non-optimal discrete Morse function for monoid posets, then use the following theorem to improve it.

DEFINITION 4.1. A *delinquent chain* in a least-increasing labelling (or in a content-lex facet order) is an increasing chain that is not lexicographically smallest on an interval. A content-lex facet order is *consistently delinquent* if the existence of a delinquent chain labeled $a_1 \cdots a_k$ implies that any chain segment labeled $b_1 \cdots b_{k'}$ for $b_1 = a_1, b_{k'} = a_k$ and $\{a_1, \ldots, a_k\} \subseteq \{b_1, \ldots, b_{k'}\}$ is also delinquent.

Denote by $e\alpha$ the earliest facet containing a cell α, and let $\mathfrak{L}(\alpha)$ be the label sequence on $e\alpha$. We say that a chain element *covers* a delinquent chain if it is in the interior of the delinquent chain, preventing the poset chain from belonging to an earlier content class.

THEOREM 4.2. *Let τ, σ, $\dim(\tau) = \dim(\sigma) + 1$, be critical cells in the same content class in a content-lex facet order which is consistently delinquent. Suppose $\mathfrak{L}(\sigma)$ differs from $\mathfrak{L}(\tau)$ by a 321-avoiding permutation π which either shifts a single group of one or more consecutive ascending labels upward, or shifts a single label downward. Then there is at most one gradient path from τ to σ.*

PROOF. Since τ is critical, every pair of consecutive labels in $\mathfrak{L}(\tau)$ is either a descent or part of a minimal delinquent chain. Gradient paths from τ to σ can never uncover any delinquent chains, since that would cause the gradient path to pass to an earlier content class, from which it could never reach σ. Since the facet order is also least-increasing, each downward step must either preserve the label sequence or eliminate a descent by deleting an element v_r from a chain τ_i of the form $v_1 < \cdots < v_{r-1} < v_r < v_{r+1} < \cdots < v_s$, causing the labels on the segments of $e\tau_i$ from v_{r-1} to v_r and from v_r to v_{r+1} to be sorted into a single ascending list. The least-increasing property Thus, any inversions present in $\mathfrak{L}(\sigma)$ must be preserved throughout the gradient path, since they can never be re-introduced. With these observations in hand, we will describe the only possible gradient path from τ to σ, showing at each stage there is only one choice for how to proceed.

Suppose $\mathfrak{L}(\sigma)$ is obtained from $\mathfrak{L}(\tau)$ by shifting a collection of consecutive, ascending labels upward. Let μ be the label to be shifted upward to the highest destination in $\mathfrak{L}(\sigma)$, and let b be the label immediately above μ in $\mathfrak{L}(\tau)$. By virtue of the Morse function construction of [**BH**], τ includes exactly the ranks where $e\alpha$ has descents as well as exactly one rank that covers each of the minimal delinquent chains, namely the lowest ranks in the J-intervals. The first gradient path downward step must eliminate a descent since it is not allowed to uncover a delinquent chain. The only choice that will not eliminate an inversion that is present in $\mathfrak{L}(\sigma)$ is to eliminate a descent between μ and the label above it, i.e. deleting an element v_r from a chain $v_1 < \cdots < v_{r-1} < v_r < v_{r+1} < \cdots < v_s$. This yields a chain σ_1 with $e\sigma_1 = F_j$ having a cone point in $F_j \cap (\cup_{i<j} F_i)$ between v_{r-1} and v_{r+1} in F_j, but no lower cone points. Thus, the subsequent upward step must insert some $v_{r'}$ above v_{r-1} and below the label μ, with the labels between v_{r-2} and $v_{r'}$ now comprising either a descent or minimal delinquent chain. In the latter case, the delinquent chain just below $v_{r'}$ must include the label b as its highest label. Labels now below b can no longer shift upward, since any such labels to be shifted upward in $\mathfrak{L}(\sigma)$ must be smaller than b. If there were instead a descent at v_{r-1}, only one label is allowed between v_{r-1} and $v_{r'}$, so either it is the label b defined above, which

does not shift upward, or it is a label smaller than b, which is now prevented from moving upward by virtue of being smaller than b.

Continuing in this fashion, there is only one viable downward step at any given stage until μ reaches its destination, namely the chain deletion which shifts μ upward, since no J-interval is ever covered by more than one chain element while μ is shifting upward. By the argument above, labels can only shift upward while they are shifting as part of a block of consecutive labels which includes μ, so all label shifting is complete once μ has reached its destination, and there is a unique way for this to happen. All that remains is to consider additional gradient path steps which preserve label sequence.

Suppose the label above μ in $\mathfrak{L}(\sigma)$ is larger than μ. Then μ must be within a delinquent chain, since σ is critical. After inserting a cone point below μ, the gradient path must take a downward step deleting a chain element above μ, since the J-interval for the delinquent chain which includes μ will now be covered by at least two chain elements. At this point, no upward step is possible, so the gradient path must have reached σ. Finally, we show that μ cannot form a descent with the label above it in $\mathfrak{L}(e\sigma)$. Otherwise, μ would be the highest label in a minimal delinquent chain labeled $\nu_1, \dots, \nu_k, \nu, \mu$, but μ could not be shifted upward to this position from below without passing to an earlier content class.

The case where a single label μ shifts downward to obtain $\mathfrak{L}(\sigma)$ from $\mathfrak{L}(\tau)$ is quite similar, so is essentially left to the reader. The first gradient path step must again eliminate a descent so as to preserve content class, and the only allowable choice is a descent between μ and the label below it. Similar reasoning to above will show that the only possible gradient path will progressively shift μ downward to its destination, concluding once μ reaches the interior of a delinquent chain. □

5. The Cohen-Macaulay property for monoid posets with quadratic Gröbner bases

Throughout this section, we assume the toric ideal $I_\Lambda = \mathrm{Ker}(\phi)$ has a quadratic Gröbner basis B. However, Section 5.1 will apply to higher degree Gröbner bases with essentially no modification needed, and large parts of Sections 5.2–5.4 will also generalize easily to higher degree Gröbner bases.

DEFINITION 5.1. Denote by $\mathrm{in}(I_\Lambda)$ the initial ideal of I_Λ with respect to the term order giving rise to B, i.e. the ideal generated by leading terms of elements of B.

This section will show that each interval $(\hat{0}, m)$ in the resulting monomial divisibility poset has the homotopy type of a wedge of spheres of top dimension. Our approach will be to construct a lex-like discrete Morse function based on a content-lex facet order in Theorem 5.2, and then to cancel all but some of the top-dimensional critical cells in Theorem 5.21.

We will employ the simple fact that any leading term of a polynomial in a toric ideal is divisible by a Gröbner basis leading term. In particular, leading terms of degree greater than the degree of the Gröbner basis will include variables that in some sense are non-essential. We will use these non-essential variables to cancel critical cells. In the case of quadratic Gröbner bases, this approach will allow us to cancel all critical cells that are not saturated chains. Later, we will use a similar (but somewhat more intricate) analysis for Gröbner bases of degree d.

5.1. A non-optimal Morse function. The first step will be to give a content-lex facet ordering.

THEOREM 5.2. *The monoid poset interval* $(\hat{0}, m)$ *has a lex-like discrete Morse function resulting from a content-lex facet order. Its minimal skipped intervals are the saturated chain segments with label sequences of the following two types:*

(1) *descents*
(2) *sequences of weakly increasing labels* $\lambda_i, \ldots, \lambda_k$ *such that* $\lambda_i \lambda_k \in \mathrm{in}(I_\Lambda)$, *but* $\lambda_{i'} \lambda_j \notin \mathrm{in}(I_\Lambda)$ *for every other pair* $i \leq i' < j \leq k$.

PROOF. The finite saturated chains on intervals in a monoid poset correspond naturally to pairs (m_i, π) where m_i is a monomial in the ring $k[z_1, \ldots, z_n]$, and π is an ordering on the content of m_i. This is equivalent to labelling saturated chains by non-commutative monomials, in $k\langle y_1, \ldots, y_n \rangle$, the viewpoint taken in [**PRS**]. Following [**PRS**], we order saturated chains in an interval $(\hat{0}, m)$ by using the monomial term order which led to the Gröbner basis B to order the factorizations $m_i \in \phi^{-1}(m)$, and then lexicographically ordering label sequences of any fixed content, with our label order given by the monomial term order applied to monomials of degree one. Lemma 3.3 confirms the crossing condition for this facet order, implying it gives rise to a lex-like discrete Morse function from a content-lex facet order. Next, we characterize its minimal skipped intervals.

Notice that a descent on the saturated chain segment $u \prec v \prec w$ implies a lexicographically smaller ascend $u \prec v' \prec w$, obtained by reversing the order in which semi-group generators are multiplied, so descents always give minimal skipped intervals. On the other hand, any label sequence $a_1, \nu_1, \nu_2, \ldots, \nu_r, a_2$ as in (2) will give rise to a minimal skipped interval because the Gröbner basis element $a_1 a_2 - b_1 b_2$ with leading term $a_1 a_2$ implies the existence of an earlier saturated chain on the interval labeled by the increasing rearrangement of the monomial $b_1 n_1 \cdots n_j b_2$; its minimality follows from the lack of descents and of Gröbner basis leading terms not requiring both a_1 and a_2.

DEFINITION 5.3. The second type of minimal skipped interval in the statement of the theorem is called a *syzygy interval.*

To see there are no other minimal skipped intervals, note that a label sequence $a_1, \ldots, a_r$ on any other minimal skipped interval must be ascending to avoid descents which would preclude its minimality; to have an earlier saturated chain on the interval $a_1, \ldots, a_r$ must be a leading term, hence divisible by a Gröbner basis leading term m. But minimality ensures a_1, a_r divide m, and the fact that the Gröbner basis is quadratic implies $m = a_1 a_r$. □

COROLLARY 5.4. *Theorem 4.2 may be applied to the above Morse function to cancel pairs of critical cells that belong to the same fibre.*

DEFINITION 5.5. A minimal skipped interval is *non-trivial* if it has height greater than one. Notice that only syzygy intervals may be non-trivial.

EXAMPLE 5.6. Consider the interval $(1, x_1^2 x_2^2 x_3 x_4)$, or equivalently, $(1, z_1 z_2 z_3 z_4)$, in the ring $k[x_1 x_2, x_1^2, x_3, x_4, x_2^2] \cong k[z_0, z_1, z_2, z_3, z_4]/(z_1 z_4 - z_0^2)$, with $\mathrm{in}(I_\Lambda) = (z_1 z_4)$. Saturated chains are labeled by indices of the generators $z_0, \ldots, z_4$. Figure 1 shows four saturated chains on this interval, the leftmost and rightmost of which will contribute critical cells. Notice that $1, 4$ labels a syzygy interval, while there

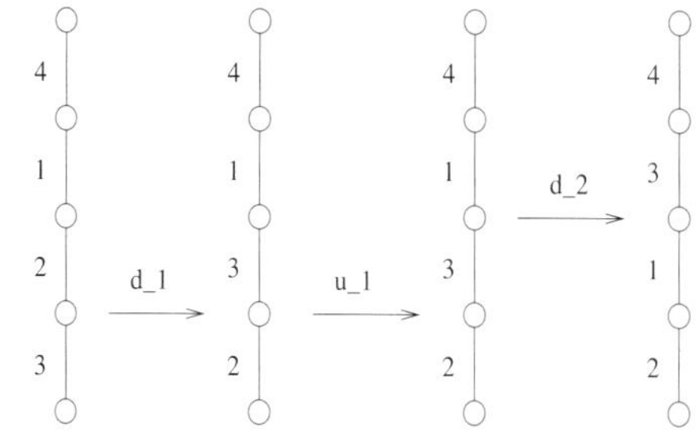

FIGURE 1. A gradient path shifting the label 3 into a syzygy interval

are descents at ranks 1 and 2 in the leftmost saturated chain, so it has a critical cell τ comprised of ranks 1,2,3. On the other hand, the label sequence $1, 3, 4$ in the rightmost saturated chain also labels a syzygy interval, and this chain has a descent at rank 1, so it has a critical cell σ comprised of ranks 1,2.

The remainder of Section 5 is devoted to cancelling pairs of critical cells by gradient path reversal, so as to eliminate all critical cells not given by saturated chains. This will require an acyclic matching on $F(\Delta)^M$ consisting of pairs of critical cells to be cancelled.

DEFINITION 5.7. We call the critical cells that remain after all this cancellation the *surviving critical cells.*

5.2. Syzygy intervals and their non-essential sets. For now we assume all monomials on our monoid poset interval are square-free. The general case is dealt with in Theorem 5.21.

REMARK 5.8. For convenience, we will refer interchangeably to a critical cell and the saturated chain which contributes it.

Figure 1 gives an example of a gradient path from a critical 2-cell to a critical 1-cell in the order complex resulting from the semigroup ring $k[z_0, z_1, z_2, z_3, z_4]/(z_1z_4 - z_0^2)$. This gradient path shifts the label z_3 to the interior of a syzygy interval, using the fact that z_3 is not essential to $z_3(z_1z_4 - z_0^2) = 0$. This is the only gradient path between these two critical cells. Our goal will be to systematically cancel many such pairs of critical cells simultaneously.

DEFINITION 5.9. Denote by $I(a_1, a_2)$ the syzygy interval with ascending labels $a_1, \lambda_1, \dots, \lambda_k, a_2$ in a saturated chain, and refer to the Gröbner basis leading term a_1a_2 with $a_1 \leq a_2$ as an *increasing leading term*, or *ILT* for short.

We will soon use ILTs to collect critical cells into Boolean algebras within $F(\Delta)^M$.

EXAMPLE 5.10. In the affine semi-group ring $k[z_1, \dots, z_6]/(z_2z_6 - z_1^2)$, consider the saturated chain F_j that is labeled $z_4z_3z_2z_5z_6$. F_j contributes the critical cell $\sigma = z_4 < z_3z_4 < z_2z_3z_4$. By Theorem 2.14, there is a unique gradient path from the critical cell $\tau = z_5 < z_4z_5 < z_3z_4z_5 < z_2z_3z_4z_5$ to σ, given by the reduced expression $s_1 \circ s_2 \circ s_3$. More generally, each $T \subseteq S = \{z_3, z_4, z_5\}$ gives rise to a critical cell $Crit(T)$ contributed by a facet F_T, as follows. F_T has label sequence

$z_{(S\setminus T)^{rev}} z_2 z_T z_6$, where z_T is the list of members of T in increasing order, and $z_{(S\setminus T)^{rev}}$ is the list of members of $S \setminus T$ listed in decreasing order. $S = \{z_3, z_4, z_5\}$ is the non-essential set of the interval. Theorem 2.14 will show that the set of critical cells $\{Crit(T) | T \subseteq S\}$ sits inside the multi-graph face poset $F(\Delta)^M$ as a Boolean algebra, depicted in Figure 2.

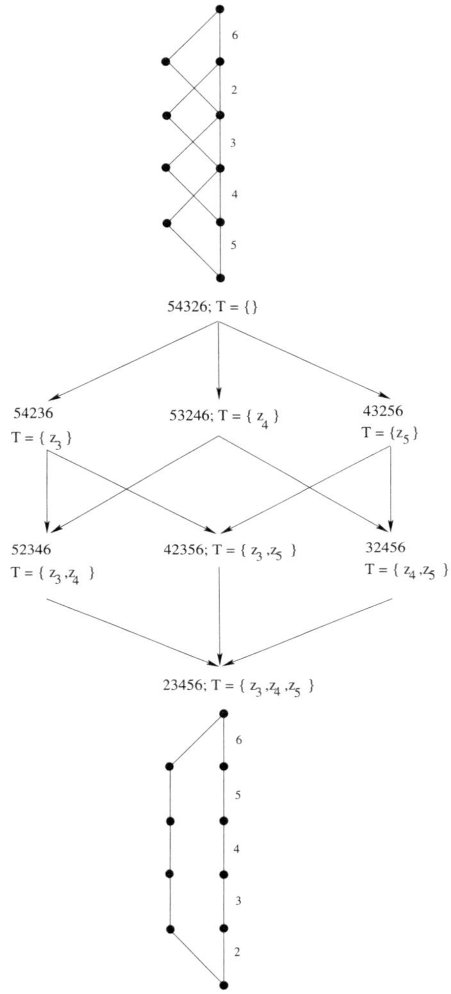

FIGURE 2. A Boolean algebra in $F(\Delta)^M$ indexed by subsets of $\{z_3, z_4, z_5\}$

This Boolean algebra has covering relations $Crit(T \cup \{z_i\}) \prec Crit(T)$ for each $T \subseteq S$ and each $z_i \in S \setminus T$.

REMARK 5.11. A gradient path cannot swap non-commuting labels (in the sense of Definition 2.11) without passing to an earlier fibre, so cells to be cancelled will agree up to allowable label commutation.

DEFINITION 5.12. A label λ in a saturated chain M is *upward-shiftable* into a syzygy interval $I(a_1, a_2)$ if it satisfies all the following conditions:

(1) $a_1 <_{monom} \lambda <_{monom} a_2$

(2) λ appears below $I(a_1, a_2)$
(3) all labels between λ and $I(a_1, a_2)$ are smaller than λ and commute with λ
(4) all labels within $I(a_1, a_2)$ commute with λ
(5) λ is not the top of some $I(\mu, \lambda)$ with either non-empty interior or such that the label ν immediately above λ would neither form a descent with μ nor be part of an ILT together with μ

Likewise, λ is *downward-shiftable* from $I(a_1, a_2)$ to just above λ' if λ commutes with all labels separating it from λ' and is larger than all such labels.

REMARK 5.13. Each label has at most one syzygy interval into which it is upward-shiftable, because the lowest such interval will separate it from all higher ones.

DEFINITION 5.14. If λ appears within $I(a_1, a_2)$, then the *topologically decreasing position* for λ below $I(a_1, a_2)$ is the highest position below $I(a_1, a_2)$ to which λ is downward-shiftable so as to obtain the label sequence for a critical cell with λ not in the interior of any ILT, if such a position exists.

In its topologically decreasing position below an ILT, λ must form descents or ILTs with the labels above and below it. Lemma 5.23 will construct gradient paths that shift labels from their topologically decreasing positions below ILTs to the interior of ILTs. In some circumstances, we will also speak of the topologically decreasing position of λ above $I(a_1, a_2)$, by which we mean the lowest position above $I(a_1, a_2)$ to which λ is upward-shiftable to yield the label sequence for a critical cell.

DEFINITION 5.15. The *non-essential set* of a syzygy interval $I(a_1, a_2)$ that appears in the label sequence $\mathfrak{L}(\sigma)$ for a critical cell σ will be a collection of labels that appear in $\mathfrak{L}(\sigma)$ either within $I(a_1, a_2)$ or in topologically decreasing positions below $I(a_1, a_2)$. This set of labels (to be defined precisely in the remainder of Section 5.2 and Section 5.3) is denoted $S(a_1, a_2)$.

To try to convey the intuition for $S(a_1, a_2)$, we now give an oversimplified definition. Section 5.3 will modify this into a much more technical definition that accomplishes exactly what is needed. Initially, let us include in $S(a_1, a_2)$ those labels that appear within $I(a_1, a_2)$ that are downward-shiftable to topologically decreasing positions below $I(a_1, a_2)$. Denote these labels by $n_1, \ldots, n_j$. Also include in $S(a_1, a_2)$ those labels $m_1, \ldots, m_r$ that are upward-shiftable into $I(a_1, a_2)$, chosen in order from highest to lowest topologically decreasing position below $I(a_1, a_2)$.

We call critical cells which are not maximal faces in the order complex $\Delta(\hat{0}, m)$ *unsaturated*. In the case of a quadratic Gröbner basis, we will match and cancel all unsaturated critical cells, using the fact that each must have one or more syzygy intervals with non-empty interior.

5.3. The matching on critical cells. In this section, we precisely define non-essential sets and show that the resulting matching on critical cells is well-defined. We also show that every critical cell which has at least one syzygy interval with non-empty interior is indeed matched and cancelled, by showing it has at least one syzygy interval with non-empty non-essential set. The fact that pairs of critical cells to be matched do indeed comprise covering relations in $F(\Delta)^M$ will be verified in Section 5.4.

DEFINITION 5.16. A label λ is *preferable* to a label μ within a label sequence if either λ is in the non-essential set of a higher syzygy interval than μ is in, or $\lambda, \mu \in S(a_1, a_2)$ with the topologically decreasing position for λ higher than for μ.

Using our oversimplified definition of non-essential set from the previous section, let $I(a_1, a_2)$ be the highest syzygy interval in a saturated chain C such that $S(a_1, a_2) \neq \emptyset$, and let λ be the label in $S(a_1, a_2)$ with highest topologically decreasing position below $I(a_1, a_2)$. The theorem below will sometimes include in a non-essential set $S(a_1, a_2)$ a single label that shifts downward into $I(a_1, a_2)$ from above. When this happens, denote this label m_0, and eliminate from $S(a_1, a_2)$ any labels below $I(a_1, a_2)$ that do not commute with m_0.

THEOREM 5.17. *Every critical cell with a syzygy interval with an internal label μ is matched. Specifically, if μ is excluded from the non-essential set of the interval, then there must be another label that allows the cell to be matched and cancelled. Moreover, the matching choices are made consistently.*

PROOF. We will typically match by shifting λ from inside $I(a_1, a_2)$ to its topologically decreasing position below $I(a_1, a_2)$, or vice versa. However, special care is needed in four circumstances described below. When λ is excluded from a non-essential set, then the matching instead shifts the label with highest preferability among those belonging to some non-essential set.

In each of these circumstances, we will show that either λ may be included in $S(a_1, a_2)$ or that there is an alternative label to λ allowing the cell to be matched. Moreover, when λ is excluded from a non-essential set, it will be excluded for all critical cells in the Boolean algebra within which our critical cell is matched.

(1) $\lambda \in I(a_1, a_2)$ shifts downward to a topologically decreasing position which is higher than some label μ with which λ does not commute; however, μ may shift upward to the interior of a syzygy interval $I(b_1, b_2)$ to obtain another critical cell when λ appears in $I(a_1, a_2)$ but not when λ has shifted downward to its topologically decreasing position.

(2) $\lambda \in I(a_1, a_2)$ shifts downward to just above a label μ with which λ does not commute to form an ILT $I(\mu, \lambda)$ which then has some label $\nu \in S(\mu, \lambda)$.

(3) $\lambda \in I(a_1, a_2)$ cannot shift downward to a topologically decreasing position without first encountering a label μ with which it does not commute.

(4) $\lambda \in I(a_1, a_2)$, but shifting λ downward causes a_2 to be in the non-essential set of a higher ILT, or more generally the shifting of all labels within $I(a_1, a_2)$ downward to topologically decreasing positions, cumulatively causes a_2 to belong to a higher non-essential set.

In the first case above, exclude λ from $S(a_1, a_2)$. When λ appears within $I(a_1, a_2)$, then $\mu \in S(b_1, b_2)$, ensuring the cell may still be matched. Matching by shifting μ will clearly give a cell which would also exclude λ from $S(a_1, a_2)$. Shifting a label μ' that is preferable to μ also gives a cell that excludes λ, either by virtue of μ, or if μ', λ do not commute, then by virtue of μ'.

In the second case, note that $\lambda \in I(a_1, a_2)$ implies either (a) ν is in the non-essential set of some $I(a_1', a_2')$ above μ, (b) ν is separated from the lowest such $I(a_1', a_2')$ by a label with which it does not commute, or (c) ν does not commute with some label in the interior of the lowest such $I(a_1', a_2')$; this follows from $\mu < \nu < \lambda < a_2$ together with the fact that μ and the label immediately above it must form either a descent or an ILT. In case 2(a), ν provides an alternative label for matching, and λ may be excluded from $S(a_1, a_2)$. Notice that the partner cell in which ν or a preferable label ν' has been shifted may also exclude λ by the following reasoning. Since λ commutes with ν, the critical cell with $\nu \in I(a_1', a_2')$ and λ immediately above μ may be matched by shifting ν downward to just above λ, instead of by shifting λ.

For 2(b), include λ in $S(a_1, a_2)$, since the cell with $\nu \in I(\mu, \lambda)$ may be matched by shifting ν upward to its topologically decreasing position above $I(\mu, \lambda)$, or by shifting a label that is preferable to ν. Notice that we this also deals with case 4, by considering it from a different viewpoint.

DEFINITION 5.18. When such a label ν is matched by such upward-shifting, we say that ν *blocks* a_2 from belonging to a higher non-essential set.

Furthermore, observe that shifting ν upward, or shifting a preferable label, still gives a critical cell which excludes λ from $S(a_1, a_2)$; this is because either $\nu \in I(\mu, \lambda)$, or ν appears above λ and would form an ascend with μ if λ were shifted upward. For 2(c), if $I(a_1', a_2')$ has non-empty interior, then this gives an alternative to λ, allowing λ to be excluded from $S(a_1, a_2)$. When this alternative label is shifted to outside $I(a_1', a_2')$, it is still preferable to λ, by the conventions from case 2(a). If there are no interior labels in $I(a_1', a_2')$, then ν does not commute with a_2'; λ is included in $S(a_1, a_2)$, noting that the cell with ν shifted into $I(\mu, \lambda)$ will be matched by shifting ν upward to just below a_1', similarly to case 2(b).

Now we turn to the third case. First notice that μ must appear in a lower ILT, either as its lowest label, or in its proper interior. If μ appears in the interior of some $I(a_1', a_2')$, then exclude λ from $S(a_1, a_2)$ and apply our argument to μ or a preferable label, proceeding downward until we find a way of matching. By virtue of case (1), the matching partner which has shifted μ or a preferable label will also have excluded λ from $S(a_1, a_2)$. If μ is the lowest label of some $I(\mu, \mu')$ with non-trivial interior, then again use a label from the interior or a preferable label for matching. On the other hand, if $I(\mu, \mu')$ has no interior, then consider the descending labels $\nu_j, \nu_{j-1}, \dots, \nu_1$ immediately above μ', up through the lowest label ν_1 in the next lowest ILT above $I(\mu, \mu')$. Let T be the subset of $\{\lambda, \nu_2, \dots, \nu_j, \mu'\}$ which consists of those μ' along with those labels which commute with μ', but not with μ. If all labels in T belong to the non-essential sets of higher ILTs, then there is a Boolean algebra of critical cells in which each $T' \subset T$ specifies which labels to leave in decreasing order immediately above μ, rather than shifted upward into the interior of various ILTs; however, the empty set is missing from this Boolean algebra unless μ forms a descent with the label just above it when all labels in T are shifts upward into interiors of ILTs.

Thus, allowing $\lambda \in S(a_1, a_2)$ and matching by shifting λ to just above μ gives nearly a complete matching on this Boolean algebra of critical cells, but there is no matching partner for the cell indexed by $T' = \{\lambda\}$. However, the cell indexed by $\{\lambda\}$ may instead be matched based on any $\nu_i \notin T$, since such a ν_i either belongs to the non-essential set of some higher ILT, or may be shifted downward into $I(\mu, \lambda)$; in the latter case, we are in a situation where ν_i blocks λ from belonging to $S(a_1, a_2)$, in the sense described above. In any event, all cells in question are matched, and it is clear that the matching partners are also matched in the same fashion.

The fourth case was already handled within the argument for the second case. $\square$

5.4. The Cohen-Macaulay Property. In this section we verify that the matching of the previous section consists of covering relations in $F(\Delta)^M$, and that these comprise an acyclic matching on $F(\Delta)^M$ with only top-dimensional surviving critical cells. We begin with an important special case which captures most of the idea.

DEFINITION 5.19. The *expanding interval* of a saturated chain C, denoted $I(a_1, a_2)$, is the highest syzygy interval with non-empty non-essential set in C..

THEOREM 5.20. *If I_Λ has a quadratic Gröbner basis and each $\phi^{-1}(m)$ is square-free, then the poset interval $(0, m)$ has a discrete Morse function whose critical cells are all saturated, implying the interval is homotopy equivalent to a wedge of spheres.*

PROOF. In a syzygy interval labeled $a_1 n_1 \dots n_j a_2$, recall that $a_1 a_2$ is a Gröbner basis leading term. Any unsaturated critical cell has at least one syzygy interval with $j > 0$. We described in the previous section how to match all such cells so that partner cells differ by exactly one in their number of minimal skipped intervals. Lemma 5.22 verifies that they in fact differ in dimension by exactly one. See Figure 1 for an example of a gradient path from one such critical cell to its matching partner.

Lemma 5.23 shows for the expanding interval $I(a_1, a_2)$ that $S(a_1, a_2)$ gives rise to a Boolean algebra of critical cells within $F(\Delta)^M$, indexed by the subsets of $S(a_1, a_2)$. $S(a_1, a_2)$ is chosen so that each $T \subseteq S(a_1, a_2)$ gives rise to a unique such cell, denoted $Crit(T)$. $Crit(T)$ is contributed by a saturated chain $M(T)$, which has exactly the labels in T inside $I(a_1, a_2)$, and each of the labels in $S(a_1, a_2) \setminus T$ shifted to its topologically decreasing position outside $I(a_1, a_2)$. All labels in $M(T)$ other than a_1, a_2 and the members of $S(a_1, a_2)$ will appear in the same relative order for all choices of $T \subseteq S(a_1, a_2)$.

Any Boolean algebra has a complete acyclic matching simply by matching by including/excluding any fixed set element. We assign each critical cell to the Boolean algebra given by the non-essential set of its expanding interval, then take a union of complete acyclic matchings on these Boolean algebras. Lemma 5.26 checks that when one critical cell is assigned to a particular Boolean algebra, then all critical cells in that Boolean algebra are assigned to it, ensuring the matching is well-defined. Section 5.3 already showed that we match all unsaturated critical cells.

The final step is to show that this union of complete matchings on Boolean algebras is an acyclic matching on $F(\Delta)^M$. By Theorem 2.10, this would imply that we may simultaneously reverse all these gradient paths to cancel all but some top-dimensional critical cells. To get acyclicity, we show two things: (1) Lemma 2.6 ensures that cycles cannot involve Boolean algebras from distinct fibres $\phi^{-1}(m)$, due to the filtration $m_1 \subseteq m_1 \cup m_2 \subseteq \dots$ based on the monomial term order $m_1, m_2, \dots$, and (2) Lemma 5.27 verifies that cycles cannot involve multiple Boolean algebras in the same fibre. Thus, we will produce a discrete Morse function whose critical cells are all top-dimensional, implying the order complex has the homotopy type of a wedge of spheres of top dimension. □

Next we deal with the possibility that not all monomials are square-free. The lemmas that follow do not use the square-free assumption, so they apply to the general case.

THEOREM 5.21. *If I_Λ has a quadratic Gröbner basis, then the monoid poset Λ has a discrete Morse function whose critical cells are all top-dimensional, implying Λ is homotopically Cohen-Macaulay.*

PROOF. The only issue left to address is repetition of labels. To this end, we adjust the definition of non-essential set and make sure surviving critical cells still do not have any syzygy intervals with non-empty interior. When multiple copies of a letter appear inside $I(a_1, a_2)$ or are upward-shiftable into it, only include one copy in $S(a_1, a_2)$ that shifts downward to below $I(a_1, a_2)$; we cannot shift more than one copy outside the interval and still get a critical cell, since consecutive identical labels not within a syzygy interval give a saturated chain rank not covered by any minimal skipped interval. Including one copy of the repeated letter in the non-essential set is enough to ensure the Boolean algebra B_n has $n \geq 1$, hence has a complete matching. A letter cannot initiate or conclude a syzygy interval and also appear in its interior, since then the syzygy interval would not be a minimal skipped interval. We may have a syzygy interval which begins and ends with the same label a_1, but then there cannot be any interior labels at all. □

LEMMA 5.22. *Critical cells that are matched differ in dimension by exactly one.*

PROOF. If none of the I-intervals are discarded in their conversion to J-intervals, then there is no issue (see Section 2.3 for definitions). When there is discardment, this means there are three or more overlapping I-intervals such that a middle one is unnecessary for covering all ranks by I-intervals, so Gröbner basis leading term elements for these intermediate I-intervals will each belong to the non-essential set of a higher I-interval, ensuring matching by shifting such an individual label to outside the collection of overlapping ILTs. This matching operation preserves the number of J-intervals from ILTs and alters by exactly one the number of J-intervals coming from descents. Thus, dimension changes by exactly one. □

LEMMA 5.23. *The critical cells indexed by subsets of $S(a_1, a_2)$, for $I(a_1, a_2)$ the expanding interval of a saturated chain $M(T)$, have the same incidences in $F(\Delta)^M$ as a Boolean algebra of subsets of $S(a_1, a_2)$. That is, there is a unique gradient path from $Crit(T)$ to $Crit(T \cup \{i\})$ for each $T \subseteq S(a_1, a_2)$ and each $i \in S(a_1, a_2) \setminus T$, and these are the only gradient paths among critical cells in $B_{S(a_1,a_2)}$.*

PROOF. Since all saturated chains in a fibre have equal content, and our facet order is content-lex, downward steps in a gradient path must sort labels. The critical cell $Crit(T)$ is obtained by arranging labels in T in increasing order within $I(a_1, a_2)$, and labels in $S \setminus T$ in unique topologically decreasing positions below $I(a_1, a_2)$ (or above $I(a_1, a_2)$, in the special circumstance that a_2 must be "blocked" from upward-shifting into another non-essential set), from which they may shift into $I(a_1, a_2)$. First we exhibit for each pair $T' = T \cup \{i\}$ with $i \in S \setminus T$, that there is a gradient path from $Crit(T)$ to $Crit(T')$. Choose r and t so that the t-th element in the chain $Crit(T)$ is just above the lowest label of $I(a_1, a_2)$, and the r-th element of $Crit(T)$ is just above the label i in $Crit(T)$, for i satisfying $T' = T \cup \{i\}$. In Example 5.24, let $i = 3, t = 3$ and $r = 1$.

EXAMPLE 5.24. Figure 1 depicts a gradient path from a critical 2-cell of rank set $\{1, 2, 3\}$ to a critical 1-cell with rank set $\{1, 2\}$, based on a ring with $z_1 z_4 \in \mathrm{in}(I_\Lambda)$. We have $S = \{2, 3\}$, $T = \emptyset$ and $T' = \{3\}$.

There is a gradient path from $Crit(T)$ to $Crit(T')$ of the form

$$d_r \circ u_r \circ d_{r+1} \circ u_{r+1} \circ \cdots \circ d_{t-1} \circ u_{t-1} \circ d_t,$$

because i commutes with all labels separating it from $I(a_1, a_2)$, i is larger than all these separating labels, and our discrete Morse function comes from a least-increasing facet order. Since the resulting permutation on labels is 321-avoiding, Theorem 4.2 ensures that this gradient path is unique, whether or not non-saturated chain segments are encountered in it.

To show that there are no other covering relations in $F(\Delta)^M$, i.e. none between other pairs of critical cells corresponding to subsets of $S(a_1, a_2)$, we use the fact that gradient paths can never introduce inversions. Thus, a gradient path from $Crit(T)$ to $Crit(T')$ would imply $T \subseteq T'$, since any $j \in T \setminus T'$, would imply an inversion (j, a_1) in $Crit(T)$ that is not present in $Crit(T')$. □

Next we verify that critical cells are indeed partitioned into Boolean algebras.

REMARK 5.25. Critical cells with label sequences of distinct content or with non-commuting labels in opposite order are assigned to distinct Boolean algebras.

LEMMA 5.26. *Whenever one critical cell is assigned to a Boolean algebra, then all critical cells in that Boolean algebra are assigned to it.*

PROOF. We must show that if $I(a_1, a_2)$ is the expanding interval for a saturated chain $M(T)$ for some $T \subseteq S(a_1, a_2)$, then the saturated chain $M(T')$ for each $Crit(T')$ with $T' \subseteq S(a_1, a_2)$ also has $I(a_1, a_2)$ as its expanding interval. Each label in a saturated chain belongs to the non-essential set of at most one syzygy interval, since it cannot pass through the lowest such syzygy interval above it to reach higher ones via a gradient path; a label is only assigned to the non-essential set of a syzygy interval below it when it cannot shift into one above it. Shifting labels belonging to $S(a_1, a_2)$ from within $I(a_1, a_2)$ to their topologically decreasing positions or vice versa cannot cause a higher non-essential set to become non-empty, by virtue of the choices made in Theorem 5.17. □

Finally, let us confirm that these complete matchings on Boolean algebras collectively give an acyclic matching on $F(\Delta)^M$.

LEMMA 5.27. *The matching on critical cells in $F(\Delta)^M$ is acyclic.*

PROOF. Lemma 2.6 ensures there are no directed cycles involving multiple fibres. Suppose there were a directed cycle C in a single fibre. Any such C must alternate upward (matching) steps with downward steps. Our matching consists of a union of complete matchings on Boolean algebras. Since the upward steps in a fixed Boolean algebra all insert the same fixed element i, each downward step must take us to a different Boolean algebra, to avoid yielding the top of an upward-oriented edge in the same Boolean algebra, from which the cycle could not have continued.

Suppose a matching step in C shifts a label μ upward from within an ILT $I(a_1, a_2)$ to above it. Then by virtue of our matching, a_2 must belong to the non-essential set of a higher ILT in the cell which has μ and all other labels within $I(a_1, a_2)$ shifted to below $I(a_1, a_2)$. Furthermore, μ must not be upward-shiftable into a higher ILT. To pass to a distinct Boolean algebra, the downward step immediately after this upward-shifting of μ must either (1) shift a_2 upward into a higher

ILT, (2) shift a label λ downward from a topologically decreasing position into the interior of an ILT, or (3) shift a label μ' upward into the interior of an ILT it then blocks (cf. Definition 5.18). (1) is impossible because a_1, μ form a non-inversion after a_2 is shifted upward, implying an ascend between consecutive commuting labels somewhere between a_1 and μ, causing the cell not to be critical. In case (3), we can never un-do this shifting of μ', since a matching step will not shift it downward from $I(a_1, a_2)$, since μ blocks a_2 from shifting upward, but μ is too large to shift below a_2 without a_1 also present. Case (2) is allowed, but eventually we still would need to shift μ downward into $I(a_1, a_2)$, at which point we would have a downward step keeping us in the same Boolean algebra, making it impossible for the cycle to continue. Thus, we can rule out upward-shifting gradient path steps in a cycle.

Next suppose there is a step that shift labels downward either creating or eliminating an ILT. Consider the lowest ILT $I(a_1, a_2)$ ever created/destroyed. It must be destroyed by a downward step shifting a_1 into the interior of a lower ILT $I(b_1, b_2)$. Eventually we have an upward (matching) step, shifting a_1 back upward from within $I(b_1, b_2)$ to below a_2. But we have already eliminated the possibility of such upward-shifting matching steps within a cycle.

Finally, if all upward (matching) steps shift labels downward from within ILTs to between them, and all downward (non-matching) steps shift labels downward from between ILTs into lower ILTs, preserving the set of ILTs at each step, then labels not initiating or concluding ILTs move progressively downward and may never return upward, making completing a cycle impossible. To be precise we create inversions between labels initiating/concluding ILTs and other labels, but we may never eliminate these inversions. □

6. Applications: minimal free cellular resolution and a finite state automaton which computes Poincaré-Betti series

This section describes the surviving critical cells in the quadratic Gröbner basis case in two ways:

(1) as the words generated by a finite state automaton, implying the generating function for Morse numbers is rational
(2) as representatives of the J'-non-stuttering, J'-commuting equivalence classes of words, as developed in [**HRW**] to count Betti numbers.

THEOREM 6.1. *The discrete Morse function of the previous section gives a minimal free cellular resolution of k as a $k[\Lambda]$-module.*

PROOF. Results of [**BW**] imply that the complex of critical cells from our Morse function supports a free cellular resolution of k as a $k[\Lambda]$-module, because our acyclic matching preserves multi-grading. Furthermore, there are no incidences among critical cells of equal multi-degree in this Morse function, because all critical cells of multidegree λ come from saturated chains with highest element λ, making gradient paths from one critical cell to another of the same multi-degree impossible, despite the fact that critical cells need not be concentrated in a single dimension. If we consider the complex obtained by tensoring the complex of critical cells with k, this implies that all its boundary maps are 0 maps. This implies that the resolution supported by the complex of critical cells is a minimal free resolution.

Alternatively, one may see that the resolution is minimal by checking that Morse numbers equal Betti numbers. Theorem 6.4 does this by constructing a bijection between the critical cells in our Morse function and the J'-non-stuttering, J'-commuting equivalence classes of words of [**HRW**]; these equivalence classes of words were shown in [**HRW**] to index a basis for Tor(k, k), because those of fixed multidegree λ index a homology basis for $(0, \lambda)$.

□

REMARK 6.2. The fact that the discrete Morse function gives a minimal free cellular resolution implies Morse numbers equal Betti numbers. Thus, the upcoming generating function for Morse numbers also computes the Poincaré-Betti series.

The next theorem will construct a finite state automaton that generates exactly the label sequences for the surviving critical cells. The list of states in this finite state automaton is far from minimal in general among all finite state automata generating this language. Specifically, we keep track of more data in each state than is strictly necessary, in order to greatly simplify the description of our automaton.

THEOREM 6.3. *The label sequences for saturated chains which contribute surviving critical cells are exactly the words of a regular language. Thus, the generating function for Morse numbers, which in this case equals the Poincaré-Betti series, is rational.*

PROOF. The alphabet for the language is the set of labels on covering relations, i.e. of generators for the monoid. For convenience, we view label sequences on saturated chains as words by reading them from top to bottom. Since all surviving critical cells are saturated chains, the dimension of each such critical cell is two less than the length of the word labelling it. Thus, the Morse number m_i counts words of length $i + 2$ in the language of label sequences. We will describe a set of states and of legal transitions between states that comprise a finite state automaton that generates exactly the language of label sequences for surviving critical cells. The existence of such an automaton will imply that the language is regular, and hence the generating function for Morse numbers is rational (cf. Section 2.6). In fact, the rational generating function may be determined from the finite state automaton (see [**BR**]). The remainder of the proof describes how to construct such an automaton.

The automaton has a unique initial state, and each time a label is read, a transition is made from one state to another state if the label sequence read so far could be the initial segment for a label sequence of a surviving critical cell. To decide which labels give valid transitions, each state must keep track of enough data about previously read labels to decide whether concatenating a newly read label will

- give a label sequence for a surviving critical cell, in which case a transition is made to a final state, or
- give a label sequence for a critical cell which is cancelled, but one where reading additional labels could again yield a surviving critical cell, in which case a transition is made to a non-final state, or
- give a label sequence not meeting either of the above forms, in which case there is no valid transition, so the word is not generated by the automaton.

Specifically, for there to be a transition labeled λ out of a state S, λ must form either a descent or an ILT with the most recently read label, and there are further constraints related to non-essential sets. The requirement about descents and ILTs is necessary because every pair of consecutive labels for a surviving critical cell must take this form.

Each state will contain the following data: the list of previously encountered ILTs and individual labels, together with the order of the most recent occurrences of these ILTs and labels. Thus, each state has associated to it a subset of the finite set of monoid generators and leading terms in our Gröbner basis, together with a permutation on the elements of this subset. Earlier occurrences of the same ILTs or individual labels are unnecessary for deciding whether all non-essential sets are empty, or else would have already caused the word to be unproducable by the automaton at an earlier stage. Thus, we have a finite list of states.

If a partial label sequence w concludes with a label μ and leads to a final state S, then the next label λ to be read gives a legal transition from S to another final state if and only the following conditions are all met:

(1) λ, μ comprise a descent or ILT
(2) λ is not in the non-essential set of any earlier ILT. That is, every previously encountered ILT $I(a_1, a_2)$ with $a_1 < \lambda < a_2$ either has (a) $\lambda\nu \in \text{in}(I_\Lambda)$ for some label ν read more recently than $I(a_1, a_2)$, (b) $\lambda a_1 \in \text{in}(I_\Lambda)$, (c) $\lambda a_2 \in \text{in}(I_\Lambda)$, or (d) λ smaller than some label ν read after $I(a_1, a_2)$
(3) If λ, μ comprise an ILT, then there is no previously encountered label μ' in its non-essential set. That is, there is no previously encountered μ' with all the following properties: (a) μ' satisfies $\lambda < \mu' < \mu$, (b) μ' is smaller than all labels read after it and before λ, (c) μ' commutes with all labels read after it, and (d) deleting μ' would cause μ to be in the non-essential set of a previously encountered ILT.

When the first and third conditions hold but the second one fails, there is still a transition to a non-final state U. However, the only legal transitions from such a non-final state U are given by labels λ' such that λ', λ form an ILT which causes λ no longer to belong to a non-essential set, i.e. when we are in one of the following circumstances:

- $\lambda' < \mu$ and $\lambda'\mu \notin \text{in}(I_\Lambda)$, because then shifting λ upward would yield a non-critical cell
- λ' would also belong to the non-essential set of some ILT once λ is shifted upward into an ILT (i.e. Theorem 5.17, case 1), implying the critical cell with ILT λ', λ is not matched by shifting λ upward.

The necessity of these constraints on allowable words is immediate from the description of critical cells and the matching to cancel them in earlier sections. These constraints on legal transitions are also sufficient to produce a surviving critical cell because any such label sequence will label a critical cell whose ILTs all have empty non-essential set, i.e. a critical cell that is not cancelled. □

Following [**HRW**], let $J = \text{in}(I_\Lambda)$, and let J' be the complement of J. Labels a, b commute if and only if $ab \notin J$, in which case we say they are *J'-commuting.* Define a *J'-commuting equivalence class* of label sequences to be a set of label sequences which agree up to J'-commutation. A label sequence is *J'-stuttering* if it has consecutive labels a_1, a_1 where $a_1^2 \notin J$. A J'-commuting equivalence class C is *J'-non-stuttering* if none of the label sequences in C are J'-stuttering.

THEOREM 6.4. *There is a bijection between the J'-non-stuttering, J'-commuting equivalence classes of a given content and the label sequences of the same content for critical cells that survive cancellation. Moreover, exactly one member of each*

J'-non-stuttering, J'-commuting equivalence class is a label sequence for a critical cell surviving cancellation.

PROOF. We will show that each J'-non-stuttering J'-commuting equivalence class contains exactly one label sequence for a critical cell surviving cancellation. First we show the existence of such a label sequence within each such J'-non-stuttering J'-commuting equivalence class by providing an algorithm which applies a series of J'-commutation relations to transform any member of such a class into the label sequence for a surviving critical cell. Then we show that each such class has at most one label sequence from a critical cell surviving cancellation. Finally, we show that J'-stuttering anywhere within a J'-commuting equivalence class of a label sequence implies that the label sequence either does not come from a critical cell or is cancelled.

The algorithm sequentially processes the labels, proceeding from smallest to largest label value, and in the case of repetition, proceeds from highest to lowest initial location for each value. The algorithm terminates because it processes a finite number of labels and will use a finite number of steps to process each label. If the label μ immediately below a label ν to be processed is smaller than ν, then the algorithm would have processed μ before ν, and we will soon see that the pair must form an ILT. In this case, we say that μ is *attached* to ν at the time ν is processed.

A label ν is processed as follows. If ν is not attached to a label immediately below it, then ν is shifted upward until ν either encounters a label with which it does not commute, reaches the top of the label sequence, or encounters a label smaller than it such that all current ILTs $I(a_1, a_2)$ above ν with $a_1 < \nu < a_2$ and $a_1\nu, \nu a_2 \notin \text{in}(I_\Lambda)$ are currently separated from ν by labels with which ν does not commute. Notice that ν will not encounter another copy of ν before reaching such a position, because the J'-commuting equivalence class is J'-non-stuttering. If ν is shifted to just below a label λ_2 with which ν does not commute, such that $\nu < \lambda_2$, then ν is now attached to λ_2. It in addition the label λ_1 previously below λ_2 had formed an ILT with λ_2, then $\nu < \lambda_1 < \lambda_2$, and we may detach λ_1 from λ_2 at the same time that we attach ν to λ_2 to form a new ILT $I(\nu, \lambda_2)$. If a label ν to be processed is attached to a label μ just below it, then the pair is shifted upward as a unit past labels larger than ν that commute with both μ and ν, with the following special rules:

- if μ, ν encounter a label $\lambda > \nu$ which commutes with ν but not with μ, then detach ν from μ, shift ν past λ, and attach λ to μ, and continue processing ν as an unattached label
- if the label immediately above ν forms either a descent or ILT with μ and ν can be shifted upward into the interior of and ILT $I(a_1, a_2)$ with $a_1 < \nu < a_2$, $a_1\nu, \nu a_2 \notin \text{in}(I_\Lambda)$, and ν commuting with all labels between ν and $I(a_1, a_2)$, then ν is detached from μ and shifted upward to above $I(a_1, a_2)$ and continues its processing.

The fact that the only ascends in the output are between non-commuting pairs ensures it labels a critical cell. To see it also is one that survives cancellation, one may check that all non-essential sets are empty. At the time μ was processed, μ could not be further shifted upward into an ILT for which it would belong to the non-essential set, or else μ would have been shifted farther upward in its processing. The fact that all smaller values had already been processed by the time μ was processed ensures that this property is preserved throughout the algorithm. We may also eliminate the possibility of non-essential set members that shift downward into ILTs, because these only arise when the top of some ILT is capable of shifting upward without its partner, but our algorithm would have actually performed this shifting, and again the fact that we process smaller labels before larger ones means this property is preserved throughout the rest of the algorithm. Thus, all non-essential sets are empty, so the algorithm indeed outputs the label sequence of a surviving critical cell.

To show that there is at most one label sequence surviving critical cell cancellation in each J'-commuting equivalence class, first note that pairs of consecutive labels that commute must appear in descending order to avoid either having an ascend not appearing within an ILT (implying the saturated chain does not contribute a critical cell) or having an ILT with non-empty interior (implying the critical cell is cancelled). Now suppose there are two label sequences in the same J'-commuting equivalence class, both from critical cells surviving cancellation. Then there must be some pair of J'-commuting labels μ_1, μ_2 with $\mu_1 < \mu_2$, such that the pair are inverted in one label sequence and not the other. One may use the intermediate value theorem to show that the non-inverted pair μ_1, μ_2 must be separated by at least one ILT $I(a_1, a_2)$ with $a_1 < \mu_1 < a_2$ and by at least one ILT $I(b_1, b_2)$ with $b_1 < \mu_2 < b_2$, such that $I(a_1, a_2)$ either equals $I(b_1, b_2)$ or occurs before $I(b_1, b_2)$. This implies either that $\mu_1 \in S(a_1, a_2)$, ensuring cancellation, or that some label not commuting with μ_1 separates μ_1 from $I(a_1, a_2)$, implying μ_2 must commute with all labels between it and a_1, in order for it to be possible to swap μ_1, μ_2. But then the critical cell with μ_2, μ_1 forming an inversion must have $\mu_2 \in S(b_1, b_2)$, and so cannot also survive cancellation.

Now we turn to the issue of non-stuttering. Any two consecutive identical labels must appear within a syzygy interval to avoid comprising an ascend which would make the cell non-critical. But we showed that any critical cell with a syzygy interval with non-empty interior is cancelled. If a label sequence ω is in the J'-commuting equivalence class of a label sequence which has consecutive identical labels, then some label λ appears more than once in ω, separated by one or more syzygy intervals. But then one of these syzygy intervals will have the lower copy of λ in its non-essential set, implying the critical cell is cancelled, unless λ does not commute with some separating label. But this label also does not commute with the other copy of λ, making it impossible for the two labels to be shifted to consecutive positions, as needed for stuttering, a contradiction. Thus, if a label sequence is J'-commuting equivalent to one with stuttering, then the label sequence does not survive cancellation. □

7. Gröbner bases of higher degree

In this section, we extend results of Section 5 from the quadratic Gröbner basis case to the degree d to prove:

THEOREM 7.1. *If I_Λ has a Gröbner basis of degree d, then $\tilde{H}_i(\Delta(\hat{0}, \lambda)) = 0$ for $i < -1 + \frac{\deg(\lambda)-1}{d-1}$, with $\deg(\lambda)$ defined as below. Hence, we have $Tor_i^{k[\Lambda]}(k,k)_\lambda = 0$ for $i < 1 + \frac{\deg(\lambda)-1}{d-1}$. Moreover, this vanishing is achieved by a free cellular resolution resulting from a discrete Morse function on Λ.*

In the standard-graded case, $\deg(\lambda)$ is given by the grading. In general, let $\deg(\lambda)$ be one more than the length of the shortest saturated chain on the poset interval $(\hat{0}, \lambda)$, i.e. the degree of the image of $\underline{x}^\lambda$ in the associated graded ring.

PROPOSITION 7.2. *Ordering saturated chains by using any monomial term order to order fibres then lexicographically ordering saturated chains within each fibre yields a content-lex facet order.*

PROOF. Syzygy intervals now must be defined to have weakly increasing labels $a_1, \dots, a_r$ such that there is a Gröbner basis leading term which divides $a_1 \cdots a_r$ and has smallest divisor a_1 and largest divisor a_r. To be a minimal skipped interval, we must also have that neither $a_2 \cdots a_r$ nor $a_1 \cdots a_{r-1}$ is divisible by a Gröbner basis leading term. Then the proof of Theorem 5.2 applies. □

The remainder of this section is concerned with cancelling pairs of critical cells to obtain a Morse function with no critical cells below dimension $-1 + \frac{\deg(\lambda)-1}{d-1}$.

DEFINITION 7.3. An *increasing leading term*, or *ILT*, is a Gröbner basis leading term, with labels arranged in weakly increasing order. We will often use the term ILT to refer to an ILT that labels a specific syzygy interval.

We will use the variables d', d'' to represent the degree of an arbitrary Gröbner basis leading term, so we always will have $d', d'' \le d$. Denote a syzygy interval with ILT $a_1 \cdots a_{d'}$ by $I(a_1, \dots, a_{d'})$. In contrast to the quadratic Gröbner basis case, now there may be several Gröbner basis leading terms specifying the same syzygy interval. This fact, that a single syzygy interval may have several ILTs beginning and ending with the same pair of labels, causes one substantial new issue to arise: the critical cells resulting from one syzygy interval may comprise several overlapping Boolean algebras, since different ILTs will give rise to different non-essential sets, as in Example 7.4.

EXAMPLE 7.4. Consider the syzygy interval labeled $a_1, \dots, a_4$ that goes with the Gröbner basis leading terms $a_1a_2a_4$ and $a_1a_3a_4$. One Boolean algebra of critical cells, based on ILT $I(a_1, a_2, a_4)$ consists of labels sequences $a_1a_2a_3a_4$ and $a_3a_1a_2a_4$, while the Boolean algebra that is associated with $I(a_1, a_3, a_4)$ consists of label sequences $a_1a_2a_3a_4$ and $a_2a_1a_3a_4$. The critical cell labeled $a_1a_2a_3a_4$ is shared by the two Boolean algebra.

Lemma 7.7 deals with such overlap by providing an acyclic matching for any such collection of overlapping Boolean algebras.

Define the non-essential set for $I(a_1, \dots, a_{d'})$, denoted $S(a_1, \dots, a_{d'})$, similarly to the $d = 2$ case, but now it may have two types of members:

(1) individual labels, in exact analogy to the non-essential set members for the $d = 2$ case, i.e. labels which either (a) appear in topologically decreasing positions below $I(a_1, \dots, a_{d'})$ (or above $I(a_1, \dots, a_{d'})$ in the exceptional circumstances discussed in Theorem 5.17) from which they may shift into $I(a_1, \dots, a_{d'})$ via a gradient path without causing $I(a_1, \dots, a_{d'})$ to cease to be a minimal skipped interval, or (b) labels that have thus shifted into the interior of $I(a_1, \dots, a_{d'})$.
(2) collections $\{b_2, \dots, b_{d'}\}$ of labels that appear either immediately above a label b_1 with which they form an ILT, or which collectively appear in the interior of one or more ILTs strictly above such a label b_1, with

$I(a_1, \ldots, a_{d'})$ serving as the highest of these ILTs. Furthermore, we require there to be a gradient path from the former to the latter which shifts the collection of labels upward into the interior of the various ILTs in order for the collection of labels to belong to $S(a_1, \ldots, a_{d'})$.

Follow the conventions of Theorem 5.17 to decide which such individual labels and collections of labels should belong to $S(a_1, \ldots, a_{d'})$, not allowing collections of labels that shift downward into $I(a_1, \ldots, a_{d'})$ from above. If there is a need for "blocking" as in Theorem 5.17, an individual label will always serve this function rather than a collection of labels. If a label $\lambda \in I(a_1, \ldots, a_{d'})$ meets the above requirements to be included individually in $S(a_1, \ldots, a_{d'})$, then it is not also included as part of a collection of labels. With these conventions, the proof that this gives a well-defined matching is identical to the proof of Theorem 5.17.

Proof of Theorem 7.1. Proposition 7.2 provides the Morse function that serves as our starting point. We will cancel all critical cells of dimension less than $-1 + \frac{\deg(\lambda)-1}{d-1}$, using a fairly similar, but somewhat more subtle, approach to the $d = 2$ argument. Notice that each critical cell of dimension less than $-1 + \frac{\deg(\lambda)-1}{d-1}$ is contributed by a saturated chain with average minimal skipped interval height greater than $d - 1$. But any minimal skipped interval of height greater than $d - 1$ is a syzygy interval with more than d labels, so it is consists of an ILT with at least one additional label interspersed. Such extra labels either allow the critical cell to be cancelled similarly to the $d = 2$ case, or in the case of a syzygy interval with multiple ILTs, Lemma 7.7 gives a matching in which all unmatched cells have average interval height at most $d-1$ for the I-intervals related to the syzygy interval. Thus, cells left unmatched must then have another syzygy interval of height greater than $d - 1$ at lower ranks. This allows us to repeat the argument until eventually reaching a syzygy interval which causes the cell to be cancelled.

The fact that there is indeed a unique gradient path from a critical cell τ to a critical cell σ for each pair τ, σ to be cancelled follows from Theorem 4.2. When a single label is shifted into a syzygy interval, the gradient path is identical to the one given in the $d = 2$ case. When a collection of labels is shifted upward into a syzygy interval, the gradient path is the one described in the proof of Theorem 4.2. Theorem 4.2 also proves the uniqueness of these gradient paths, whether shifting a single label or a collection of labels. As before, we give a complete acyclic matching on each Boolean algebra of critical cells, as long as it is not part of a collection of overlapping Boolean algebras. Similarly to the $d = 2$ case, we match all cells in this Boolean algebra by including/excluding a single non-essential set member ν from the interior of the syzygy interval. It is convenient to choose ν to be the individual label with highest topologically decreasing position outside the syzygy interval, if there is such a label, and otherwise to choose the collection of labels with highest topologically decreasing position outside the syzygy interval. Lemma 7.7 provides the matching for collections of overlapping Boolean algebras.

With these choices, acyclicity is similar to the $d = 2$ case, since Lemma 7.7 will verify acyclicity of the matching on a single collection of overlapping Boolean algebras resulting from several ILTs on a single syzygy interval. Applying results of [**BW**], the desired resolution is immediate from this acyclic matching. □

Next we prove Lemma 7.7 in the special case of degree $d = 3$. In this case we deduce a stronger result than for general d, but the proof is also much simpler than in general, but gives the flavor of the upcoming proof for degree d.

DEFINITION 7.5. A pair of ILTs $a_1 \dots a_r$ and $b_1 \dots b_s$ with $a_1 \le b_1, a_r \le b_r$ are *concatenating* if either (1) $a_1 < b_1$ and $a_r = b_i$ for some $1 \le i < s$, or (2) $a_r < b_s$ and $b_1 = a_i$ for some $1 < i \le r$.

LEMMA 7.6. *Let I_Λ be a toric ideal with Gröbner basis with leading terms all of degree at most 3. Let $I = I(a_1, \dots, a_{d'})$ be a syzygy interval, given by one or more ILTs, each of which gives rise to a Boolean algebra of critical cells. Then this collection of overlapping Boolean algebras has an acyclic matching which matches all critical cells with average interval height at most 2 for I together with any descents coming from labels shifted out of I.*

PROOF. Order the ILTs $M_1, \dots, M_k$. If some M_i has degree 2, then the Boolean algebra for each M_j is contained in the Boolean algebra for M_i, so we use the complete matching on a single Boolean algebra. Otherwise, we have labels $\lambda_1, \dots, \lambda_k$ such that for $1 \le i \le k$, $M_i = a_1 \lambda_i a_{d'}$, for fixed initial and final labels $a_1, a_{d'}$. Let Δ_i be the set of critical cells in the Boolean algebra $B^{(i)}$ for M_i which are not shared with any earlier Boolean algebra $B^{(i')}$ for $i' < i$. Notice that Δ_i consists of exactly those critical cells in $B^{(i)}$ which have $\lambda_1, \dots, \lambda_{i-1}$ all shifted to topologically decreasing positions outside I. Thus, Δ_i has the structure of a Boolean algebra, resulting from all other labels in the non-essential set for $I(a_1, \lambda_i, a_{d'})$, so this has a complete acyclic matching unless this set is empty. But when the set is empty, then I consists of only the three labels $a_1, \lambda, a_{d'}$, as well as labels essential to concatenating ILTs, so matching is not necessary. In the case of concatenating ILTs, the average interval height is still at most 2. □

The situation gets much more complex when labels other than a_1 and $a_{d'}$ may divide more than one of the Gröbner basis leading terms specifying ILTs on the syzygy interval.

LEMMA 7.7. *Suppose that a single expanding interval I has multiple ILTs. Then the resulting collection of overlapping Boolean algebras has an acyclic matching such that all unmatched cells have minimal skipped interval average height at most $d-1$.*

PROOF. Choose a total order $M_1, \dots, M_k$ on the ILTs for $I(a_1, \dots, a_{d'})$. Thus, each M_i is a Gröbner basis leading term with smallest divisor a_1, largest divisor $a_{d'}$ and with divisors of intermediate value, all of which appear as labels that can shift in/out of the syzygy interval. Thus, any shifting of labels in/out of the syzygy interval still gives a syzygy interval as long as at least one of these ILTs appears entirely within the syzygy interval. Each ILT $I(M_i)$ has its own non-essential set, denoted $S(M_i)$, giving rise to its own Boolean algebra of critical cells. Denote by Δ_r the collection of critical cells in the Boolean algebra given by M_r which are not shared with any of the earlier Boolean algebras given by $M_1, \dots, M_{r-1}$.

We will provide an acyclic matching on each such Δ_r. Note that Δ_r consists of those subsets of $S(M_r)$ which shift enough labels to outside the syzygy interval $I(M_r)$ so that the label sequence on $I(M_r)$ is not divisible by any of the monomials $M_1, \dots, M_{r-1}$. If there is any label in $S(M_r)$ that does not divide any of the monomials $M_1, \dots, M_{r-1}$, then we obtain a complete acyclic matching on Δ_r by including/excluding one such label in $I(M_r)$. Next we consider the case where each member of $S(M_r)$ does divide some earlier M_i.

Fix an ordering $\lambda_1, \dots \lambda_l$ on the elements of $S(M_r)$. It will be convenient in the next section if we order them from highest to lowest topologically decreasing position outside $I(M_r)$. Now apply the following matching procedure to each critical cell in Δ_r:

(1) match the cell by including/excluding λ_1 from $I(M_r)$ unless shifting λ_1 to inside $I(M_r)$ yields a cell in an earlier Δ_i,
(2) if the cell is not yet matched, then match by including/excluding λ_2 from $I(M_r)$, unless this yields a matching partner which was already matched at the first step or which belongs to an earlier Δ_i
(3) continue inductively, matching the cell by including/excluding λ_i from $I(M_r)$ if the cell was not already matched based on any of the labels $\lambda_1, \dots, \lambda_{i-1}$ and the partner cell based on shifting λ_i also does not belong to an earlier Δ_i and is not already matched based on any earlier label $\lambda_{i'}$ with $i' < i$.

Notice that a cell cannot be matched based on the label λ_i if either (a) $\lambda_i \notin I(M_r)$ and shifting λ_i into $I(M_r)$ gives a cell in an earlier Boolean algebra, or (b) shifting λ_i in or out of $I(M_r)$ gives a cell previously matched. Thus, any unmatched critical cell that has exactly the labels $M = \{\mu_1, \dots, \mu_k\}$ shifted to outside $I(M_r)$ will have the property that each such μ_i is necessary outside $I(M_r)$ either to avoid overlap with an earlier Boolean algebra or in order for some $\nu \in I(M_r)$ coming earlier than μ_i also not to allow matching. That is, in the latter case there must be some $\mu_{i'} < \nu < \mu_i$ with $\mu_{i'} \notin I(M_r)$, such that $\mu_{i'}$ "covers" multiple ILTs (see Definition 7.8), some of which could also be covered by ν, and the rest of which are also covered by μ_i.

DEFINITION 7.8. A label μ *covers* an earlier ILT M_j if μ divides the Gröbner basis leading term specifying M_j.

Assign to each $\mu_i \in M$ either an ILT $M_{i'}$ which it exclusively covers, or an ILT $M_{i''}$ that it would exclusively cover if the earliest forbidden $\nu \in I(M_r)$ were shifted to outside $I(M_r)$, or which it would exclusively cover after some number of iterations of this reasoning, i.e. an ILT which makes it impossible to shift μ_i from outside $I(M_r)$ to inside $I(M_r)$ as a matching step. Call this ILT which is assigned to μ_i the *indexing ILT* of μ_i.

If we can show that every label in $S(M_r)$ belongs to one of the k indexing ILTs, this will imply $|T| \leq k\cdot(d-2)$, as desired. Suppose some $\lambda \in S(M_r)$ is not in any of the indexing ILTs, and choose the label λ of this form which comes earliest in our ordering on labels in $S(M_r)$. First note that $\lambda \in I(M_r)$, since otherwise λ would belong to its own indexing ILT. We will show next that the cell with λ shifted to outside $I(M_r)$ is not matched based on a label of higher precedence than λ. Since shifting λ to outside $I(M_r)$ also cannot give a cell belonging to an earlier Δ_i, we will be able to conclude that the critical cell will be matched based on λ. Thus, any unmatched cell will satisfy $|T| \leq k \cdot (d-2)$.

Now we prove the claim that the cell may be matched by shifting λ to outside $I(M_r)$. When λ is shifted to outside $I(M_r)$, each μ_i of higher precedence which appears outside $I(M_r)$ cannot be shifted to inside $I(M_r)$ as a matching step, by virtue of its indexing ILT, since λ cannot cover this indexing ILT. Likewise any

λ_i of higher precedence which appears within $I(M_r)$ in the critical cell cannot be shifted to outside $I(M_r)$ without rendering some $\mu_{i'}$ unnecessary for covering its indexing ILT, since otherwise we would have matched based on the smallest λ_i which did not have this property; in particular, this means that λ_i must belong to the indexing ILT for $\mu_{i'}$ in the critical cell. Shifting λ to outside $I(M_r)$ does not change this relationship, so the cell with λ shifted to outside $I(M_r)$ also cannot match by shifting λ_i. Thus, λ is the first label allowing matching for both cells, so both are indeed matched by shifting λ.

In the case where all elements of $S(M_r)$ are individual labels that shift to topologically decreasing positions outside $I(M_r)$, this yields the following upper bound on average interval height for this portion of the interval system, using the fact that total height is one less than the total number of labels involved, and that k is non-negative:

$$\frac{\text{total height}}{\text{no. of intervals}} = \frac{d-1+|T|}{k+1} \leq$$

$$\frac{d-1+k\cdot(d-2)}{k+1} = (d-2) + \frac{1}{k+1} \leq d-1.$$

Let us now handle the more general case, where some non-essential set members are collections of labels. All labels belonging to such collections will contribute individually to the bound $|T| \leq k \cdot (d-2)$ when the labels appear within $I(M_r)$, because each label contributes individually to monomial degree. When such a collection of labels appears outside the ILT, it would increase the number of intervals k by one, but would increase the total height by as much as $d-1$, seemingly invalidating the above computation of average interval height. However, the highest label in the newly created ILT must also form a descent with the label immediately above it, and we may use this descent rather than the new ILT in order to compute the above bound, since the descent will not also be counted in a similar computation for any other syzygy interval. We may safely ignore the newly created ILT in the bound computation, since it also has height at most $d-1$.

Acyclicity will follow from the Cluster Lemma of [**Jo**] (see Lemma 2.6), using the filtration of subcomplexes $B_1 \subseteq B_1 \cup B_2 \subseteq \cdots \subseteq B_1 \cup \cdots \cup B_k$ where $B_1 \cup \cdots \cup B_i$ is the union of Boolean algebras given by $M_1, \ldots, M_i$. All we need to do is show that the matching on each $B_i \setminus (B_1 \cup \cdots \cup B_{i-1})$ is acyclic. But if there were a cycle, let μ_i be the highest precedence label to be inserted as a matching step in the cycle. This would necessitate a downward step in the cycle shifting μ_i back into the interior of $I(M_r)$, but this would be preceded and followed by matching steps inserting labels of lower precedence than μ_i. This contradicts our greedy matching procedure, because it would instead make the downward edge a matching edge inserting μ_i, since this has higher precedence than the matching step of either endpoint. Thus, there are no cycles. □

8. Rationality of Morse number generating function

In this section we describe a finite state automaton that generates exactly the language of label sequences for surviving critical cells, in the case of a Gröbner basis of degree d. The existence of such a generating function again implies the language is regular, and hence that the generating function for Morse numbers is a rational function which gives upper bounds on the terms in the Poincaré-Betti series. In contrast, for $d \geq 3$ the Poincaré-Betti series is not always rational. The generating function for Morse numbers does come close enough to the Poincaré-Betti series to achieve the vanishing of Betti numbers described by Theorem 7.1. Due to the similarity of the finite state automaton to the one given in the quadratic Gröbner basis case, less detail is provided here than in Section 6.

The states in the automaton keep track of the set of previously encountered ILTs and individual labels, in their order of most recent appearance. Reading label sequences from top to bottom, the following are the legal transitions from one state to another.

(1) a single label λ that is larger than its predecessor, i.e. which forms a descent with the label above it. For the transition to be to a final state, we require the further property that λ is separated from each previously encountered ILT $I(a_1, \ldots, a_{d'})$ which satisfies $a_1 < \lambda < a_{d'}$ and $\lambda a_2 \cdots a_{d'}, \lambda a_1 \cdots a_{d'-1} \notin \text{in}(I_\Lambda)$. by a label with which λ does not commute.
(2) a single label which forms an ILT together with its predecessor, exactly as in the $d = 2$ case
(3) a collection $\{a_1, \ldots, a_{d'-1}\}$ of labels with $d' > 2$, which together with the most recently encountered label $a_{d'}$ form an ILT $I(a_1, \ldots, a_{d'})$ such that (a) the labels $\{a_2, \ldots, a_{d'}\}$ cannot all simultaneously shift upward into the interior of higher ILTs to yield a critical cell which does not have any of the labels $\{a_2, \ldots, a_{d'}\}$ individually as members of any non-essential set, and (b) no label above $I(a_1, \ldots, a_{d'})$ may shift downward into $I(a_1, \ldots, a_{d'})$ by a gradient path to yield a critical cell. Such a transition leads to a final state.
(4) a collection of labels that collectively complete an ILT, with an allowable collection of interspersed labels. Allowable collections are those that arise as a result of concatenating ILTs, as described below, and those which may be within the ILT in a surviving critical cell when there are multiple ILTs on the same syzygy interval. In this case the transition is to a non-final state, and we will justify below that there are only a finite number of these transitions.

The point is to use non-final states for label sequences for critical cells that are cancelled, if the concatenation of additional labels may yield a critical cell that is not cancelled. to be cancelled. The "concatenating" ILTs mentioned in the fourth type of transition come from situations such as the following example.

EXAMPLE 8.1. Consider a label sequence $abcde$ where abd and cde are each Gröbner basis leading terms. The label c cannot be shifted out of the ILT $I(a, b, d)$ to yield a critical cell, because the ascend (d, e) would no longer be part of a minimal skipped interval, so $c \notin S(a, b, d)$ for the label sequence $abcde$, though it would belong to $S(a, b, d)$ in the label sequence $abcd$.

Specifically, a pair of ILTs $M_1 = a_1 \cdots a_r$ and $M_2 = b_1 \cdots b_s$ are *concatenating* if either (a) a_r divides M_2 with $a_r \neq b_s$ and $a_1 < b_1$, or (b) b_1 divides M_1 with $b_1 \neq a_1$ and $a_r < b_s$.

The fourth type of transition also accommodates the matching procedure of Lemma 7.7.

PROPOSITION 8.2. *The automaton has finitely many states and transitions.*

PROOF. There is a finite list of possible ILTs, even when we consider all possible label interspersions that could still allow the cell not to be cancelled, i.e. from concatenating ILTs and from multiple ILTs on a single syzygy interval. This follows from the fact that the semi-group ring is finitely generated, and that each Gröbner basis leading term has finite degree, so labels occurring in the interior of an ILT with multiplicity greater than the Gröbner basis degree will always allow critical cell cancellation. The transitions out of a state are limited by the finite list of labels. □

PROPOSITION 8.3. *Word length equals critical cell dimension shifted by two.*

PROOF. Any label sequence which has more I-intervals than J-intervals will be cancelled, unless there are two concatenating ILTs such that their concatenation contains another ILT, causing three or more overlapping I-intervals in which one is discarded, in such a way that no labels may be shifted from the interior of any of these ILTs without making the cell non-critical. But in the case of this type of concatenation, where two ILTs share labels and cover a third ILT, this means we can use just the labels in these two ILTs for labelling the transitions in the finite state automaton, so we get the correct word length. □

Using the observations and propositions above, it is not hard to generalize the automaton from the $d = 2$ case to obtain:

THEOREM 8.4. *The surviving critical cells are labeled by the words of a regular language, with word length measuring cell dimension, shifted by two. Thus, the generating function for Morse numbers is a rational generating function which is determined by the given finite state automaton.*

9. Some remarks and open questions

REMARK 9.1. When a variable does not appear in any syzygies, then it may be "factored out" before starting our analysis, similar to the situation with computing Tor groups directly. Specifically, if some z_i does not appear in any generators of the toric ideal I_Λ for $k[z_1, \ldots, z_n]/I_\Lambda$, then the partial order Λ is the product of an infinite chain together with the poset of monomials ordered by divisibility in $k[z_1, \ldots, \hat{z}_i, \ldots, z_n]/I_\Lambda$. Thus, any finite interval is the product of a finite chain together with a monoid poset interval $(0, \lambda)$ for the ring $k[z_1, \ldots, \hat{z}_i, \ldots, z_n]/I_\Lambda$; the order complex of such an interval is the suspension of the join of the order complexes for the two terms in the product, so the suspension of the join of a simplex (i.e. the order complex of a chain) with the order complex $\Delta(0, \lambda)$.

REMARK 9.2. We sometimes have gradient paths which reverse a decreasing sequence of labels of length $d > 2$ to produce an ILT, in which case the permutation on labels is not 321-avoiding. We have not matched and cancelled any such pairs of critical cells. Theorem 2.14 shows there are at most two gradient paths between a pair of critical cells related by such a reversal for lexicographic discrete Morse functions; the proof of Theorem 2.14 generalizes to those facet orders which satisfy the crossing condition, so in particular to content-lex facet orders.

THEOREM 9.3. *Suppose I_Λ has degree at most three. Then for each critical cell τ in our complex Δ^M of critical cells after cancellation, $\partial(\tau)$ is a linear combination of critical cells of content strictly earlier than τ.*

PROOF. Suppose τ, σ, $\dim(\tau) = \dim(\sigma)+$, are surviving critical cells with equal content and there is a gradient path from τ to σ. Then τ, σ each have no syzygy intervals with non-empty non-essential set. Any gradient path from τ to σ must sort labels, but in such a way that σ still has no syzygy intervals with non-empty non-essential set. This can only be accomplished by reversing three or more descending labels to form a new ILT. This ILT must come from a Gröbner basis leading term of degree exactly three, since pairs of labels comprising degree 2 leading terms cannot be swapped without passing to an earlier content class. Lemma 7.6 ensures that the three or more labels must occur in a single string of descending labels within τ, to avoid τ being cancelled by virtue of a syzygy interval with non-empty non-essential set.

The ILT to be created cannot come from a Gröbner basis leading term of degree greater than three, both because of the assumptions of our theorem, and also because this would decrease critical cell dimension by more than one, implying σ could not be in the image of the boundary map applied to τ. Theorem 2.14 shows there are at most two gradient paths reversing three labels, resulting from the Coxeter relation $s_i s_{i+1} s_i = s_{i+1} s_i s_{i+1}$ being applied at the conclusion of a reduced expression. But one may easily check that one will indeed get two gradient paths when we reverse three labels as required for σ in the boundary of τ, and furthermore, that these will be oriented so that the two ways in which σ is incident to τ will cancel. Thus, σ will appear with coefficient 0 in the boundary of τ. □

REMARK 9.4. The following example shows that the Morse function bound on which Tor groups vanish is sharp. Consider

$$k[\Lambda] = k[z_1, \ldots, z_{2d}]/(z_1 \cdots z_d - z_{d+1} \cdots z_{2d}),$$

or equivalently,

$$k[z_1 z_2, z_3 z_4, \ldots, z_{2d-1} z_{2d}, z_1 z_{d+1}, z_2 z_{d+2}, \ldots, z_d z_{2d}].$$

This clearly has a Gröbner basis of degree d and none of lower degree. The interval $(1, z_1 \cdots z_d)$ in Λ is disconnected.

QUESTION 9.5. Is there a nice description of the gradient paths between surviving critical cells? This would be needed for a completely explicit description of the boundary maps in our resolution, since these are sums over such gradient paths.

QUESTION 9.6. Is it possible to improve our discrete Morse function into one that would provide a combinatorial proof of the following theorem? If an affine semi-group ring is standard graded, and its toric ideal of syzygies has a Gröbner basis of degree d, then its $(d-1)$-st Veronese is Koszul.

EXAMPLE 9.7. There is only one circumstance in which critical cells could skip more than $2d-3$ consecutive elements of a saturated chain, and this only may happen in the $d > 3$ case. Namely, if there are distinct Gröbner basis leading terms with the same initial and final labels, this may result in overlapping Boolean algebras of critical cells, with cells with large syzygy intervals not necessarily cancelled.

QUESTION 9.8. In [**HRW**], Tor groups related to quotients of affine semi-group rings by monomial ideals are translated to homology of certain relative complexes $\Delta(\lambda, A)$, where λ specifies a monoid poset interval and A is a graphic subspace arrangement. Does our Morse function translate to this setting to provide useful new information?

Acknowledgments

The authors thank Phil Hanlon, Mel Hochster, Ezra Miller and Vic Reiner for helpful conversations.

References

[An] D. Anick, On the homology of associative algebras, Trans. Amer. Math. Soc. **296** (1986), 641-659.

[Av] L. Avramov, Infinite free resolutions, in Elias, J. (ed.) et al., *Six lectures on commutative algebra*, Basel: Birkhuser. Prog. Math. **166**, 1-118 (1998)

[BH] E. Babson and P. Hersh, *Discrete Morse functions from lexicographic orders*, Trans. Amer. Math. Soc. to appear.

[BW] E. Batzies and V. Welker, *Discrete Morse theory for cellular resolutions*, J. Reine Angew. Math. **543** (2002), 147–168.

[BS] D. Bayer and B. Sturmfels, *Cellular resolutions of monomial modules*, J. Reine Angew. Math. **502** (1998), 123–140.

[BR] J. Berstel and C. Reutenauer, *Les se'ries rationnelles et leurs languages. (French) [Rational power series and their languages]*, Etudes et Recherches en Informatique. [Studies and Research in Computer Science], Masson, Paris, 1984. 132 pp.

[BiH] L. Billera and G. Hetyei, *Linear inequalities for flags in graded partially ordered sets*, J. Combin. Theory Ser. A **89** (2000), 77–104.

[BjW] A. Björner and M. Wachs, *On lexicographically shellable posets*, Trans. Amer. Math. Soc. **277, No. 1** (1983), 323–341.

[BjWe] A. Björner and V. Welker, Segre and Rees products of posets with ring theoretic applications, J. pure and appl. Alg. to appear. Preprint (2003).

[Ch] M. Chari, On discrete Morse functions and combinatorial decompositions, Discrete Math. **217** (2000) 101-113.

[Ei] D. Eisenbud, *Commutative Algebra with a View Towards Algebraic Geometry*, Springer-Verlag, New York, 1995.

[ERT] D. Eisenbud, A. Reeves and B. Totaro, *Initial ideals, Veronese subrings and rates of algebras*, Adv. Math **109** (1994), 168–187.

[Fo] R. Forman, *Morse theory for cell complexes*, Adv. Math. **134** (1998), 90–145.

[Fr] R. Fröberg, Koszul algebras, in Dobbs, David E. (ed.) et al., *Advances in commutative ring theory*. Lect. Notes Pure Appl. Math. **205**, 337-350, Marcel Dekker, New York, 1999.

[HHS] P. Hanlon, P. Hersh and J. Shareshian, *A $GL(n,q)$-analogue of the partition lattice*, Preprint 2003.

[He1] P. Hersh, *Lexicographic shellability for balanced complexes*, J. Algebraic Combinatorics **17** (2003), no. 3, 225–254.

[He2] P. Hersh, *On optimizing discrete Morse functions*, Preprint 2003.

[HRW] J. Herzog, V. Reiner and V. Welker, *The Koszul property in affine semigroup rings*, Pacific J. Math, **186** (1998), 39–65.

[Jo] J. Jonsson, *On the topology of simplicial complexes related to 3-connected and Hamiltonian graphs*, J. Comb. Theory, Ser A, **104** (2003), 169-199.

[LS] O. A. Laudal and A. Sletsjøe, *Betti numbers of monoid algebras. Applications to 2-dimensional torus embeddings*, Math. Scand. (1985), 145–162.

[MacL] S. MacLane, *Homology*, Springer-Verlag, 1975.

[PRS] I. Peeva, V. Reiner and B. Sturmfels, *How to shell a monoid*, Mathematische Annalen, **310** (1998), 379–393.

[RS] J.-E. Roos and B. Sturmfels, *A toric ring with irrational Poincaré-Betti series*, C. R. Acad. Sci. Paris Ser. I. Math. **326** (1998), no. 2, 141–146.

Department of Mathematics, Indiana University, Bloomington, Rawles Hall, Bloomington, IN 47405, USA

E-mail address: `phersh@indiana.edu`

Philipps-Universität Marburg, Fachbereich Mathematik und Informatik, D-35032 Marburg, Germany

E-mail address: `welker@mathematik.uni-marburg.de`

Contemporary Mathematics
Volume **374**, 2005

Integer programming, duality and superadditive functions

Jean B. Lasserre

ABSTRACT. Given $A \in \mathbb{Z}^{m\times n}$, $b \in \mathbb{Z}^m, c \in \mathbb{R}^n$, we consider the integer program $\mathbb{P}_d$: $\max\{c'x | Ax = b; x \in \mathbb{Z}^n_+\}$ which has a well-known abstract dual optimization problem stated in terms of superadditive functions. Using a linear program $\mathbb{Q}$ equivalent to $\mathbb{P}_d$ that we have introduced recently, we show that its dual $\mathbb{Q}^*$ can be interpreted as a simplified and tractable form of the abstract dual, and identifies a subclass of superadditive functions, sufficient to consider in the abstract dual. This class of superadditive functions is also used to characterize the integer hull of the convex polytope $\{x \in \mathbb{R}^n \,|\, Ax = b; x \geq 0\}$.

1. Introduction

Let $A \in \mathbb{Z}^{m\times n}, b \in \mathbb{Z}^m, c \in \mathbb{R}^n$ and consider the integer program

$$(1.1) \qquad \mathbb{P}_d \rightarrow \quad \max\{\, c'x \quad | \quad Ax = b; \quad x \geq 0; \quad x \in \mathbb{Z}^n\},$$

where the convex polyhedron $\mathrm{P} := \{x \in \mathbb{R}^n \mid Ax = b;\ x \geq 0\}$ is compact.

Related to $\mathbb{P}_d$ is the *dual* optimization problem

$$(1.2) \qquad \min_{f\in\Gamma}\{\, f(b) \mid \quad f(A_j) \geq c_j,\ j = 1,\ldots,n\}$$

where Γ is a certain set of functions $f : \mathbb{R}^m \rightarrow \mathbb{R}$ that are *superadditive* and such that $f(0) = 0$; see e.g. Gomory and Johnson [**1**], Jeroslow [**2**], Johnson [**3**], and Wolsey [**10**] who considers the particular inequality case $\{Ax \leq b, x \geq 0\}$ of (1.1) (add m nonnegative *slack* variables to retrieve the equality case). The dual problem (1.2) is rather conceptual in nature, but from it one still retrieves several concepts already available in standard linear programming (LP) duality (see [**10**, p. 175] still for the particular case $\{Ax \leq b, x \geq 0\}$). More importantly, the fundamental and basic Gomory (fractional) *cuts* for integer programs, which are crucial for the efficiency of today 's most powerful codes for solving $\mathbb{P}_d$, have an interpretation in terms of superadditive functions f in (1.2). For instance see Wolsey [**9**, §7]. Therefore, besides its theoretical interest, any insight on the dual problem (1.2) is of potential interest as it could provide useful information for deriving efficient cuts in standard solving procedures for $\mathbb{P}_d$. Moreover, problem (1.2) which is non-linear, can be transformed into an equivalent finite LP. For instance, in the particular

---

1991 *Mathematics Subject Classification.* Primary 54C40, 14E20; Secondary 46E25, 20C20.
Key words and phrases. Integer programming; duality.

inequality case $\{Ax \leq b, x \geq 0\}$ of (1.1) (with $A \in \mathbb{N}^{m\times n}, b \in \mathbb{N}^m$), introducing the $s := \prod_{i=1}^m (b_i + 1)$ variables $\{\pi(\lambda)\}$ with $\lambda \in \mathbb{N}^m$ and $\lambda \leq b$, (1.2) is equivalent to

$$(1.3) \qquad \left\{\begin{array}{llll} \min\limits_{\pi} & \pi(b) & & \\ \text{s.t.} & \pi(\lambda) + \pi(\mu) & \leq \pi(\lambda+\mu), & 0 \leq \lambda+\mu \leq b \\ & \pi(\lambda) & \leq \pi(\mu), & 0 \leq \lambda \leq \mu \leq b \\ & \pi(A_j) & \geq c_j, & j = 1, \ldots, n \\ & \pi(0) & = 0. & \end{array}\right.$$

The first set of constraints simply states that $\pi : \prod_{j=1}^n \{0, \ldots, b_j\} \to \mathbb{R}$ is a monotone superadditive function, whereas the second set of constraints is just that of (1.2); see Wolsey [**9**, §2].

Contribution: The goal of this paper is to relate duality results obtained in Lasserre [**4, 5, 6**] with the abstract dual problem (1.2), and to provide some insights for the latter problem. Namely, we present a dual problem $\mathbb{P}_d^*$ of $\mathbb{P}_d$ in the form (1.2), where a subclass of superadditive functions is identified and also yields a finite LP, simpler than (1.3). In fact, we obtain our result the other way around. From the discrete Farkas lemma proposed in [**5**], we have also obtained a linear programming problem $\mathbb{Q}$ *equivalent* to $\mathbb{P}_d$ (which also yields a simple characterization of the *integer hull* of the convex polytope P; see Lasserre [**6**]). It turns out that the LP dual $\mathbb{Q}^*$ of this equivalent LP, can be put in the form (1.2), for a particular subclass of superadditive functions. In view of the importance of Gomory cuts for solving $\mathbb{P}_d$ and their interpretation in terms of superadditive functions in (1.2), we hope that the new insights on $\mathbb{P}_d$ and (1.2) provided by the equivalent linear program $\mathbb{Q}$ and its dual $\mathbb{Q}^*$ (or equivalently $\mathbb{P}_d^*$), will prove to be useful in solving $\mathbb{P}_d$. Finally the integer hull of P is also characterized in terms of finite family of superadditive functions.

2. Notation and preliminary results

Let $\mathbb{N}$ denote the natural numbers or, equivalently, the positive cone $\mathbb{Z}_+$ of the integer numbers $\mathbb{Z}$. For a vector $b \in \mathbb{R}^m$ and a matrix $A \in \mathbb{R}^{m\times n}$, denote by b' and $A' \in \mathbb{R}^{n\times m}$ their respective transpose. Let $\mathbb{R}[x_1, \ldots, x_n]$ be the ring of real-valued polynomials in the variables $x_1, \ldots, x_n$. A polynomial $f \in \mathbb{R}[x_1, \ldots, x_n]$ is written

$$x \mapsto f(x) = \sum_{\alpha\in\mathbb{N}^n} f_\alpha x^\alpha = \sum_{\alpha\in\mathbb{N}^n} f_\alpha x_1^{\alpha_1} \cdots x_n^{\alpha_n},$$

for *finitely many* real coefficients $\{f_\alpha\}$, in the (usual) basis of monomials.

Given a matrix $A \in \mathbb{Z}^{m\times n}$, let $A_j \in \mathbb{Z}^m$ denote its j-th column (equivalently, the j-th row of A'); then z^{A_j} stands for

$$z^{A_j} := z_1^{A_{1j}} \cdots z_m^{A_{mj}} = \mathrm{e}^{\langle A_j, \ln z\rangle} = \mathrm{e}^{(A' \ln z)_j},$$

and if $A_j \in \mathbb{N}^m$ then z^{A_j} is a monomial of $\mathbb{R}[z_1, \ldots, z_m]$.

2.1. The abstract dual of $\mathbb{P}_d$. Let $A \in \mathbb{N}^{m\times n}, b \in \mathbb{N}^m$ and let $\mathbb{P}_d$ be the integer problem (1.1). Denote by $\max \mathbb{P}_d$ the optimal value of the integer program (1.1) so that $\max \mathbb{P}_d = -\infty$ whenever $Ax = b$ has no solution $x \in \mathbb{N}^n$; observe that $\max \mathbb{P}_d < +\infty$ whenever $A \in \mathbb{N}^{m\times n}$.

Consider the *value function* $f^* : \mathbb{N}^m \to \mathbb{R} \cup \{-\infty\}$, defined by

$$(2.1) \qquad b \mapsto f^*(b) := \max \mathbb{P}_d, \qquad b \in \mathbb{N}^m.$$

Actually, f^* can extended to $\mathbb{R}^m$ by $f(x) = -\infty$ for all $x \notin \mathbb{N}^m$. It follows that $f^*(0) = 0$ and f^* is superadditive. Indeed, if $f^*(x), f^*(y) > -\infty$ then we immediately have $f^*(x+y) \geq f^*(x) + f^*(y)$. This is because if $u, v \in \mathbb{N}^n$ are respective optimal solutions of $\mathbb{P}_d$ with $f^*(x) = c'u$ and $f^*(y) = c'v$, then $u + v \in \mathbb{N}^n$ and $A(u+v) = x + y$, so that $f^*(x+y) \geq c'(u+v) = f^*(x) + f^*(y)$.

If $f^*(x) = -\infty$ or $f^*(y) = -\infty$ (or both), then either $f^*(x+y) > -\infty$ or $f^*(x+y) = -\infty$; in both cases $f^*(x+y) \geq f^*(x) + f^*(y)$.

Next, consider the abstract dual problem (1.2) where Γ is the set of functions $f : \mathbb{N}^m \to \mathbb{R} \cup \{-\infty\}$ such that $f(0) = 0$ and f is superadditive. To prove that (1.2) is indeed a dual problem for $\mathbb{P}_d$, we need to show that the *weak duality* property, that is, for any two feasible solutions $x \in \mathbb{N}^n$, $f \in \Gamma$ of $\mathbb{P}_d$ and (1.2) respectively, we have $f(b) \geq c'x$.

Thus, let $f \in \Gamma$ be any feasible solution of (1.2). Then, if $x \in \mathbb{N}^n$ is any feasible solution of $Ax = b$,

$$\begin{aligned} f(b) = f(\sum_{j=1}^n A_j x_j) &\geq \sum_{j=1}^n f(A_j x_j) \quad \text{[by superadditivity]} \\ &\geq \sum_{j=1}^n f(A_j) x_j \quad \text{[by superadditivity]} \\ &\geq \sum_{j=1}^n c_j x_j = c'x, \end{aligned}$$

and so, $f(b) \geq c'x$, that is, the weak duality property holds.

Strong duality requires that both problems $\mathbb{P}_d$ and its dual $\mathbb{P}_d^*$ have an optimal solution with same optimal value. It holds because $f^*(b) = \max \mathbb{P}_d = c'x^*$ for any optimal solution $x^* \in \mathbb{N}^n$ of $\mathbb{P}_d$; notice that whenever $Ax = b$ has no solution $x \in \mathbb{N}^n$, we also have $f^*(b) = \max \mathbb{P}_d = -\infty$.

Let $\mathrm{P} := \{x \in \mathbb{R}^n_+ \mid Ax = b\}$. As already mentioned, the dual problem (1.2) is rather conceptual, as the set Γ is very large and not practical. For instance, Γ contains the value function f^* in (2.1), which is a bit too much of a requirement if one wishes to solve $\mathbb{P}_d$ for a single value b of the right-hand-side. To make a parallel with linear programming, one may also define the dual of the LP

$$(2.2) \qquad \mathbb{P} \to \quad \max \{ c'x \mid Ax = b; \quad x \geq 0; \quad x \in \mathbb{R}^n \},$$

to be (1.2) replacing now Γ with the set Γ' of *concave* functions $f : \mathbb{R}^n \to \mathbb{R} \cup \{-\infty\}$. The (large) set Γ' would also contain the value function associated with $\mathbb{P}$. However, by considering the subclass of affine functions of Γ', one obtains the usual (and tractable) LP dual $\mathbb{P}^*$ of $\mathbb{P}$.

Similarly, the maximal valid inequalities that define the *integer hull* P_1 of the convex polyhedron P, are *Chvátal-Gomory cuts* formed from linear combinations and rounding of the facet inequalities that describe P. They are obtained from a subclass of superadditive functions that satisfy the constraints of (1.2); see e.g. Nemhauser and Wolsey [**7**]. In §4 we also provide a characterization of P_1 in terms of a finite family of superadditive functions.

2.2. A class of superadditive functions. Let $\mathcal{D} \subset \mathbb{N}^m$ be a finite set such that

$$(2.3) \qquad 0 \in \mathcal{D}; \quad \text{and} \quad \alpha \in \mathcal{D} \quad \Rightarrow \quad \beta \in \mathcal{D} \quad \forall \beta \leq \alpha,$$

and let $\Delta_{\mathcal{D}}$ be the set of functions $\pi : \mathbb{N}^m \to \mathbb{R} \cup \{+\infty\}$, such that

$$\pi \in \Delta_{\mathcal{D}} \quad \text{if } \pi(0) = 0 \quad \text{and} \quad \pi(\alpha) \begin{cases} < +\infty & \text{if } \alpha \in \mathcal{D} \\ = +\infty & \text{if } \alpha \notin \mathcal{D}. \end{cases} \tag{2.4}$$

Next, given $\pi \in \Delta_{\mathcal{D}}$ let $f_\pi : \mathbb{N}^m \to \mathbb{R} \cup \{+\infty\}$ be defined as

$$x \mapsto f_\pi(x) := \inf_{\alpha \in \mathcal{D}} \{\pi(\alpha + x) - \pi(\alpha)\}, \qquad x \in \mathbb{N}^m. \tag{2.5}$$

Observe that $f_\pi \in \Delta_{\mathcal{D}}$ whenever $\pi \in \Delta_{\mathcal{D}}$, that is,

$$f_\pi(0) = 0 \quad \text{and} \quad f_\pi(x) = +\infty \quad \text{only if } x \notin \mathcal{D}.$$

Indeed, $f_\pi(0) = 0$ follows from the definition (2.5) of f_π. Next, if $x \notin \mathcal{D}$ then so does $\alpha + x$ for every $\alpha \in \mathbb{N}^m$; thus $\pi(\alpha + x) = +\infty$ and from (2.5) $f_\pi(x) = +\infty$ because $\pi(\alpha) < \infty$ whenever $\alpha \in \mathcal{D}$. Next, let $x \in \mathcal{D}$ so that $\pi(x) < +\infty$. From (2.5), $f_\pi(x) \leq \pi(x) - \pi(0) = \pi(x) < +\infty$.

LEMMA 2.1. *For every* $\pi \in \Delta_{\mathcal{D}}$ *:*
(i) $f_\pi \leq \pi$, *and* f_π *is superadditive.*
(ii) *If* $\pi \in \Delta_{\mathcal{D}}$ *is superadditive then* $\pi = f_\pi$.

PROOF. (i) We have just seen that $f_\pi \leq \pi$. Next, let $\pi \in \Delta_{\mathcal{D}}$, and let $x, y \in \mathbb{N}^m$ be fixed, arbitrary. First consider the case where $x+y \in \mathcal{D}$ so that $x, y \in \mathcal{D}$. Observe that if $\alpha \in \mathcal{D}$ and $\alpha + x + y \notin \mathcal{D}$ then $\pi(\alpha + x + y) - \pi(\alpha) = +\infty$; hence

$$\inf_{\alpha \in \mathcal{D}} \{\pi(\alpha + x + y) - \pi(\alpha)\} = \inf_{\alpha + x + y \in \mathcal{D}} \{\pi(\alpha + x + y) - \pi(\alpha)\}.$$

Therefore, we have:

$$\begin{aligned} f_\pi(x+y) &= \inf_{\alpha \in \mathcal{D}} \{\pi(\alpha + x + y) - \pi(\alpha)\} \\ &= \inf_{\alpha + x + y \in \mathcal{D}} \{[\pi(\alpha + x + y) - \pi(\alpha + x)] + [\pi(\alpha + x) - \pi(\alpha)]\} \\ &\geq \inf_{\alpha + x \in \mathcal{D}} \{\pi(\alpha + x + y) - \pi(\alpha + x)\} + \inf_{\alpha \in \mathcal{D}} \{\pi(\alpha + x) - \pi(\alpha)\} \\ &= f_\pi(y) + f_\pi(x), \end{aligned}$$

where we have used that $\alpha + x + y \in \mathcal{D}$ implies $\alpha + x \in \mathcal{D}$ (see (2.3)).

If $x+y \notin \mathcal{D}$, then $\pi(\alpha+x+y) = +\infty$ for every $\alpha \in \mathcal{D}$, and so from (2.5) it follows that $f_\pi(x+y) = +\infty$. And thus in this case, we also have $f_\pi(x+y) \geq f_\pi(x) + f_\pi(y)$.

(ii) We already have $f_\pi \leq \pi$. On the other hand, if π is superadditive

$$\pi(\alpha + x) - \pi(\alpha) \geq \pi(x) \quad \forall \alpha \in \mathcal{D},$$

and so,

$$f_\pi(x) = \inf_{\alpha \in \mathcal{D}} \{\pi(\alpha + x) - \pi(\alpha)\} \geq \pi(x), \quad \forall x \in \mathbb{N}^m,$$

which combined with $f_\pi \leq \pi$ yields the desired conclusion. □

3. Duality and superadditivity

In this paper we treat the case (1.1) where the constraint matrix A in $Ax = b$, has only nonnegative integer entries. This involves no essential loss of generality, for in §6 we show that the general case case of integer matrix $A \in \mathbb{Z}^{m \times n}$ (and P compact) reduces to the case $A \in \mathbb{N}^{m \times n}$ after a slight modification of the initial problem (introducing an additional constraint and an additional variable).

3.1. A discrete Farkas lemma. In this section we present a discrete analogue of the continuous Farkas lemma. However, for a better comparison with the continuous case, we first briefly recall the following (nonstandard) form of the continuous Farkas lemma.

Denote by $\mathbb{R}[z_1, \dots, z_m]$ the ring of real polynomials in the m variables $z_1, \dots, z_m$.

THEOREM 3.1 (Continuous Farkas lemma). *Let $A \in \mathbb{R}^{m \times n}, b \in \mathbb{R}^m$. Then the following two satements are equivalent.*

(a) *The linear system $Ax = b$ has a nonnegative solution $x \in \mathbb{R}^n$.*

(b) *The (linear) polynomial $\lambda \mapsto b'\lambda$ can be written*

$$b'\lambda = \sum_{j=1}^{n} Q_j(\lambda)(A'_j\lambda), \qquad \lambda \in \mathbb{R}^m, \tag{3.1}$$

for some polynomials $\{Q_j\}_{j=1}^n \subset \mathbb{R}[\lambda_1, \dots, \lambda_m]$, all with nonnegative coefficients.

PROOF. Let $0 \leq x \in \mathbb{R}^n$ satisfy $Ax = b$. Then,

$$b'\lambda = x'A'\lambda = \sum_{j=1}^{n} x_j(A'_j\lambda),$$

so that (3.1) holds with $Q_j \equiv x_j \geq 0$, for all $j = 1, \dots, n$. Conversely, assume that (3.1) holds for some polynomials Q_j's with nonnegative coefficients. Evaluating the gradients of polynomials in both sides of (3.1) at the point $\lambda = 0$, yields

$$b = \sum_{j=1}^{n} Q_j(0)A_j = \sum_{j=1}^{n} A_j x_j,$$

wih $x_j = Q_j(0) \geq 0$ for all $j = 1, \dots, n$. □

It also follows that if (3.1) holds then necessarily, the Q_j's are constant polynomials, say $Q_j \equiv x_j \geq 0$ (which provide a nonnegative solution $x \in \mathbb{R}^n$ to $Ax = b$). One also retrieves easily that if $A'\lambda \geq 0$ then $b'\lambda \geq 0$.

This nonstandard (continuous) Farkas lemma has been introduced to better compare with the discrete Farkas lemma below, obtained in Lasserre [**5**].

THEOREM 3.2 (Discrete Farkas lemma). *Let $A \in \mathbb{N}^{m \times n}, b \in \mathbb{N}^m$. Then the following two satements are equivalent.*

(a) *The linear system $Ax = b$ has a solution $x \in \mathbb{N}^n$.*

(b) *The polynomial $z \mapsto z^b - 1$ can be written*

$$z^b - 1 = \sum_{j=1}^{n} Q_j(z)(z^{A_j} - 1), \qquad z \in \mathbb{R}^m, \tag{3.2}$$

for some polynomials $\{Q_j\}_{j=1}^n \subset \mathbb{R}[z_1, \dots, z_m]$, all with nonnegative coefficients. Moreover, the total degree of each polynomial Q_j can be bounded by $b^* := \sum_{j=1}^m b_j - \min_{k=1,\dots,n} \sum_{i=1}^m A_{ik}$.

As the degree of each Q_j is bounded, let $q \geq 0$ be the vector of (nonnegative) coefficients of all the polynomials Q_j's. Then, checking the existence of such polynomials $\{Q_j\}$ in (3.2) reduces to solving a linear system

$$\mathrm{M}q = r, \quad q \geq 0, \tag{3.3}$$

for some matrix M and vector r with only 0 and ± 1 entries. The constraints $\mathrm{M}q = r$ state that the polynomials $z \mapsto z^b - 1$ and $z \mapsto \sum_j Q_j(z)(z^{A_j} - 1)$ are identical by equating their respective coefficients; see Lasserre [**5**] for more details.

3.2. A linear program equivalent to $\mathbb{P}_d$. We use the discrete Farkas lemma to construct a linear program equivalent to $\mathbb{P}_d$, using a method developed in Lasserre [**6**]. In fact, from the proof of Theorem 3.2 in [**6**], each polynomial Q_j in (3.2) may be restricted to contain only monomials z^α with $\alpha \in \mathbb{N}^m$ such that $\alpha \leq b - A_j$. Indeed, if $x \in \mathbb{N}^n$ solves $Ax = b$ then (3.2) holds with $Q_j \equiv 0$ whenever $x_j = 0$, $Q_1 = (1 + z^{A_1} + \ldots + z^{A_1(x_1-1)})$ if $x_1 \neq 0$ and for $j > 1$,

$$z \mapsto Q_j(z) := z^{\sum_{k=1}^{j-1} A_k x_k}\left[1 + z^{A_j} + \ldots + z^{A_j(x_j-1)}\right], \qquad \text{if } x_j \neq 0. \tag{3.4}$$

And so, the monomials z^α of Q_j satisfy $\alpha \leq \sum_{k=1}^{j} A_k x_k - A_j \leq b - A_j$.

Therefore, in the constraints $\mathrm{M}q = r$ which states that the polynomials $z \mapsto z^b - 1$ and $z \mapsto \sum_j Q_j(z)(z^{A_j} - 1)$ are identical, we only need to equate their respective coefficients of same monomials z^α for those $\alpha \in \mathbb{N}^m$ that satisfy $\alpha \leq b$. This is because, as each Q_j contains only monomials z^α with $\alpha \leq b - A_j$, each polynomial $Q_j(z)(z^{A_j} - 1)$ contains only monomials z^β with $\beta \leq b$.

Hence, in the LP (3.3), the vector q and the matrix M can be taken in $\mathbb{R}^s$ and $\mathbb{R}^{p \times s}$, respectively, where :

- $p = \prod_{i=1}^m (b_i + 1)$, i.e., the number of monomials z^α with $\alpha \leq b$.
- $s = \sum_{j=1}^n s_j$ with $s_j = \prod_{i=1}^m (b_i - A_{ij} + 1)$ for all $j = 1, \ldots, n$ (the number of monomials z^α with $\alpha - A_j \leq b$.

Note that the matrix M is *totally unimodular* because it is a network matrix (each column has only two nonzero entries $+1$ and -1).

Define the row vectors $\mathrm{e}_{s_j} := (1, \ldots, 1) \in \mathbb{R}^{s_j}$, for every $j = 1, \ldots, n$, and let $\mathrm{E} \in \mathbb{N}^{n \times s}$ be the n-block diagonal matrix, whose each diagonal block is a row vector e_{s_j}, that is,

$$\mathrm{E} := \begin{bmatrix} \mathrm{e}_{s_1} & 0 & \ldots & 0 \\ 0 & \mathrm{e}_{s_2} & 0 & \ldots \\ \ldots & \ldots & \ldots & \ldots \\ 0 & \ldots & 0 & \mathrm{e}_{s_n} \end{bmatrix}. \tag{3.5}$$

Given $c \in \mathbb{R}^n$, let $\widehat{c} \in \mathbb{R}^s$ be the vector $\mathrm{E}'c$. Then, in the same manner as we did in Lasserre [**6**], we have :

THEOREM 3.3. *Let $A \in \mathbb{N}^{m \times n}, b \in \mathbb{N}^m, c \in \mathbb{R}^n$. Let $\mathrm{M} \in \mathbb{R}^{p \times s}, r \in \mathbb{R}^p, \mathrm{E} \in \mathbb{R}^{n \times s}$ be as in (3.3) and (3.5), and let $\widehat{c} := \mathrm{E}'c$. Then :*

(a) *The optimal value* $\max \mathbb{P}_d$ *of the integer program* $\mathbb{P}_d$ *is the same as the optimal value* $\max \mathbb{Q}$ *of the linear program*

$$\mathbb{Q} \quad \to \max \{ \widehat{c}'q \mid \; \mathrm{M}q = r; \quad q \geq 0\}. \tag{3.6}$$

(b) *With every optimal vertex $q^* \in \mathbb{R}^s$ of the linear program $\mathbb{Q}$, is associated an optimal solution $x^* := \mathrm{E}\, q^* \in \mathbb{N}^n$ of the integer program $\mathbb{P}_d$.*

The proof is exactly the same as in [**6**].

3.3. The link with superadditivity. To relate the above linear program $\mathbb{Q}$ with superadditive functions and with the abstract dual (1.2), we proceed as follows. As we have just seen, we may restrict to the case $A \in \mathbb{N}^{m\times n}, b \in \mathbb{N}^m$. Let $\mathcal{D} \subset \mathbb{N}^m$ be the set

$$\mathcal{D} := \prod_{j=1}^{m} \{0, 1, \ldots, b_j\}. \tag{3.7}$$

In view of the simple form of the matrix M of the linear program $\mathbb{Q}$ in (3.6), its LP dual is easy to state. Namely, it is the linear program

$$\left\{ \begin{array}{lll} \min_{\gamma} & \gamma(b) - \gamma(0) & \\ \text{s.t.} & \gamma(\alpha + A_j) - \gamma(\alpha) & \geq c_j,\ \alpha + A_j \in \mathcal{D};\ j = 1, \ldots, n. \end{array} \right. \tag{3.8}$$

Clearly, by the change of variable $\pi(\alpha) := \gamma(\alpha) - \gamma(0)$, $\alpha \in \mathcal{D}$, the above linear program also reads

$$\mathbb{Q}^* \left\{ \begin{array}{lll} \min_{\pi} & \pi(b) & \\ \text{s.t.} & \pi(\alpha + A_j) - \pi(\alpha) & \geq c_j, \quad \alpha + A_j \in \mathcal{D};\ j = 1, \ldots, n, \\ & \pi(0) & = 0. \end{array} \right. \tag{3.9}$$

with optimal value denoted $\min \mathbb{Q}^*$.

Now, extend π to $\mathbb{N}^m$ by $\pi(\alpha) = +\infty$ whenever $\alpha \notin \mathcal{D}$. Then with $\Delta_{\mathcal{D}}$ as in (2.4), the linear program $\mathbb{Q}^*$ is equivalent to the optimization problem

$$\rho_1 := \left\{ \begin{array}{lll} \min_{\pi \in \Delta_{\mathcal{D}}} & \pi(b) & \\ \text{s.t.} & \pi(\alpha + A_j) - \pi(\alpha) & \geq c_j, \quad \alpha \in \mathcal{D};\ j = 1, \ldots, n, \end{array} \right. \tag{3.10}$$

that is, $\min \mathbb{Q}^* = \rho_1$.

THEOREM 3.4. *Let $A \in \mathbb{N}^{m\times n}, b \in \mathbb{N}^m, c \in \mathbb{R}^n$ and let $\mathbb{Q}^*$ be the linear program defined in (3.8) or (3.9), so that $\min \mathbb{Q}^* = \max \mathbb{P}_d$. Consider the optimization problem*

$$\mathbb{P}_d^* : \quad \rho_2 := \left\{ \begin{array}{lll} \inf_{\pi \in \Delta_{\mathcal{D}}} & f_\pi(b) & \\ \text{s.t.} & f_\pi(A_j) & \geq c_j \quad j = 1, \ldots, n, \end{array} \right. \tag{3.11}$$

where $f_\pi : \mathbb{N}^m \to \mathbb{R}$ is the superadditive function defined in (2.5), for every $\pi \in \Delta_{\mathcal{D}}$. If $\mathbb{P}_d$ is solvable, i.e., if $\max \mathbb{P}_d > -\infty$, then

$$\max \mathbb{P}_d = \min \mathbb{Q}^* = \rho_2 = f_{\pi^*}(b) \quad \textit{for some } \pi^* \in \Delta_{\mathcal{D}}. \tag{3.12}$$

PROOF. First, by Theorem 3.3, $\max \mathbb{Q} = \max \mathbb{P}_d$, and by linear programming duality, $\max \mathbb{Q} = \min \mathbb{Q}^*$. In particular, if $\max \mathbb{P}_d > -\infty$,

$$\max \mathbb{P}_d = \max \mathbb{Q} = \min \mathbb{Q}^* = \rho_1.$$

Next, $\rho_1 \geq \rho_2$. Indeed, let $\pi \in \Delta_{\mathcal{D}}$ be an admissible solution of (3.10). Then the associated superadditive function f_π defined in (2.5) is feasible in (3.11) because, clearly, for all $j = 1, \ldots, n$,

$$\pi(\alpha + x) - \pi(\alpha) \geq c_j \quad \forall \alpha \in \mathcal{D} \quad \Rightarrow \quad f_\pi(A_j) \geq c_j,$$

and $\rho_1 \geq \rho_2$ follows because $\pi \geq f_\pi$. In addition, $f_\pi(b) = \pi(b)$ because $\pi(b + \alpha) - \pi(\alpha) = +\infty$ whenever $\pi \in \Gamma$ and $0 \neq \alpha \in \mathcal{D}$ (because then $b + \alpha \notin \mathcal{D}$).

Now, let $x \in \mathbb{N}^n$ be a feasible solution for $\mathbb{P}_d$, i.e., $Ax = b$, and with $\pi \in \Delta_{\mathcal{D}}$, let f_π be a feasible solution in (3.11). Then

$$\begin{aligned} f_\pi(b) = f_\pi(\sum_{j=1}^n A_j x_j) &\geq \sum_{j=1}^n f_\pi(A_j x_j) \quad \text{[by superadditivity]} \\ &\geq \sum_{j=1}^n f_\pi(A_j) x_j \quad \text{[by superadditivity]} \\ &\geq \sum_{j=1}^n c_j x_j = c'x. \end{aligned}$$

This proves that $\rho_2 \geq \max \mathbb{P}_d$, hence $\rho_2 = \max \mathbb{P}_d$, which completes the proof. □

Thus the dual problem of $\mathbb{P}_d$ has either the superadditive formulation $\mathbb{P}_d^*$ in (3.11) or the equivalent LP formulation $\mathbb{Q}^*$ in (3.9). We discuss futher the interpretation of the linear program $\mathbb{Q}^*$ in §4.

3.4. Characterizing the integer hull. We end up with a characterization of the integer hull of the polytope P via superadditive functions.

Let M, r, and E be as in (3.3) and (3.5) respectively. When P is compact (i.e. P is a polytope) then the integer hull P_1 of $\mathrm{P} \cap \mathbb{Z}^n$ is defined by:

$$\mathrm{P}_1 = \{x \in \mathbb{R}^n \mid r'\pi^k + x'\lambda^k \geq 0\}, \tag{3.13}$$

where $\{(\pi^k, \lambda^k)\}_{k=1}^l =: \mathrm{G} \subset \mathbb{R}^p \times \mathbb{R}^n$ is a set of generators of the convex polyhedral cone of $\mathbb{R}^p \times \mathbb{R}^n$, defined by

$$\{(\pi, \lambda) \in \mathbb{R}^p \times \mathbb{R}^n \mid \mathrm{M}'\pi + \mathrm{E}'\lambda \geq 0\}. \tag{3.14}$$

For more details see Lasserre [**6**].

Recall that a vector $\pi \in \mathbb{R}^p$ may be viewed as a function $\pi : \mathcal{D} \to \mathbb{R}$, with $\mathcal{D}$ as in (3.7), and can be extended to $\mathbb{N}^m$ by setting $\pi(x) = +\infty$ if $x \notin \mathcal{D}$. Therefore, with a vector $\pi \in \mathbb{R}^p$ obtained from a generator $(\pi, \lambda) \in \mathrm{G}$, we may and will associate a superadditive function f_π as in (2.5).

THEOREM 3.5 (Characterization of the integer hull). *Let $A \in \mathbb{N}^{m \times n}, b \in \mathbb{N}^m$, and let M, r, and E be as in (3.3) and (3.5) respectively. Let $\mathrm{G} \subset \mathbb{R}^p \times \mathbb{R}^n$ be a set of generators of the convex cone defined in (3.14). Then the integer hull P_1 of the convex polytope $\mathrm{P} = \{x \in \mathbb{R}^n \mid Ax = b;\ x \geq 0\}$ is defined by*

$$\mathrm{P}_1 = \{x \in \mathbb{R}^n \mid \sum_{j=1}^n f_\pi(A_j) x_j \leq f_\pi(b), \quad (\pi, \lambda) \in \mathrm{G}\} \tag{3.15}$$

where

- *f_π is the superadditive function obtained from π in (2.5), and*
- *$\lambda_j = -f_\pi(A_j)$ for all $j = 1, \ldots, n$.*

PROOF. In view of the definition of M, r and E, (3.14) reads

$$\pi(\alpha + A_j) - \pi(\alpha) + \lambda_j \geq 0, \quad \alpha + A_j \in \mathcal{D}, \quad j = 1, \ldots, n,$$

where the vector $\pi \in \mathbb{R}^p$ is viewed as a function $\pi : \mathcal{D} \to \mathbb{R}$, with $\mathcal{D}$ as in (3.7). Therefore, every generator $(\pi, \lambda) \in \mathrm{G}$ satisfies

$$-\lambda_j = \min_{\alpha + A_j \in \mathcal{D}} \{\pi(\alpha + A_j) - \pi(\alpha)\}, \quad j = 1, \ldots, n.$$

Extend $\pi : \mathbb{N}^m \to \mathbb{R} \cup \{+\infty\}$, by $\pi(x) = +\infty$ if $x \notin \mathcal{D}$. Then, equivalently

$$-\lambda_j = \min_{\alpha \in \mathcal{D}} \{\pi(\alpha + A_j) - \pi(\alpha)\} = f_\pi(A_j), \quad j = 1, \ldots, n,$$

where f_π is the superadditive function obtained from π in (2.5). Therefore, as $\pi' r = \pi(b) - \pi(0)$, and $\lambda_j = -f_\pi(A_j)$ for all $j = 1, \ldots, n$, we finally obtain that

$$\mathrm{P}_1 = \{x \in \mathbb{R}^n \mid \sum_{j=1}^{n} f_\pi(A_j)\, x_j \le f_\pi(b), \quad (\pi, \lambda) \in \mathrm{G}\},$$

that is, (3.15) holds. □

So, as in Nemhauser and Wolsey [**7**], the integer hull P_1 is characterized in (3.15) via a *finite* family of superadditive functions. However, the superadditive functions in (3.15) are *not* obtained from linear combinations and rounding of the hyperplanes that define P, but rather from the set of generators G of the convex cone (3.14).

4. Discussion

We now consider the usefulness of the dual $\mathbb{P}_d^*$ of the integer program $\mathbb{P}_d$ and its equivalent linear program $\mathbb{Q}^*$ presented in §3.3. Observe that in contrast with the LP (1.3), the LP $\mathbb{Q}^*$ in (3.9) does *not* optimize over the superadditive functions $\pi \in \Delta_{\mathcal{D}}$ with $\Delta_{\mathcal{D}}$ as in (2.4). In general, the constraints in (3.9) do *not* define a superadditive function π. But the function f_π obtained from π in (2.5), is superadditive and satisfies the constraint of the abstract dual (1.2). However, given a feasible solution π in (3.9), one does not need to define f_π at all points $\alpha \in \mathcal{D}$. We only need its value at the points A_j, $j = 1, \ldots, n$. Finally note that the LP (3.9) is simpler than the analogue for the case $Ax = b$ of the LP (1.3) considered in Wolsey [**9**] for the particular case $\{Ax \le b, x \ge 0\}$, that is,

$$(4.1) \qquad \left\{ \begin{array}{lllll} \min\limits_{\pi} & \pi(b) & & & \\ \text{s.t.} & \pi(\lambda) + \pi(\mu) & \le \pi(\lambda + \mu), & 0 \le \lambda + \mu \le b \\ & \pi(A_j) & \ge c_j, & j = 1, \ldots, n \\ & \pi(0) & = 0, & \end{array} \right.$$

whose constraints state that $\pi : \mathcal{D} \to \mathbb{R}$ is superadditive, with $\pi(0) = 0$, and $\pi(A_j) \ge c_j$, for all $j = 1, \ldots, n$. The LP (4.1) and the LP dual (3.9) have the same number of variables. On the other hand, with $p := \prod_{j=1}^m (b_j + 1)$, the LP (4.1) has $O(p^2)$ constraints in contrast with only $O(np)$ constraints for the LP dual (3.9). However, the reduction in size is not so important as (3.9) remains an LP of potentially very large size.

EXAMPLE 4.1. Consider the following simple illustrative example where $A := [2, 3] \in \mathbb{N}^{1 \times 2}$ and $b := 5$ so that $Ax = b$ has only one solution $x^* = (1, 1)$. Let the cost vector be $c = [c_1, c_2]$, and $\mathcal{D} := \{0, 1, \ldots, b\}$. The dual problem (3.9) reads

$$\mathbb{Q}^* \text{ (or } \mathbb{P}_d^*) \qquad \left\{ \begin{array}{lllll} & \min\limits_{\pi} \ \pi(5) & & & \\ \text{s.t.} & \pi(2) & \ge c_1; & \pi(3) & \ge c_2 \\ & \pi(3) - \pi(1) & \ge c_1; & \pi(4) - \pi(1) & \ge c_2 \\ & \pi(5) - \pi(3) & \ge c_1; & \pi(5) - \pi(2) & \ge c_2 \\ & \pi(4) - \pi(2) & \ge c_1, & \pi(0) & = 0, \end{array} \right.$$

with optimal value $\min \mathbb{Q}^* = \pi^*(5) = c_1 + c_2$, and optimal solution

$$\pi^*(1) = c_2 - c_1,\ \pi^*(2) = c_1,\ \pi^*(3) = c_2,\ \pi^*(4) = \max[2c_1, 2c_2 - c_1].$$

The superadditive function $f_{\pi^*} : \mathbb{N}^2 \to \mathbb{R}$ defined in (2.5) (with $\pi^*(x) = +\infty$ if $x > 5$) satisfies $f_{\pi^*}(5) = c_1 + c_2$.

If $b = 1$ instead of $b = 5$, the system $Ax = b$ has no solution $x \in \mathbb{N}^n$. As now $\mathcal{D} = \{0, 1\}$, the LP dual (3.9) reads

$$\min \mathbb{Q}^* \;=\; \min_{\pi} \{\, \pi(1) \mid \quad \pi(0) = 0\} \;=\; -\infty,$$

because $\alpha + A_j \notin \mathcal{D}$ for every $\alpha \in \mathcal{D}$, which is consistent with $\max \mathbb{P}_d = -\infty$.

5. Conclusion

We now summarize the main results of this paper in Table 1 which compares the continuous and discrete optimization problems $\mathbb{P}$ and $\mathbb{P}_d$, in the case of a non-negative constraint matrix $A \in \mathbb{N}^{m\times n}$ (and so, with $\mathcal{D}$ as in (3.7)). We have also included the expressions of the convex polytope P and its integer hull P_1

TABLE 1. Problems $\mathbb{P}, \mathbb{P}_d$ and their respective duals

Primal linear program $\mathbb{P}$	**Primal integer program** $\mathbb{P}_d$
$\mathbb{P}: \begin{array}{lll} \max & c'x & \\ \text{s.t.} & Ax & = b \\ & x & \in \mathbb{R}^n_+ \end{array}$	$\mathbb{P}_d: \begin{array}{lll} \max & c'x & \\ \text{s.t.} & Ax & = b \\ & x & \in \mathbb{Z}^n_+ \end{array}$
Dual $\mathbb{P}^*$	**Dual** $\mathbb{Q}^*$ **(or** $\mathbb{P}^*_d$**)**
$\mathbb{P}^*: \begin{array}{lll} \min\limits_{\lambda\in\mathbb{R}^m} & b'\lambda & \\ \text{s.t.} & A'\lambda & \geq c. \end{array}$	$\mathbb{Q}^*: \begin{array}{ll} \min\limits_{\pi\in\Delta_{\mathcal{D}}} & \pi(b) \\ \text{s.t.} & \pi(\alpha + A_j) - \pi(\alpha) \geq c_j, \\ & \alpha + A_j \in \mathcal{D};\ j = 1, \ldots, n, \\ & \pi(0) = 0. \end{array}$
Polytope P	**Integer hull** P_1
$\mathrm{P}: \begin{array}{ll} \sum\limits_{j=1}^{n} A_j x_j & = b \\ x & \geq 0 \end{array}$	$\mathrm{P}_1: \ \sum\limits_{j=1}^{n} f_\pi(A_j) x_j \ \leq f_\pi(b), \quad (\pi, \lambda) \in \mathrm{G}$

In Table 1 one may also use the superadditive formulation $\mathbb{P}^*_d$ in (3.11), instead of $\mathbb{Q}^*$. However, we have chosen the LP formulation $\mathbb{Q}^*$ in (3.9), to better highlight the difference with the LP dual $\mathbb{P}^*$ of $\mathbb{P}$. Notice that if π is taken to be linear in $\mathbb{P}^*_d$, i.e. $\pi(\alpha) = \pi'\alpha$ for all $\alpha \in \mathcal{D}$ (for some vector $\pi \in \mathbb{R}^m$), then one retrieves exactly the LP dual $\mathbb{P}^*$ of $\mathbb{P}$.

As already mentioned, we hope that the insights on $\mathbb{P}_d$ and $\mathbb{P}_d^*$ obtained by this approach will help in solving $\mathbb{P}_d$ more efficiently, e.g. by defining appropriate and efficient Gomory cuts from $\mathbb{P}_d^*$, to be used in software packages.

6. Appendix : The general case $A \in \mathbb{Z}^{m\times n}$

We now assume that $A \in \mathbb{Z}^{m\times n}$, i.e., A may have negative entries. We will assume that the convex polyhedron P is compact.

Let $\alpha \in \mathbb{N}^n, \beta \in \mathbb{N}$ be such that for all $j = 1, \ldots, m$,

$$\widehat{b}_j := b_j + \beta \geq 0; \quad \widehat{A}_{jk} := A_{jk} + \alpha_k \geq 0; \quad k = 1, \ldots, n. \tag{6.1}$$

As P is compact we have

$$\max_{x\in\mathbb{N}^n} \{\sum_{j=1}^n \alpha_j x_j \mid Ax = b\} \leq \max_{x\in\mathbb{R}^n; x\geq 0} \{\sum_{j=1}^n \alpha_j x_j \mid Ax = b\} =: \rho^*(\alpha) < \infty.$$

Given $\alpha \in \mathbb{N}^n$, the scalar $\rho^*(\alpha)$ is easily calculated by solving a LP problem. Let $\widehat{A} \in \mathbb{N}^{m\times n}, \widehat{b} \in \mathbb{N}^m$ be as in (6.1) with $\beta \geq \rho^*(\alpha)$, and let $\mathrm{e}_m \in \mathbb{R}^m$ be a vector of ones. The integer program $\mathbb{P}_d$ is equivalent to the integer program

$$\widehat{\mathbb{P}}_d \quad \left\{ \begin{array}{lll} \max\limits_{x,u} & c'x & \\ \text{s.t.} & \widehat{A}x + \mathrm{e}_m u & = \widehat{b} \\ & \alpha' x + u & = \beta \\ & (x, u) & \in \mathbb{N}^n \times \mathbb{N}. \end{array} \right. \tag{6.2}$$

It is straightforward to check that $\max \mathbb{P}_d = \max \widehat{\mathbb{P}}_d$ and if $x^* \in \mathbb{N}^n$ is an optimal solution of $\mathbb{P}_d$ then so is (x^*, u^*) for $\widehat{\mathbb{P}}_d$ (with $u^* = \beta - \alpha' x^* \in \mathbb{N}$). As the matrix of the constraints of $\widehat{\mathbb{P}}_d$ has nonnegative entries, we are back to to the case analyzed in §3.1, with matrix and vector

$$\left[\begin{array}{cc} \widehat{A} & \mathrm{e}_m \\ \alpha' & 1 \end{array} \right] \in \mathbb{N}^{(m+1)\times(n+1)}, \quad \left[\begin{array}{c} \widehat{b} \\ \beta \end{array} \right] \in \mathbb{N}^{m+1},$$

in lieu of A and b, respectively.

References

[1] R.E. Gomory, E.L. Johnson, *The group problem and subadditive functions*, in : Mathematical Programming, T.C. Hu and S.M. Robinson, eds., Academic Press, New York, 1973.

[2] R. Jeroslow, *An introduction to the theory of cutting-planes*, Ann. Discrete Math. **5** (1979), 71–95.

[3] E.L. Johnson, *Integer Programming: Facets, Subadditivity, and Duality for Group and Semi-group Problems*, Society for Industrial and Applied Mathematics, Philadelphia, 1980.

[4] J.B. Lasserre, *Duality and a Farkas lemma for integer programs*, in : *Optimization : Structure and Applications*, E. Hunt and C.E.M. Pearce (eds.), Kluwer Academic Publishers, in press.

[5] J.B. Lasserre, *A discrete Farkas lemma*, Discr. Optim. **1** (2004), 67–75.

[6] J.B. Lasserre, *The integer hull of a convex rational polytope*, Discr. Comp. Geom. **32** (2003), 129–139.

[7] G.L. Nemhauser, L.A. Wolsey, *Integer and Combinatorial Optimization*, Wiley, John & Sons, 1988.

[8] A. Schrijver, *Theory of Linear and Integer Programming*, John Wiley & Sons, Chichester, 1986.

[9] L.A. Wolsey, *Integer programming duality: Price functions and sensitivity analysis*, Math. Prog. **20** (1981), 173–195.

[10] L.A. Wolsey, *Integer Programming*, John Wiley & Sons, Inc., New York, 1998.

LAAS-CNRS, 7 AVENUE DU COLONEL ROCHE, 31 077 TOULOUSE CÉDEX 4, FRANCE
E-mail address: lasserre@laas.fr

Contemporary Mathematics
Volume **374**, 2005

The Cayley Trick and Triangulations of Products of Simplices

Francisco Santos

Abstract. We use the Cayley Trick to study polyhedral subdivisions of the product of two simplices. For arbitrary (fixed) $l \geq 2$, we show that the numbers of regular and non-regular triangulations of $\Delta^l \times \Delta^k$ grow, respectively, as $k^{\Theta(k)}$ and $2^{\Omega(k^2)}$.

For the special case of $\Delta^2 \times \Delta^k$, we relate triangulations to certain class of lozenge tilings. This allows us to compute the exact number of triangulations up to $k = 15$, show that the number grows as $e^{\beta k^2/2 + o(k^2)}$ where $\beta \simeq 0.32309594$ and prove that the set of all triangulations is connected under geometric bistellar flips. The latter has as a corollary that the toric Hilbert scheme of the determinantal ideal of 2×2 minors of a $3 \times k$ matrix is connected, for every k.

We include "Cayley Trick pictures" of all the triangulations of $\Delta^2 \times \Delta^2$ and $\Delta^2 \times \Delta^3$, as well as one non-regular triangulation of $\Delta^2 \times \Delta^5$ and another of $\Delta^3 \times \Delta^3$.

Introduction

The *polyhedral Cayley Trick* gives a canonical bijection between *mixed subdivisions* of the Minkowski sum $\sum_{i=1}^k P_i$ of several polytopes $P_1, \dots, P_k \in \mathbb{R}^d$ and all polyhedral subdivisions of a certain polytope $\mathcal{C}(P_1, \dots, P_k) \subset \mathbb{R}^d \times \mathbb{R}^{k-1}$ called the *Cayley embedding* of $P_1, \dots, P_k$. The correspondence was first developed by Sturmfels [**19**] for the case of coherent subdivisions, and then generalized to all subdivisions by Huber et al. [**10**].

Originally, the trick was devised as a way of understanding and computing fine (i.e., minimal with respect to refinement) mixed subdivisions, taking advantage of the much deeper knowledge and specific software that exists for triangulations. But the trick can be also used in reverse, to understand triangulations of the $d+k-1$-dimensional polytope $\mathcal{C}(P_1, \dots, P_k)$ in terms of a d-dimensional object. This is what we do here.

2000 *Mathematics Subject Classification.* 52B20; 52B11.

This paper was finished in the fall of 2003, while I was a member of the Mathematical Sciences Research Institute (Berkeley, CA), supported by MSRI and the Spanish Ministry of Education. My research is also supported by grant BFM2001-1153 of the Spanish Ministry of Science and Technology.

Specially interesting is the case when all the P_i's are copies of a simplex Δ^{l-1}. Then, the Cayley Trick relates polyhedral subdivisions of $\Delta^{k-1} \times \Delta^{l-1}$ to mixed subdivisions of the dilation $k\Delta^{l-1}$. If, moreover, we fix $l = 3$, then the mixed subdivisions we have to study are essentially the same as lozenge tilings of a triangle of size k. Using this interpretation we prove:

THEOREM 1. (1) *The graph of flips between triangulations of $\Delta^2 \times \Delta^k$ is connected (Theorem 4.4) and it has diameter $\Theta(k^2)$ (Corollary 4.5).*

(2) *(Theorem 4.9) The number of triangulations of $\Delta^2 \times \Delta^k$ grows as*

$$e^{\frac{3}{\pi}L\left(\frac{\pi}{3}\right)k^2+o(k^2)},$$

where $L(x)$ is the Lobachevsky function

$$L(x) = -\int_0^x \log|2\sin t|dt.$$

(3) *The number of triangulations of $\Delta^2 \times \Delta^{k-1}$, for $k = 1, \ldots, 16$ is $k!$ times the number shown in Table 1.*

k	lozenge tilings of $k\Delta^2$
1	1
2	3
3	18
4	187
5	3135
6	81462
7	3198404
8	186498819

k	lozenge tilings of $k\Delta^2$
9	15952438877
10	1983341709785
11	355891356876534
12	91655826195854811
13	33726014269095727260
14	17665249123640876125464
15	13130399067698641838496272
16	13813411778618644581617635925

TABLE 1. Tilings of $k\Delta^2$ into k triangles and $\binom{k}{2}$ lozenges. Times $k!$, this is the number of triangulations of $\Delta^2 \times \Delta^{k-1}$

The number $\frac{3}{\pi}L\left(\frac{\pi}{3}\right) \simeq 0.323$ that appears in part (2) of this statement is the maximum asymptotic normalized entropy of lozenge tilings of a planar region, as computed in [**4**]. The exact number of triangulations of $\Delta^2 \times \Delta^k$ that appears in part (3) had previously been computed only up to $k = 5$ [**6, 15**]. Part (1) of the statement is interesting for two reasons. On the one hand, there are not many examples where the graph of flips is known to be connected. Essentially, only the case of dimension at most 2 (classical), codimension at most 3 [**1**] and that of cyclic polytopes [**14**]. On the other hand, since all triangulations of products of simplices are unimodular, the graph of flips has a very direct interpretation in toric algebraic geometry; see Theorem 2 below.

Besides the results in Theorem 1, the Cayley Trick allows to picture triangulations of $\Delta^2 \times \Delta^k$ as 2-dimensional objects. We include pictures of all (non-isomorphic) triangulations of $\Delta^2 \times \Delta^2$ and $\Delta^2 \times \Delta^3$ (Figures 5 and 14). Also, of non-regular triangulations of $\Delta^2 \times \Delta^5$ and $\Delta^3 \times \Delta^3$ and of a non-regular coarse subdivision of $\Delta^2 \times \Delta^7$ (Figures 6 and 7). $\Delta^2 \times \Delta^5$ and $\Delta^3 \times \Delta^3$ are the minimal products of simplices that have non-regular triangulations. To the best of our knowledge, our subdivision of $\Delta^2 \times \Delta^7$ is the first known non-regular coarse subdivision of a product of simplices.

Subdivisions and triangulations of $\Delta^{k-1} \times \Delta^{l-1}$ are interesting from several perspectives. They have been studied for their own sake in [**3, 6**]and [**8**, Sect. 7.3.D] and they have been used as building blocks to find efficient triangulations of high dimensional cubes [**9, 12**] or to find disconnected flip-graphs [**16, 18**]. Also, $\Delta^{k-1} \times \Delta^{l-1}$ is an example of a *totally unimodular* polytope; that is, a lattice polytope all the simplices of which have the same volume. This implies it is equidecomposable [**2**], i.e., all its triangulations have the same f-vector. From the f-vector of any of its triangulations one can recover, for example, its Erhart polynomial.

From a more algebraic point of view, the toric ideal associated to the vertex set of $\Delta^{k-1} \times \Delta^{l-1}$ is a fundamental object. It is the determinantal ideal generated by the 2×2-minors of a $k \times l$-matrix, and the variety associated to it is the *Segre embedding* of $\mathbb{CP}^{k-1} \times \mathbb{CP}^{l-1}$ in $\mathbb{CP}^{kl-1}$. The study of this ideal has connections to enumeration, sampling and optimization for contingency tables and transportation problems (cf. [**20**, Chapter 5]).

In this context, part (1) of Theorem 1 has the following algebro-geometric application: Every lattice point set defines a toric (binomial) ideal and a toric Hilbert scheme. Contrary to standard Hilbert schemes, toric Hilbert schemes are sometimes non-connected [**18**]. In fact, for a totally unimodular polytope, connectivity of the corresponding toric Hilbert scheme is equivalent to connectivity of the graph of triangulations (cf. [**11**] and Theorem 10.13 in [**20**]). Hence:

THEOREM 2. *The toric Hilbert scheme of the determinantal ideal of 2×2 minors of a $3 \times k$ matrix is connected.*

Another recent source of interest in triangulations of $\Delta^{k-1} \times \Delta^{l-1}$ comes from tropical geometry. Develin and Sturmfels [**5**] have proved that the combinatorial types of point configurations with k points in the tropical space of dimension $l-1$ are in bijective correspondence to the *regular* subdivisions of $\Delta^{k-1} \times \Delta^{l-1}$. Via this relation, we prove:

THEOREM 3 (Corollary 5.5). *For every fixed $l \geq 2$, the number of regular subdivisions of $\Delta^l \times \Delta^k$ grows as $2^{\Theta(k \log k)}$ while the number of non-regular subdivisions grows as $2^{\Omega(k^2)}$.*

The structure of the paper is as follows: After a first section with preliminaries on the Cayley Trick, Section 2 shows how to study mixed subdivisions from a purely geometric point of view. The results hold for arbitrary mixed subdivisions, but have a specially simple form for dilations of a simplex. Sections 3 and 4 contain our main results, on triangulations of $\Delta^2 \times \Delta^{k-1}$. The first explores the relation between mixed subdivisions of $k\Delta^2$ and lozenge tilings, and the second uses it to prove Theorem 1. Finally, Section 5 describes the relation between regular subdivisions of $\Delta^l \times \Delta^k$ and tropical polytopes in l-space, and proves Theorem 3.

1. The Cayley Trick

Except for Theorem 1.5, the contents of this section are special cases of results from [**10**].

1.1. Polyhedral subdivisions. Let P be a polytope in $\mathbb{R}^d$. A *cell* of P is any sub-polytope of the same dimension as P and whose vertices are a subset of those of P. A *polyhedral subdivision* of P is any family of cells which cover P and intersect properly, meaning that $B \cap B'$ is a common face of both B and B' for every pair of cells B and B' in the subdivision. This definition is a special case of the definition of polyhedral subdivisions of a point configuration [**7, 6, 8, 22**].

Polyhedral subdivisions form a poset under the refinement relation

$$S \leq S' \qquad \Longleftrightarrow \qquad \forall B \in S \ \ \exists B' \in S' \ : \ B \subseteq B'.$$

The minimal elements in this poset, i.e., those subdivisions whose cells are all simplices, are the *triangulations* of P. The unique maximal element in the poset is the *trivial subdivision*, which has the whole polytope P as its only cell. The subdivisions which refine only the trivial one are called *coarse*.

A polyhedral subdivision is called *regular* (or, sometimes, coherent) if it can be obtained as the orthogonal projection of the lower facets of a $d+1$-dimensional polytope. Equivalently, S is regular if there is a height function $h : \text{vertices}(P) \to \mathbb{R}$ such that for every cell $B \in S$ the points $\{(v, h(v)) : v \in B\} \subset \mathbb{R}^{d+1}$ lie in a hyperplane that passes strictly below all other points $\{(w, h(w)) : w \notin B\}$.

1.2. Mixed subdivisions. Let $P_1, \dots, P_k \subset \mathbb{R}^d$ be convex polytopes. We in principle do not need to assume them to be full-dimensional, but we require their Minkowski sum to be. This Minkowski sum is defined as:

$$\{ x_1 + \cdots + x_k : x_i \in P_i \}.$$

DEFINITION 1.1. A *Minkowski cell* of the Minkowski sum $\sum_{i=1}^k P_i$ is any full-dimensional polytope $B = \sum_{i=1}^k B_i$, where each B_i is a (perhaps not full-dimensional) polytope with vertices among those of P_i. A *mixed subdivision* of $\sum_{i=1}^k P_i$ is any family S of Minkowski cells which cover $\sum P_i$ and intersect properly as Minkowski sums, meaning that for any two cells $B = \sum B_i$ and $B' = \sum B_i'$ in S and for every $i \in \{1, \dots, k\}$, the polytopes B_i and B_i' intersect properly (their intersection is a face of both).

REMARK 1.2. We have intentionally decided to slightly abuse notation in these definitions and in the rest of the paper, to simplify it and to make the geometry more apparent. Indeed, although we speak of "mixed subdivisions of $P = \sum P_i$", the concept depends not only on the polytope P but also on the particular Minkowski decomposition of it that we are given. The same occurs in the definition of proper intersection of B and B'. To resolve this ambiguity, *every time we write a Minkowski sum, the expression $\sum_{i=1}^k Q_i$ should be formally understood as an ordered k-tuple of polytopes $(Q_1, \dots, Q_k)$.* Only in sentences like "a family of Minkowski cells covers $\sum P_i$" we are referring to the underlying polytope resulting from the sum.

In particular, if there are different ways of obtaining a certain subpolytope of P as a sum of subpolytopes from the P_i's, we consider these as different Minkowski cells and we have to specify which of them we are using in a particular mixed subdivision. For example, part (a) of Figure 1 shows a mixed subdivision of the Minkowski sum of two equal squares. Below the figure, each of the six Minkowski cells is expressed as a sum of sub-polytopes of the squares $a_1a_2a_3a_4$ and $b_1b_2b_3b_4$. Clearly, the exchange of all the a's and b's in these expressions would still provide a (different) mixed subdivision, with the same picture. On the other hand, considering $B_4 = a_4 + b_1b_3b_4$ and $B_5 = a_1a_2a_3 + b_4$ would not provide a mixed subdivision, since these two Minkowski cells do not intersect properly.

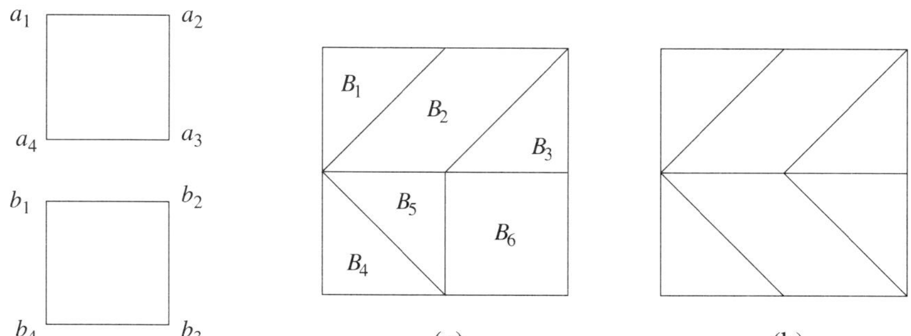

(a) (b)

$$B_1 = a_1a_2a_4 + b_1, \quad B_2 = a_2a_4 + b_1b_2, \quad B_3 = a_2a_3a_4 + b_2,$$
$$B_4 = a_4 + b_1b_3b_4, \quad B_5 = a_4 + b_1b_2b_3, \quad B_6 = a_3a_4 + b_2b_3.$$

FIGURE 1. (a) A mixed subdivision of the Minkowski sum of two equal squares. (b) A subdivision which is not mixed

Part (b) of the figure shows another reason why the "labeling" of cells as Minkowski sums is important. There, the same Minkowski sum of two equal squares is decomposed into cells which intersect properly as polytopes and which individually can be considered Minkowski cells, but which cannot be labeled in a way that makes them intersect properly in the Minkowski sense. In other words, the picture is not compatible with any mixed subdivision.

REMARK 1.3. Our treatment of mixed subdivision is formally different, but equivalent to, the one in [**10**]. There, in order to keep track of the Minkowski decompositions of cells, instead of speaking of the sum of polytopes $\sum P_i$ the authors speak of the sum of point sets $\sum A_i$, where each A_i is the vertex set of P_i. $\sum A_i$ is considered as a multiset, where different "points" (different Minkowski sums) may have identical geometric coordinates.

For the reader familiar with the theory of fiber polytopes (cf. Chapter 10 in [**22**]), mixed subdivisions of $\sum P_i$ can also be described as the subdivisions which are Π-induced, where Π is the natural projection

$$\Pi : \Delta^{\#\,\text{vertices}(P_1)-1} \times \cdots \times \Delta^{\#\,\text{vertices}(P_k)-1} \to P_1 + \cdots + P_k.$$

This is Lemma 2.4 in [**10**].

As in the case of subdivisions, mixed subdivisions form a poset. A Minkowski cell $B = \sum B_i$ is smaller than (or contained in) another one $B' = \sum B'_i$ if for every i we have $B_i \subseteq B'_i$. We write $B \leq B'$ if this happens. This induces the following refinement relation among mixed subdivisions:

$$S \leq S' \iff \forall B \in S \;\; \exists B' \in S' \; : \; B \leq B'.$$

The unique maximal element in this poset is again the trivial subdivision with only one Minkowski cell $B = \sum P_i$. The minimal elements are called *fine mixed subdivisions*. In Proposition 2.3 we will see that fine mixed subdivisions are characterized in terms of what Minkowski cells they use. This is analogous to the fact that "fine polyhedral subdivisions" (i.e., triangulations) of a polytope are the polyhedral subdivisions whose cells are simplices. As in the case of subdivisions, we call *coarse mixed subdivisions* those which only refine the trivial one. They do not have an easy intrinsic characterization, as far as we know.

A mixed subdivision of $P_1, \dots, P_k$ is called *coherent* if it can be obtained as the orthogonal projection of the lower facets of a $d+1$-dimensional Minkowski sum $\tilde{P}_1 + \cdots + \tilde{P}_k \subset \mathbb{R}^{d+1}$, where each $\tilde{P}_i$ orthogonally projects to the corresponding P_i. Equivalently, S is regular if there are height functions h_i : vertices$(P_i) \to \mathbb{R}$ such that for every cell $B = \sum B_i \in S$ the points

$$\{(v_1, h_1(v_1)) + \cdots + (v_k, h_k(v_k)) : v_i \in B_i \text{ for all } i\} \subset \mathbb{R}^{d+1}$$

lie in a hyperplane that passes strictly below all other points

$$\{(w_1, h_1(w_1)) + \cdots + (w_k, h_k(w_k)) : w_i \notin B_i \text{ for some } i\}.$$

1.3. The Cayley Trick. As before, let $P_1, \dots, P_k$ be polytopes in $\mathbb{R}^d$. Let $e_1, \dots, e_k$ be an affine basis in $\mathbb{R}^{k-1}$ and let $\mu_i : \mathbb{R}^d \to \mathbb{R}^d \times \mathbb{R}^{k-1}$ be the inclusion $\mu_i(x) = (x, e_i)$. We define

$$\mathcal{C}(P_1, \dots, P_k) := \mathrm{conv}(\cup_{i=1}^k \mu_i(P_i)).$$

We call $\mathcal{C}(P_1, \dots, P_k)$ the *Cayley embedding* of $P_1, \dots, P_k$. We will primarily be interested in the Cayley embedding of k copies of the same polytope P. Clearly, it equals $P \times \Delta^{k-1}$, where Δ^{k-1} is the simplex with vertices $e_1, \dots, e_k$.

The Cayley Trick (see [**10**] for a full exposition) is a poset isomorphism between polyhedral subdivisions of the Cayley embedding and mixed subdivisions of the Minkowski sum. More precisely, observe that for any choice of affine coordinates $(\lambda_1, \dots, \lambda_k)$ in the simplex Δ^{k-1}, the intersection of $\mathcal{C}(P_1, \dots, P_k)$ with the affine subspace $\mathbb{R}^d \times \{\sum \lambda_i e_i\}$ equals the "weighted Minkowski sum" $\sum \lambda_i P_i$. Choosing $\lambda_1 = \cdots = \lambda_k = 1/k$ gives a (scaled) copy of the standard Minkowski sum. What the Cayley Trick says is that:

THEOREM 1.4 (Cayley Trick [**10**, Theorem 3.1]).

(1) *For any polyhedral subdivision of* $\mathcal{C}(P_1, \dots, P_k)$, *intersecting its cells with* $\mathbb{R}^d \times \{\sum \frac{1}{k} e_i\}$ *we get a mixed subdivision of* $\frac{1}{k} \sum P_i$.
(2) *This correspondence is a poset isomorphism between polyhedral subdivisions of* $\mathcal{C}(P_1, \dots, P_k)$ *and mixed subdivisions of* $\sum P_i$, *and bijects regular subdivisions of* $\mathcal{C}(P_1, \dots, P_k)$ *to coherent mixed subdivisions of* $\sum P_i$.

Part (1), and hence one direction of part (2), are straightforward: every full-dimensional cell in a subdivision of $\mathcal{C}(P_1, \dots, P_k)$ is itself a Cayley embedding $\mathcal{C}(B_1, \dots, B_k)$ of certain subpolytopes $B_i \subseteq P_i$, and hence its intersection with $\mathbb{R}^d \times \{\sum \frac{1}{k} e_i\}$ is a Minkowski cell in $\frac{1}{k} \sum P_i$. The other direction, that every mixed subdivision arises in this way, can be easily proved using different values of $\lambda = (\lambda_1, \dots, \lambda_k) \in \Delta^{k-1}$: The "Minkowski intersection property" of the cells in a mixed subdivision S of $P_1 + \cdots + P_k$ guarantees that S induces a mixed subdivision of $\sum \lambda_i P_i$ for every λ (by just replacing each Minkowski cell $B = \sum B_i$ by its weighted version $\lambda B := \sum \lambda_i B_i$) and the cells $\mathcal{C}(B_1, \dots, B_k) = \cup_{\lambda \in \Delta^{k-1}} \lambda B$ for $B \in S$ form the desired polyhedral subdivision of $\mathcal{C}(P_1, \dots, P_k)$.

1.4. The case $P_1 = \cdots = P_k$. When all the polytopes P_i are copies of a single polytope $P \subset \mathbb{R}^d$, then $\mathcal{C}(P, \ldots, P) = P \times \Delta^{k-1}$. Hence, the Cayley Trick is a bijection between subdivisions of the product $P \times \Delta^{k-1}$ and mixed subdivisions of the dilation kP of P. The main topic of this paper is to take advantage of this fact in order to study triangulations of $P \times \Delta^{k-1}$, via fine mixed subdivisions of $P + \cdots + P$. Here comes a first result in this direction.

THEOREM 1.5. *In the action of the affine symmetry group of Δ^{k-1} on triangulations of $P \times \Delta^{k-1}$, every orbit has exactly $k!$ elements (that is, the action is free). In particular, the number of triangulations of $P \times \Delta^{k-1}$ is divisible by $k!$.*

PROOF. The action of the permutation group, when regarded in the corresponding mixed subdivision of kP, amounts to a mere permutation of the Minkowski summands of every Minkowski cell, without affecting the cell itself (as a polytope). In particular, if some permutation sends a mixed subdivision S to itself, it will send every Minkowski cell of S to itself. Hence, it suffices to prove that for every fine mixed subdivision S and every permutation σ there is a Minkowski cell $B = \sum B_i$ that is not invariant under reordering of the summands (that is to say, in which $B_i \neq B_{\sigma(i)}$ for some i).

Finding such cells is easy. Take any i such that $i \neq \sigma(i)$ and let $B = \sum B_i$ be a cell in which B_i is full-dimensional. These are easy to find in the Cayley embedding: they are the ones with a full-dimensional (in P) intersection with the face $P \times \{e_i\}$ of $P \times \Delta^{k-1}$, where e_i is the ith vertex of Δ^{k-1}. □

Theorem 1.5 can be proved directly (and easily) in the world of triangulations of $P \times \Delta^{k-1}$. What we want to emphasize is that the proof is much more transparent using the Cayley Trick. In particular, we have not found this result in the bibliography about triangulations of the product of two simplices [**6, 3**].

2. The labeling of a mixed subdivision

2.1. General case. We said earlier (and illustrated with Figure 1) that there may be different ways in which a given polyhedral subdivision of $P_1 + \cdots + P_k$ can be labeled as a mixed subdivision. In this section we address the problem of what additional information is needed to make the mixed subdivision unique. We start with a straightforward, but useful, observation:

LEMMA 2.1. *Let $I \subset \{1, \ldots, k\}$ be a subset of indices. Let S be a mixed subdivision of $P_1 + \cdots + P_k$. Then, the following is a mixed subdivision of $\sum_{i \in I} P_i$:*

$$S|_I := \{\sum_{i \in I} B_i : \sum_{i \in \{1, \ldots, k\}} B_i \in S\}.$$

PROOF. To see that the cells in $S|_I$ cover $\sum_{i \in I} P_i$ one can use a limiting process: for each $\lambda = (\lambda_1, \ldots, \lambda_k) \in (\mathbb{R}_+)^k$ the cells $\{\lambda B : B \in S\}$ form a mixed subdivision of $\sum_{i \in \{1, \ldots, n\}} \lambda_i P_i$. If we fix $\lambda_i = 1$ for the indices in S and make all other λ's go to zero, the Minkowski sum tends to $\sum_{i \in I} P_i$. That the cells intersect properly in the Minkowski sense is straightforward. □

This result can be easily understood in the Cayley embedding: $\mathcal{C}(P_i : i \in I)$ is a face of $\mathcal{C}(P_1, \ldots, P_k)$ and, certainly, every polyhedral subdivision of $\mathcal{C}(P_1, \ldots, P_k)$ induces a subdivision of it.

THEOREM 2.2. *Let S be a polyhedral subdivision of $P_1 + \cdots + P_k$ and suppose that a labeling of it as a mixed subdivision exists. Assume further that all the P_i have the same dimension. If we know:*

(1) *The subdivisions $S_i := S|_{\{i\}}$ induced by S in each of the individual P_i's, and*
(2) *Which cell of S collapses to each full-dimensional cell of each S_i,*

then the whole mixed subdivision can be recovered from that data (that is to say, S is the only mixed subdivision compatible with that information and there is an algorithm to recover the whole labeling).

PROOF. We work with one i at a time. That is to say, we are going to fix i and show how to recover the i-th summand B_i of a certain cell $B \in S$. We argue by induction on the codimension of B_i. The hypotheses give us the case of codimension zero.

So, suppose that we already know the i-th summand of all cells for which that summand has dimension $d' + 1$. Let B_i be a d'-dimensional cell in S_i. Our goal is to determine all maximal cells of S whose i-th component is precisely B_i:

– First, if a cell C of S has been determined to have as i-th summand a cell $C_i \in S_i$ that contains B_i (hence as a face), then the faces of C in any of the directions defined by the (relatively open) normal cone) of B_i in C_i will have B_i as their i-th component.

– Second, if a full-dimensional cell C' has been determined by the previous rule to have a face with i-th summand equal to B_i then its i-th summand contains B_i as a face. In particular, it either has been determined already or equals B_i.

Our claim is that applying these two rules we can recover all the cells whose i-th summand is B_i. A way to see it is in the process that gives S_i as the limit of mixed subdivisions of $\lambda(P_1 + \cdots + P_k) + (1 - \lambda)P_i$ then λ goes to zero. When λ is very close to zero, all the cells whose i-th summand is B_i become very close to B_i itself and certainly we can go from any of them to one that becomes close to the afore-mentioned C_i traversing only cells whose i-th summand is B_i. □

2.2. Fine and pure subdivisions. Lemma 2.1 makes it easy to show that fine mixed subdivisions can be characterized in terms of what Minkowski cells they are allowed to use:

PROPOSITION 2.3. *A mixed subdivision S is fine if and only if for every Minkowski cell $B = \sum B_i$ in S, the B_i's are all simplices and $\dim(B) = \sum \dim(B_i)$ (in other words, the B_i's lie in independent affine subspaces).*

PROOF. The "if" direction is straightforward: If all the cells are as claimed then no mixed cell can be properly contained in one of S and, hence, S is minimal. For the "only if" direction, a direct proof can be given but it is simpler to use the Cayley Trick. The Cayley Trick implies that fine mixed subdivisions are those for which the associated polyhedral of $\mathcal{C}(P_1, \ldots, P_k)$ is a triangulation. In other words, those that use only Minkowski cells $B = \sum B_i$ whose associated Cayley cells $\mathcal{C}(B_1, \ldots, B_k)$ are simplices. This condition is easily seen to be equivalent to the one in the statement. □

We will refer to the special mixed cells described in Proposition 2.3 as *fine mixed cells.*

COROLLARY 2.4. *For a fine mixed subdivision, the information in (1) and (2) of Theorem 2.2 can be recovered if we know, for each i, what maximal cells of S have a full-dimensional i-th summand.*

PROOF. In a fine mixed subdivision, if a cell has a full-dimensional summand then all the other summands are 0-dimensional (points). Hence, the information we are given is what maximal cells of S form the subdivision S_i, for every i, except we are not told how to arrange them to subdivide P_i. If we show how to do that, then Theorem 2.2 gives the rest. It is actually enough to find out, for every cell B of S, what is (up to translation) the ith summand B_i of that cell. If we know this, we know how to write $B = B_i + C_i$ for each cell (where the C_i is the sum of all the other summands) and we can scale down the C_i components of all cells to recover (in the limit) the subdivision P_i.

Suppose then that we have identified (up to translation) all the i-th summands of dimension greater than a certain d'. Exactly as in Theorem 2.2, it is then possible to propagate along S all the d' dimensional faces of those summands, hence getting all the i-th summands of dimension d'. □

Figure 2 illustrates this result. The figure shows all the triangulations of the 3-dimensional cube (the Cayley embedding of two equal squares), pictured as mixed subdivisions. Only one representative modulo the symmetries of the square and modulo the exchange of the labels 1 and 2 is shown. In each picture, the two triangles labeled 1 are the triangulation of the bottom square and the two triangles labeled 2 are the triangulation of the top square of the cube. Knowing that information is enough to recover the mixed subdivision labeling (and hence, the corresponding triangulation of the 3-cube).

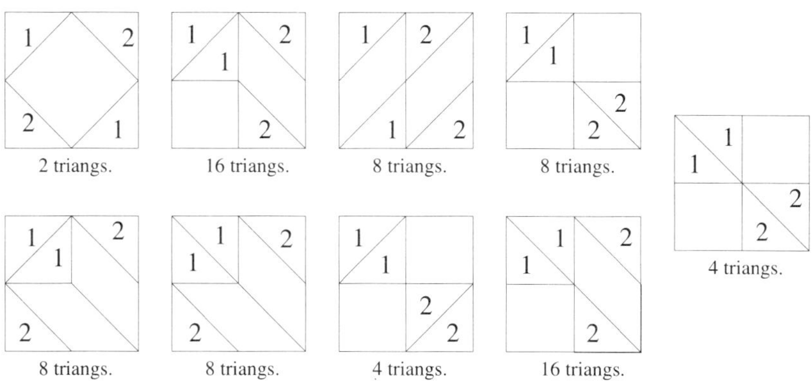

FIGURE 2. The 74 triangulations of a regular cube

It may be of interest also to define *pure mixed subdivisions* as those for which every Minkowski cell $B = \sum B_i$ satisfies $\dim(B) = \sum \dim(B_i)$. If all the P_i's are simplices then pure is equivalent to fine. In general, pure mixed subdivisions form a lower ideal in the poset of mixed subdivisions and contain all the fine ones. Their Minkowski cells are (combinatorially) products of their summands. Corollary 2.4 and Theorem 1.5 hold for pure mixed subdivisions, with the same proofs.

2.3. The product of two simplices. The product $\Delta^p \times \Delta^q$ of two simplices can be considered a Cayley embedding in two ways: $q+1$ copies of Δ^p or $p+1$ copies of Δ^q. It is easy to conclude from the previous results that in this case the "Minkowski labeling" of a subdivision can totally be neglected.

LEMMA 2.5. *Let $B \subset \mathbb{R}^d$ be a polytope and $\Delta \subset \mathbb{R}^d$ be a d-simplex. If there is a way of writing B as a Minkowski sum of (positive dimensional, not necessarily distinct) faces of Δ, then this way is unique, modulo reordering.*

PROOF. We argue by induction on the dimension of Δ. The case where Δ is a segment is trivial: B is the sum of as many copies of Δ as indicated by its length.

For the inductive step, let Δ^0 be a facet of Δ, and let p_0 be the opposite vertex. For every polytope P in $\mathbb{R}^d$ let P^0 denote the face of P in the direction of Δ^0 (that is, the face containing the outer normal vector to Δ^0 in its relatively open outer normal cone) and let P^1 denote the face in the opposite direction (that is, the face whose outer normal cone contains the opposite vector).

Since B can be written as a Minkowski sum of faces of Δ, every face of it can. In particular, the faces B^0 and B^1. Moreover, any decomposition $B = F_1 + \cdots + F_k$ will restrict to decompositions $B^0 = F_1^0 + \cdots + F_k^0$ and $B^1 = F_1^1 + \cdots + F_k^1$. By inductive hypothesis, the decompositions of B^0 and B^1 are unique, so we can assume we know them, except perhaps for the ordering and for the fact that some F_i^α may be points and we cannot recover them.

For every i, F_i^1 must be either equal to F_i^0 or to the point p_0. This allows us to assume that we have matched each F_i^0 to its F_i^1 and that we can recover F_i, with the only exception of the summands where both F_i^0 and F_1^i are points (and hence F_i is a segment containing p_0). But after we subtract from B all the summands which are not segments, what remains is a parallelotope (Minkowski sum of linearly independent segments) whose Minkowski decomposition is straightforward, and unique. □

THEOREM 2.6. *Let Δ be a simplex. Let S be a polyhedral subdivision of $k\Delta$ and assume that every cell B can be written as a Minkowski sum of faces of Δ. Then:*

(1) *S can be labeled as a mixed subdivision.*
(2) *The labeling is unique, modulo reordering of the k summands.*

PROOF. Uniqueness follows immediately from Theorem 2.2 and Lemma 2.5. Indeed, the S_i's are known because the simplex has only the trivial subdivision, and which cells collapse to full-dimensional in $S|_{\{i\}}$ is also known (modulo reordering of the factors): those which have a full-dimensional summand in their unique decompositions as Minkowski sum of faces of Δ.

To prove existence, let us show how to find the ith summand of all cells of S. First, identify the cells that have the full simplex Δ as one of the summands in their (unique, by Lemma 2.5) Minkowski decomposition. There will be exactly k such cells, counted with multiplicity if some have Δ as a repeated summand. Assign the numbers 1 to k to them arbitrarily.

Once this is done, fix an index $i \in \{1, \ldots, k\}$. As in the previous results, starting from that cell we can conclude what the ith summand of every other cell should be, starting with those where this summand has codimension 1, then 2, etc. The only difficulty is to show that this assignment is globally consistent, meaning that the ith summand obtained for a given cell B is independent of the path that led from the cell with a full-dimensional ith summand to B.

To show consistency we use the following idea: assume without loss of generality that $i = k$, and in each cell of S shrink by a factor of λ the k-th summand obtained by the previous method. Since being a polyhedral subdivision is a local property, this produces a polyhedral subdivision of $(k-1)\Delta + \lambda\Delta$. In the limit where $\Delta = 0$ we get a polyhedral subdivision of $(k-1)\Delta$ all of whose cells are Minkowski sums and, by induction on k, the labelings are consistent (and unique). □

3. Mixed subdivisions of $k\Delta^2$ and lozenge tilings

The triangulations of $\Delta^1 \times \Delta^{k-1}$ are well-understood. Their number is $k!$ and they form one only orbit under the action of the affine symmetry group S_k of Δ^{k-1} (see [**8**, Section 7.3.C]). All this can be easily derived from the Cayley Trick: The Minkowski sum of k copies of a segment is just a segment k times longer, and the fine mixed subdivisions of it are the $k!$ ways of placing the segments one after another (in other words, the fine mixed subdivisions differ only by the "labeling"). This and the next section are devoted to the next case, $\Delta^2 \times \Delta^{k-1}$.

By the Cayley Trick, subdivisions (resp., triangulations) of $\Delta^2 \times \Delta^{k-1}$ are in bijection with the mixed subdivisions (resp., fine mixed subdivisions) of the Minkowski sum of k copies of the triangle Δ^2. Let us denote T_k this Minkowski sum, which we think of as an equilateral triangle of size k. Hence $T_1 = \Delta^2$ is a triangle of unit size, whose vertices we denote a, b and c.

3.1. Fine mixed subdivisions of T_k.

Fine mixed cells must be sums of faces from the summand triangles and the sum of dimensions of the faces involved must be 2. This leaves two possibilities: either one of the k triangles plus vertices from the other ones or the sum of two non-parallel edges of two triangles plus vertices from the other ones. In other words, they are the upward triangles and lozenge tiles in the next definition:

DEFINITION 3.1. Let T_k be an equilateral triangle in the plane with side length k. Consider it tiled into k^2 equilateral triangles of side length 1, $(k^2+k)/2$ of them parallel to T_k (we call them *upward triangles*) and $(k^2-k)/2$ of them opposite to T_k (*downward triangles*). See Figure 3 where $k = 4$.

A *lozenge* in T_k is the union of a pair of adjacent triangles in the tiling (one upward and one downward). A *lozenge tiling* (or *rhombus tiling*) of T_k is a decomposition of T_k into $(k^2-k)/2$ lozenges and k upward triangles.

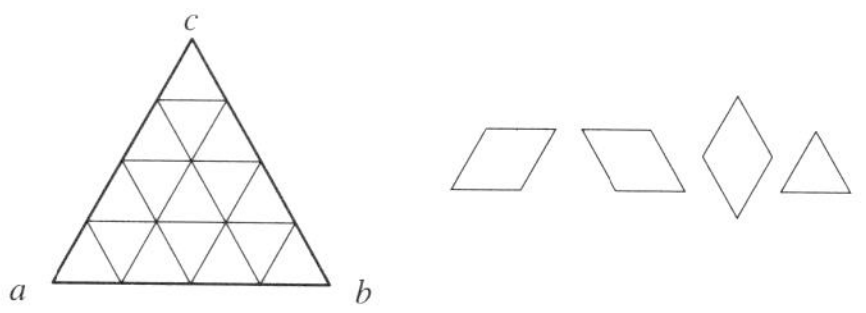

(a) (b)

FIGURE 3. (a) The triangular tiling of T_4. (b) The tiles allowed in a lozenge tiling of T_k

The terms "lozenge" and "rhombus" tilings are synonyms in the literature but they are used normally in a sense different from ours; they refer to tilings by only lozenges, of a shape containing as many upward as downward triangles. For example, a classical object of study is the set of lozenge tilings of the centrally symmetric hexagon with sides of lengths a, b, c, a, b and c. They are in bijection with plane partitions that fit into an $a \times b \times c$ box. A classical result of MacMahon gives the number of them. See, for instance, [**13**], for more information on this subject. Theorem 2.6 implies:

THEOREM 3.2. *Every lozenge tiling of T_k admits a labeling as a mixed subdivision of k copies of a triangle. Moreover, in any such a labeling,*

(1) *Each of the k copies of Δ^2 appears exactly once as a summand in one of the upward triangles of the tiling.*
(2) *Specifying an assignment of the k copies of Δ^2 to the k upward triangles uniquely determines the labeling of the mixed subdivision.*

There is a more direct way of proving Theorem 3.2, that explicitly tells how to get the labelings. First, a simple counting argument shows that there are exactly k upward triangles in every lozenge tiling: in the triangular tiling of T_k there are k more upward than downward triangles. After assigning the numbers 1 to k to the k upward triangles arbitrarily, we can define the ith *zone* of the lozenge tiling as the union of the ith upward triangle plus the lozenges that are obtained from it by parallel sweep of its three edges along the tiling. Figure 4 shows the four zones in a certain lozenge tiling of T_4. The name "zone" is borrowed from a somewhat similar concept in zonotopal tilings.

The complement of (the relative interior of) any of the zones consists of three regions, one containing each of the three vertices of T_k. We label the three regions as a, b or c depending on the vertex of T_k they contain. Every lattice point of T_k lies in exactly one of the three regions. Then, the ith summand of a cell B in the tiling equals the convex hull of the vertices of T_1 corresponding to the (closed) regions of the complement of the ith zone intersected by B.

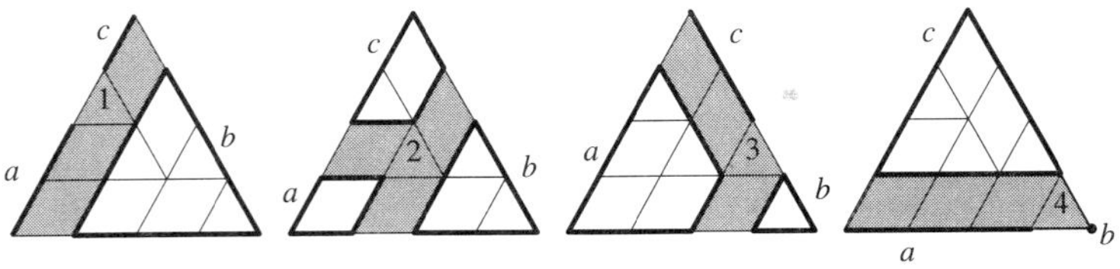

FIGURE 4. The four zones in a certain lozenge tiling of T_4

As an example, the central triangle in figure 4 gets labeled as

$$\{b\} + \{a, b, c\} + \{a\} + \{c\}$$

COROLLARY 3.3. *There is a bijection between lozenge tilings of T_k and orbits of triangulations of $\Delta^2 \times \Delta^{k-1}$ by the action of the symmetry group of Δ^{k-1}. In particular, the number of triangulations of $\Delta^2 \times \Delta^{k-1}$ equals $k!$ times the number of lozenge tilings of T_k.*

As a first application of this result, we can "draw" all the triangulations of $\Delta^2 \times \Delta^2$, that is to say, all mixed subdivisions of T_3. Modulo the symmetries of the triangle, there are 5 lozenge tilings of T_3, displayed in Figure 5. Each represents as many S_3-orbits of triangulations as lozenge tilings in its orbit modulo the symmetries of the triangle[1]. Hence, the number of lozenge tilings of T_3 is $3+6+1+2+6=18$ and the number of triangulations of $\Delta^2 \times \Delta^2$ is $18 \cdot 6 = 108$. The reader can compare Figure 5 with Figure 39 in [**8**] (page 150), where a different representation of the triangulations of $\Delta^2 \times \Delta^2$ is used. There, the vertices of $\Delta^{k-1} \times \Delta^{l-1}$ are represented as a $k \times l$ grid, and each simplex of a triangulation is represented by marking some squares in the grid. Incidentally, comparing the two figures the reader can easily detect an error in the adjacency graph of one of the triangulations of [**8**, Fig. 39].

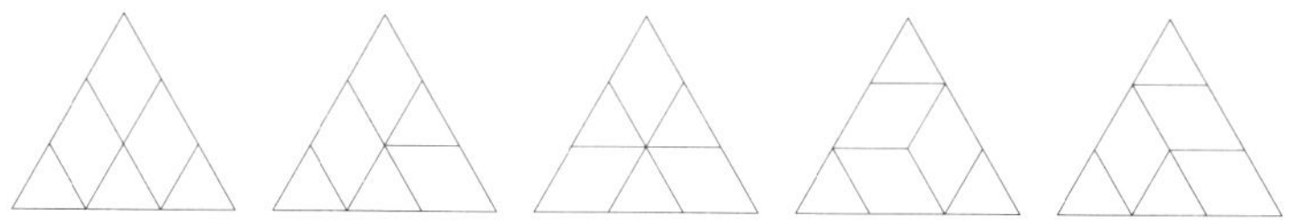

FIGURE 5. The 108 triangulations of $\Delta^2 \times \Delta^2$

3.2. Non-fine mixed subdivisions of T_k. What should a non-fine mixed subdivision look like? In the first place, it must be a polyhedral subdivision in the usual sense, with vertex set contained in the $\binom{k+2}{2}$ lattice points of T_k. Second, it must possess fine mixed refinements, hence each cell must be a convex union of lozenge tiles and upward triangles. We consider any such union as a hexagon with three pairs of parallel edges, each pair parallel to one edge of T_k. Hexagons may degenerate to have some edges of length zero.

LEMMA 3.4. *Let B be a convex union of triangles of the triangular tiling. Then, the following properties are equivalent:*

(1) *B is a Minkowski sum of faces (perhaps with repetition) of the unit upward triangle.*
(2) *B can be tiled by lozenges and upward triangles.*
(3) *B contains at least as many upward triangles as downward triangles.*
(4) *In each of the three pairs of parallel sides of B the one with the same outer normal as a side of the unit upward triangle is at least as long as the opposite one.*

PROOF. If B has a pair of opposite sides of positive length, then reducing both by one unit does not affect whether B satisfies any of the properties. Hence, there is no loss of generality in assuming that B has no pair of opposite sides, hence it is a triangle (or a point). If it is an upward triangle of any size (or a point) then the four conditions hold; if it is a downward triangle then none of them does. □

As in the fine case, Theorem 2.6 has the following consequence. Observe, however, that the orbits of the permutation group may now have cardinality smaller than $k!$ (an extreme example of this is the trivial subdivision).

THEOREM 3.5. *A polygonal subdivision of T_k can be labeled as a mixed subdivision if and only if all the cells are in the conditions of Lemma 3.4. The labeling is unique modulo the action of the permutation group.*

---

[1]Observe that in this sentence there are of two different S_3-actions: one on the labels, the other on T_3. They correspond to the symmetries in the two factors of $\Delta^2 \times \Delta^2$.

A direct way of getting the Minkowski labeling in this case is as follows: We define the excess of a cell as the difference between upward and downward triangles contained in it. Our assumption is that all cells have non-negative excess. Actually, the excess of a cell has an interpretation in any of the four settings of Lemma 3.4: it is the number of upward triangles in a Minkowski decomposition, the number of upward triangles in a lozenge tiling, and the difference in length between any of the three pairs of opposite edges.

The total excess in the tiling is clearly k. Let us distribute the numbers 1 through k to the different cells, giving a cell as many numbers as its excess. In much the same way as we did for lozenge tilings, we can define the k zones of the polyhedral subdivision: The i-th zone contains the cell to which we assigned the label i and then three arms, obtained as the cells adjacent to it in the directions towards the three edges of T_k, and then the ones adjacent to these, and so on. The main difference with the zones in a lozenge tiling is that now, as we travel along an arm, the edges that we cross may increase in length from one cell to the next. Also, the ith and jth zones may coincide (if the ith and jth excess reside in the same cell) or an arm of one zone be contained in the other zone.

As in the case of lozenge tilings, the definition of i-th zone classifies cells into seven types: the one labeled i, the ones in the three arms and the ones in the three regions of the complement of the zone (some of the last six types may be empty). This classification says whether the i-th summand in the mixed cell expression of a given cell is going to be $\{a, b, c\}$ (if the cell is labeled i), $\{a, b\}$, $\{a, c\}$ or $\{c, b\}$ (if the cell lies in one of the three arms, depending on the edge of T_k they are heading to) or just $\{a\}$, $\{b\}$, or $\{c\}$ (if the cells in the complement of the zone).

As an example, the right part of Figure 6 shows a valid polygonal subdivision of T_8. The cells labeled 1 to 8 have excess 1. The two unlabeled cells have excess zero. The dots, arrow and shading in the figure are there for later use. In order to illustrate the above concepts, Table 2 shows the mixed subdivision labeling (i.e., the Minkowski decomposition) of the ten cells. Cells 1 to 8 appear first in the list, then the shaded hexagon and finally the parallelogram. The eight columns of summands correspond to the eight zones.

4. Subdivisions of $\Delta^2 \times \Delta^{k-1}$

In this section we list several properties of triangulations and subdivisions of $\Delta^2 \times \Delta^{k-1}$ which can be derived from representing them as lozenge tilings.

4.1. Non-regular subdivisions of $\Delta^2 \times \Delta^{k-1}$. In order for a mixed subdivision to be coherent it has first to be regular as a subdivision in the standard sense (i.e., the projection of the lower hull of a polytope in one dimension more). Hence:

PROPOSITION 4.1. *The tilings of Figure 6 represent a non-regular triangulation of $\Delta^5 \times \Delta^2$ and a coarse non-regular subdivision of $\Delta^2 \times \Delta^7$, respectively.*

PROOF. The proof of non-regularity is sketched in the picture in both cases. In the left, if a lifting existed, there would be no loss of generality (by addition of an affine function to all the heights) in assuming that the three neighbors of the central point get height zero. The central point must then get a negative height that we denote $-a$. We let b, c, d, e, f and g denote the heights of certain boundary points, as shown in the figure. From these heights some others can be deduced, and

$\{a,b,c\}$ +	$\{a,c\}$ +	$\{a,c\}$ +	$\{a\}$ +	$\{a\}$ +	$\{a\}$ +	$\{a,b\}$ +	$\{a\}$.
$\{b\}$ +	$\{a,b,c\}$ +	$\{a,c\}$ +	$\{a\}$ +	$\{a\}$ +	$\{a\}$ +	$\{b\}$ +	$\{a,b\}$.
$\{b\}$ +	$\{b\}$ +	$\{a,b,c\}$ +	$\{a\}$ +	$\{a,b\}$ +	$\{a\}$ +	$\{b\}$ +	$\{b\}$.
$\{b\}$ +	$\{b\}$ +	$\{b,c\}$ +	$\{a,b,c\}$ +	$\{b\}$ +	$\{a,b\}$ +	$\{b\}$ +	$\{b\}$.
$\{b\}$ +	$\{b,c\}$ +	$\{c\}$ +	$\{a,c\}$ +	$\{a,b,c\}$ +	$\{a\}$ +	$\{b\}$ +	$\{b\}$.
$\{b,c\}$ +	$\{c\}$ +	$\{c\}$ +	$\{c\}$ +	$\{c\}$ +	$\{a,b,c\}$ +	$\{b\}$ +	$\{b\}$.
$\{c\}$ +	$\{c\}$ +	$\{c\}$ +	$\{a,c\}$ +	$\{a,c\}$ +	$\{a\}$ +	$\{a,b,c\}$ +	$\{a\}$.
$\{c\}$ +	$\{c\}$ +	$\{c\}$ +	$\{c\}$ +	$\{c\}$ +	$\{a,c\}$ +	$\{b,c\}$ +	$\{a,b,c\}$.
$\{bc\}$ +	$\{c\}$ +	$\{c\}$ +	$\{a,c\}$ +	$\{a,c\}$ +	$\{a\}$ +	$\{b\}$ +	$\{a,b\}$.
$\{b\}$ +	$\{b,c\}$ +	$\{c\}$ +	$\{c\}$ +	$\{b,c\}$ +	$\{a,b\}$ +	$\{b\}$ +	$\{b\}$.

TABLE 2. Minkowski decomposition of the cells in the non-regular subdivision of $\Delta^2 \times \Delta^7$ of Figure 6 (right)

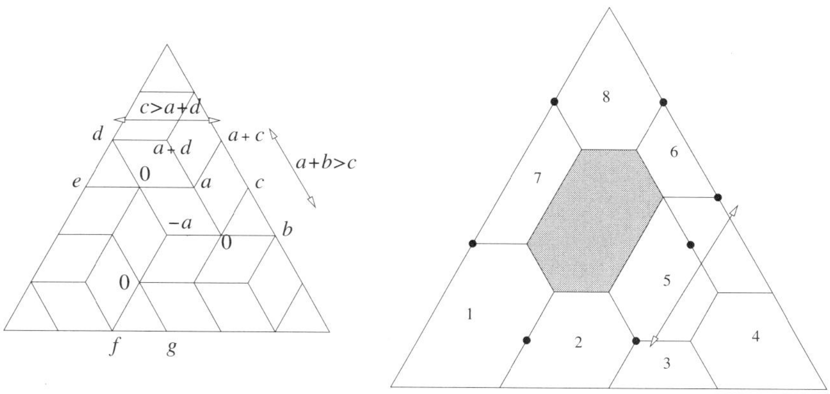

FIGURE 6. A non-regular triangulation of $\Delta^5 \times \Delta^2$ (left) and a coarse non-regular subdivision of $\Delta^2 \times \Delta^7$ (right)

in particular the figure shows how to conclude that $b > d$. The same arguments applied cyclically show that $d > f$ and $f > b$, which is impossible.

For the picture on the right, there is no loss of generality in assuming height zero for all the vertices of the shaded cell. Then, the seven marked points can easily be proved to get all the same height, but this contradicts convexity at the edge between region 5 and its adjacent parallelogram. That the right picture represents a coarse subdivision follows from the fact that it is coarse as a subdivision of T_8 in the standard sense. □

Non-regular triangulations of the product of two simplices were first constructed by de Loera [**6**], for $\Delta^3 \times \Delta^3$. He also proved that all triangulations of $\Delta^2 \times \Delta^k$ are regular, up to $k = 4$. Later, Sturmfels [**20**], constructed a non-regular triangulation of $\Delta^2 \times \Delta^5$, hence concluding that $\Delta^k \times \Delta^l$ has non-regular triangulations if and only if $(k-1)(l-1) \geq 4$. In particular, the non-regular lozenge tiling of T_6

that we show is smallest possible. As for our second example, to the best of our knowledge it is the first known coarse non-regular subdivision of the product of two simplices. Observe that coarse subdivisions of polytopes in general, and of products of simplices in particular, are not well-understood objects.

The Cayley Trick can also be used to picture non-regular triangulations of $\Delta^3 \times \Delta^3$. Figure 7 is our attempt to do so. The picture shows (an explosion of) the

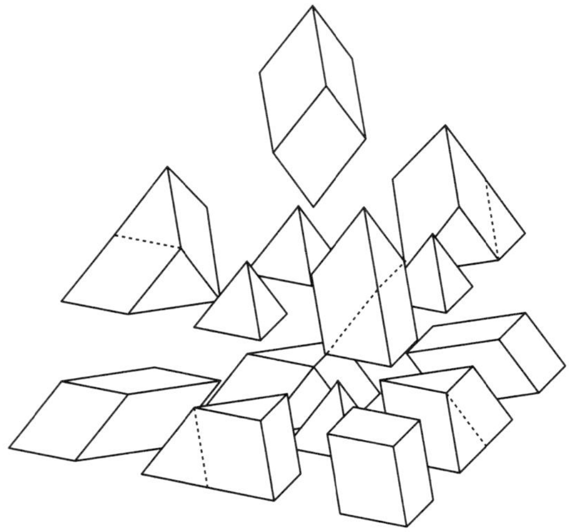

FIGURE 7. A regular subdivision of $\Delta^3 \times \Delta^3$ and a non-regular refinement of it

coherent mixed subdivision of $4\Delta^3$ produced by the following lifting matrix. Each row represents the lifting of one of the four copies of Δ^3:

$$\begin{pmatrix} 0 & 1 & 1 & 1 \\ 1 & 0 & 1 & 1 \\ 1 & 1 & 0 & 1 \\ 1 & 1 & 1 & 0 \end{pmatrix}.$$

This subdivides $4\Delta^3$ into 14 cells: four parallelepipeds in the four corners of $4\Delta^3$; four tetrahedra incident to the center of the four facets of $4\Delta^3$; and six "Minkowski sums of two triangles" along the six edges of $4\Delta^3$. Another way of describing this subdivision is that it is obtained by cutting $4\Delta^3$ with the four planes through its centroid and parallel to its facets.

This mixed subdivision is not fine, because the six special cells along edges of $4\Delta^3$ can be refined into two triangular prisms each, as Figure 8 shows. If the six cells are refined in the particular "skew" way sketched by dashed lines in Figure 7, then the fine mixed subdivision obtained is not coherent. (It is actually a non-regular subdivision; the proof is easy and left to the reader). Hence, it corresponds to a non-regular triangulation of $\Delta^3 \times \Delta^3$.

It is interesting to observe that $\Delta^{k-1} \times \Delta^{l-1}$ has non-regular triangulations if and only if there is a matroid on $k + l$ elements and of rank k which is not representable over the reals. This was first noticed in [**20**], where the subdivisions in Figure 7 and the left part of Figure 6 were related, respectively, to the Vamos and the non-Pappus matroids.

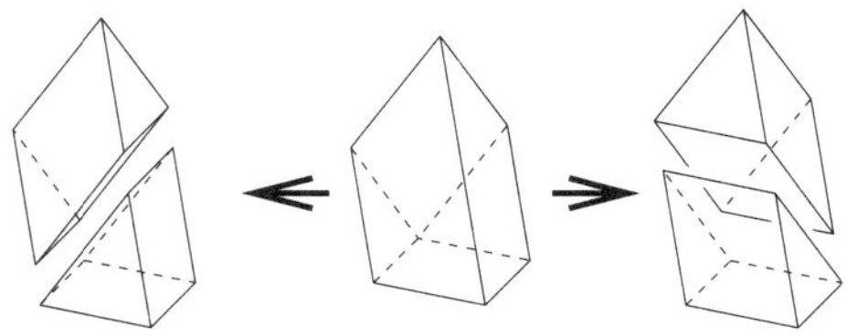

FIGURE 8. Refining the non-fine cells of Figure 7

4.2. Lozenge flips versus bistellar flips. A basic concept to understand the set of all triangulations of a polytope is that of geometric bistellar flip. Roughly speaking, it is the minimum possible difference (the "elementary move") between two triangulations. One simple definition, (see [**17**]) is that two triangulations differ by a bistellar flip if and only if they are the only two refinements of a certain polyhedral subdivision. A more explicit definition that says what the difference between the two triangulations has to be for this to happen is contained, for example, in [**8, 7**]. We do not need it here.

DEFINITION 4.2. We say that two lozenge tilings of T_k differ by a *lozenge flip* if one can be obtained from the other by one of the four substitutions of tiles shown in Figure 9. More precisely, the first three will be called *trapezoid flips* and the last one a *hexagon flip.*

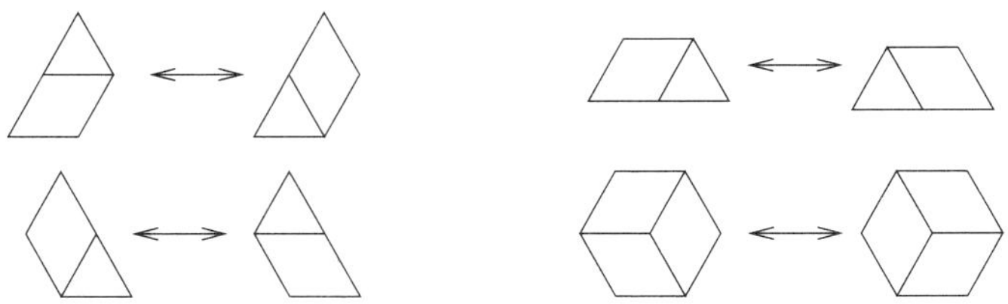

FIGURE 9. The four types of lozenge flips

PROPOSITION 4.3. *Let L and L' be two labeled lozenge tilings of T_k, corresponding to two triangulations S and S' of $\Delta^2 \times \Delta^{k-1}$. Then, S and S' differ by a bistellar flip if and only if the following three properties hold:*

1. *L and L' differ by a lozenge flip.*
2. *The labeling of upward triangles is the same in L and L' (except for the displacement of the triangle in case of a trapezoid flip).*
3. *For a trapezoid flip, the arm of the triangle affected by the flip in the direction of the big edge of the trapezoid does not change with the flip (it is only translated).*

PROOF. Let us look at what a polyhedral subdivision of T_k has to look like in order to admit only two lozenge refinements. First, all cells must be convex and individually admit only two lozenge refinements. The possibilities are a hexagon as the one in a hexagon flip, a trapezoid, or the union of two parallel lozenges with a common edge. If a hexagon arises, then its refinements are independent of the refinement of any other cell, which means that no other refinable cell can be present. Hence, the two tilings differ by a hexagon-flip. If a trapezoid or union of two lozenges arises, then the edge (or edges) of length two in that cell must be propagated up to the boundary of T_k on one side and to a trapezoid on the other side. The flip is a trapezoid flip and satisfies the arm condition in the statement. That the labels must be the same in the two lozenge tilings is trivial. □

As an example, the two lozenge tilings on the right part of Figure 10 represent triangulations which differ by a bistellar flip, because they are the two refinements of the subdivision in the bottom-right. The two lozenge tilings on the left differ by a lozenge flip, but not by a bistellar flip.

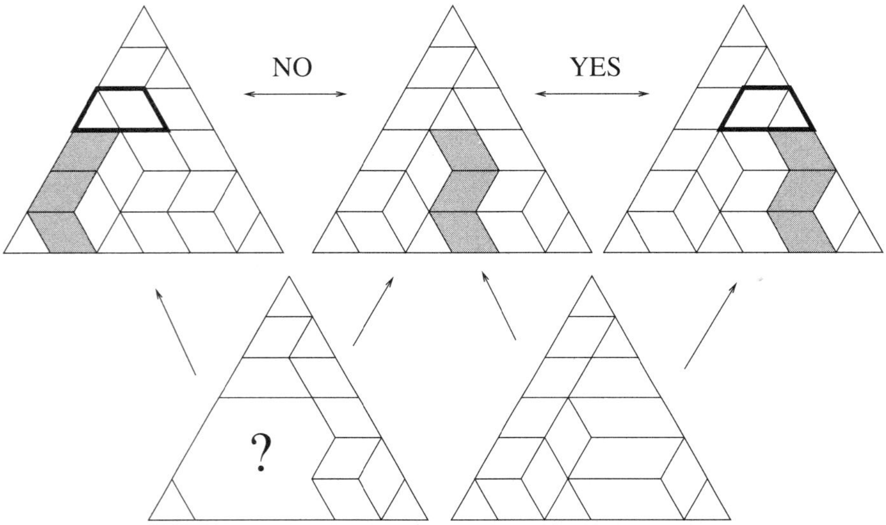

FIGURE 10. A trapezoid flip is not always a bistellar flip

THEOREM 4.4. (1) *The set of all (labeled) lozenge tilings of* T_k *is connected under trapezoid flips.*

(2) *The set of triangulations of* $\Delta^2 \times \Delta^{k-1}$ *is connected under geometric bistellar flips.*

PROOF. Any triangle not on the bottom row of a lozenge tiling is adjacent to a lozenge below it. Performing the trapezoid flip there produces a triangle one level lower. With this idea, we can eventually arrive to a lozenge tiling with all its triangles on the bottom row, and there is only one such tiling. We have not taken care of labels, but once we have the tiling with all triangles in the bottom row there is a sequence of three trapezoid flips which exchanges two consecutive triangles. Hence, any permutation of the labels can be implemented as a sequence of trapezoid flips, too. This proves part (1).

For part (2), we proceed similarly. Not all trapezoid flips are bistellar flips, but we can prove that unless all the triangles are in the bottom row there must be some trapezoid flip which decreases the height of the triangle involved and which is a geometric bistellar flip. To see this, start a triangle i with maximum height in the tiling. This implies that all the rows above that triangle are tiled with vertical lozenges, as in Figure 11. Let us consider the trapezoid flip that would decrease the height of i and, more specifically, at the downward looking triangle next to the lozenge below i, in the direction of the big side of the trapezoid. There are two possibilities for the lozenge containing that downward triangle: if it is a vertical lozenge, then the trapezoid flip at i is a geometric bistellar flip, and we are done. If it is not, then above it there is another triangle at the same height as j, and we get a trapezoid flip looking in the same direction and "closer to the boundary". Repeating this process we must eventually arrive at a trapezoid flip which is itself a bistellar flip. □

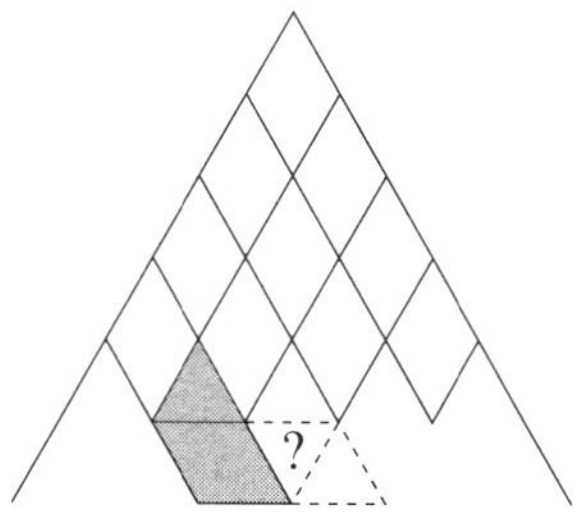

FIGURE 11. Every lozenge tiling has *some* trapezoid flip which is a bistellar flip

Further analysis of the above proof gives bounds on the diameter:

COROLLARY 4.5. *The graph of lozenge tilings of T_{k+1} and the graph of triangulations of $\Delta^2 \times \Delta^k$ both have diameter in $\Theta(k^2)$.*

More precisely, the graph of unlabeled lozenge tilings has diameter at least $\binom{k}{2}$ and the graph of triangulations of $\Delta^2 \times \Delta^k$ has diameter at most $5\binom{k}{2}$.

PROOF. For a quadratic lower bound, observe that at least $\binom{k}{2}$ lozenge flips are needed to go from the lozenge tiling with all triangles on the bottom to the lozenge tiling with all triangles on one side. This is so because each lozenge flip changes the height of only one triangle and only by one unit.

For an upper bound, the process in the proof of Theorem 4.4 shows how to go from any lozenge tiling to the one with all triangles in the bottom by a sequence of $\sum(\text{height}_T(i)) \leq \binom{k}{2}$ bistellar flips. As mentioned there, once we are in that lozenge tiling we can permute any two labelings with three times the number of pairs of indices which are ordered differently, that is, at most $3\binom{k}{2}$ bistellar flips. With another $\binom{k}{2}$ flips we can go back to the second lozenge tiling. □

The constants in the previous statement can surely be improved. For example, instead of going to the tiling with all triangles on the bottom, we can choose to go to the tiling with all triangles on one side, which provides a different definition of height. For each triangle, the sum of its three heights is clearly $k-1$, so that with respect to one of the three sides we get

$$\sum(\text{height}_T(i)) + \sum(\text{height}_{T'}(i)) \leq 2k(k-1)/3$$

instead of the $2\binom{k}{2} = k(k-1)$ used in the proof.

QUESTION 4.6. Is there a triangulation of $\Delta^2 \times \Delta^{k-1}$ with less than $2k-2$ bistellar flips? Observe that $2k-2$ is the dimension of the corresponding secondary polytope, hence it is a lower bound for the number of flips of every regular triangulation. Also, it is easy to prove that every lozenge tiling has at least $2k-2$ lozenge flips: The k upward triangles have a total of $3k$ sides and at most $k+3$ of them are in the boundary. This implies there are at least $2k-3$ trapezoid flips and that there are exactly that number if and only if three triangles are at the corners and the others are on the boundary. In this case, the lozenges produce a lozenge tiling of a simply connected region in the standard sense. Since lozenge tilings of a simply connected region are connected by hexagon flips, there has to be at least one hexagon flip.

4.3. Counting lozenge tilings. The number of lozenge tilings of T_k can be computed in the following way. Let k be fixed, and let S be a subset of $\{1, 2 \ldots, k\}$. We classify the lozenge tilings of T_k according to what triangles they have in the bottom line. More precisely, let

- $f_k(S)$ denote the number of lozenge tilings of T_k which have triangles exactly in the positions of the bottom line given by S.
- $g_k(S)$ denote the number of lozenge tilings of T_k which have triangles at least in the positions of the bottom line given by S.

Clearly,

$$g_k(S) = \sum_{S \subseteq S'} f_k(S'). \tag{1}$$

But, moreover,

PROPOSITION 4.7. *Let $S = \{s_1, \ldots, s_j\} \neq \emptyset$, where $1 \leq s_1 < \cdots < s_j \leq k$. If $j = 1$, then $f_k(S) = g_{k-1}(\emptyset)$. If $j > 1$, then:*

$$f_k(S) = \sum_{\substack{s_1 \leq s'_1 < s_2 \\ \vdots \\ s_{j-1} \leq s'_{j-1} < s_j}} g_{k-1}(\{s'_1, \ldots, s'_{j-1}\}) \tag{2}$$

PROOF. Between every two triangles of the bottom row there must be one and only one vertical lozenge. Once we fixed the positions $s'_1, \ldots, s'_{j-1}$ of these vertical lozenges, the ways to complete the lozenge tiling are exactly the same as the lozenge tilings of T_{k-1} containing triangles in (at least) the positions $s'_1, \ldots, s'_{j-1}$ of the bottom row. □

Table 3 shows all the values of $f_k(S)$ and $g_k(S)$ with $k = 1, 2, 3$, as well as the values of $f_4(S)$, computed using the recursive equations (1) and (2). Adding all the entries of $f_4(S)$ we get the number of lozenge tilings of T_4, which is $g_4(\emptyset) = 187$. Hence, the number of triangulations of $\Delta^2 \times \Delta^3$ is $187 \times 4! = 4488$.

The numbers shown in Table 1 in the introduction were computed with an implementation of these recursive formulas in Maple. The computation is clearly exponential in time, since we need to compute 2^k values of $f_k(S)$ and $g_k(S)$ for each k. In practice, the computation of each value took about five times the previous one: 21 seconds for $k = 10$ and 70 hours for $k = 16$. By Corollary 3.3, multiplying the kth number by $k!$ we get the number of triangulations of $\Delta^{k-1} \times \Delta^2$. A direct approach

S	$\emptyset$	1	2	1, 2	3	1, 3	2, 3	1, 2, 3
$f_1(S)$	0	1						
$g_1(S)$	**1**	1						
$f_2(S)$	0	1	1	1				
$g_2(S)$	**3**	2	2	1				
$f_3(S)$	0	3	3	2	3	$2+2$ $=4$	2	1
$g_3(S)$	**18**	10	8	3	10	5	3	1
$f_4(S)$	0	18	18	10	18	$10+8$ $=18$	8	3
$f_4(S \cup \{4\})$	18	$10+8+10$ $=28$	$8+10$ $=18$	$3+5$ $=8$	10	$5+3$ $=8$	3	1

TABLE 3. The number of triangulations of $\Delta^2 \times \Delta^3$ computed by hand. It is 4! times the sum of entries in the last two rows

allowed Jesús de Loera and Jörg Rambau [**6, 15**] to compute these numbers of triangulations only up to $k = 4$ and $k = 6$ respectively.

4.4. The asymptotic number of lozenge tilings. Let l_k denote the number of lozenge tilings of T_k. It is easy to show that l_k is in $e^{\Theta(k^2)}$:

- Since a lozenge tiling can be specified by which of the three upward neighbors of each of the $(k^2 - k)/2$ downward triangles forms a lozenge with it, $l_k \leq 3^{(k^2-k)/2} < 3^{\frac{k^2}{2}}$.
- Assume k is a multiple of 3. T_k can be tiled into $3\binom{k/3}{2} = \frac{k^2-3k}{6}$ hexagons plus k boundary trapezoids (see Figure 12), each of which can independently be refined in two ways. Hence, $l_k \geq 2^{(k^2+3k)/6}$.

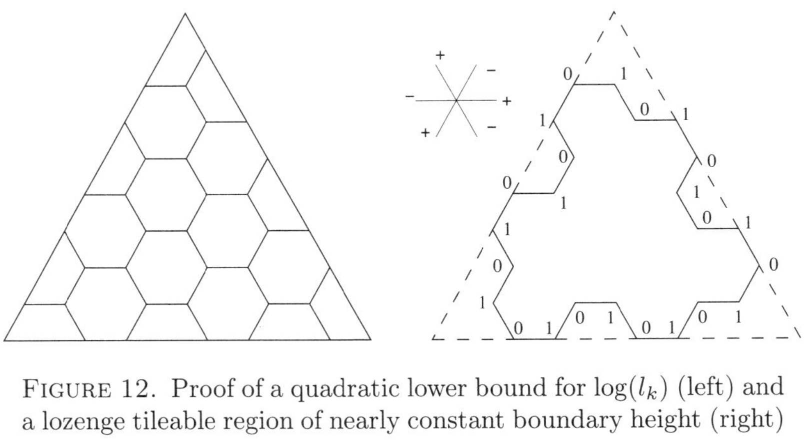

FIGURE 12. Proof of a quadratic lower bound for $\log(l_k)$ (left) and a lozenge tileable region of nearly constant boundary height (right)

In particular, it is asymptotically not relevant to distinguish between labeled and unlabeled lozenge tilings. We can as well think of l_k as the number of triangulations of $\Delta^2 \times \Delta^{k-1}$.

The property that the logarithm of the number of tilings is proportional to the area for dilations of a given shape is well-known in the context of usual lozenge tilings, as follows from the following result of Cohn, Kenyon and Propp [**4**] (they consider mostly the case of domino tilings, i.e., perfect matchings in a sub-region of the square grid, but the case of lozenge tilings arises as a particular case in which certain edges are forbidden in the matching). Let R^* be a simply-connected region in the plane, and let $(R_n)_{n\in\mathbb{N}}$ be a sequence of lozenge-tileable (in the standard sense) simply-connected regions such that R_n/n converges to R^*. Let $f_n : \partial R_n \to \mathbb{R}$ be the boundary height function of the region R_n (defined below). Assume that, after scaling it down, f_n/n converges to a certain function $f^* : \partial R^* \to \mathbb{R}$.

LEMMA 4.8 (Cohn et al. [**4**]). *In the above conditions,*

(1) *The logarithm of the number of lozenge tilings of R_n divided by the area of R_n (measured in lozenge tiles) converges to a constant that depends only on R^* and f^*.*
(2) *This constant is maximized if $f^* = 0$. In this case it equals*

$$\frac{3}{\pi} L\left(\frac{\pi}{3}\right) \simeq 0.32306594.$$

Here, $L(x)$ is the Lobachevsky function, defined as

$$L(x) = -\int_0^x \log|2\sin t|\, dt.$$

The constant in part (1) (and its specific instance in part (2)) is computed as an integral of the average (in a well-defined sense) extension of the boundary height function f^* to the interior of R^*.

The boundary height function of a simply connected union R of triangles of the regular triangular tiling is defined as follows: choose alternating signs for the six directions of edges in the tiling. Starting at any particular boundary vertex, give height zero to that vertex and then propagate the height along the boundary cycle of R, increasing or decreasing the height by one depending on the direction of the edge traversed. A simply connected region is lozenge-tileable if and only if the height becomes 0 again when you return to the starting point [**21**]. The right part of Figure 12 is an example of a tileable region with nearly constant boundary function.

The following statement says that the asymptotic entropy per unit tile is the same in our lozenge tilings of T_k as in classical lozenge tilings of a simply connected region with nearly constant boundary height function. The proof we give is essentially glued from personal communications to the author by J. Propp, H. Cohn and, specially, David Wilson:

THEOREM 4.9.

$$l_k = e^{\beta\frac{k^2}{2}\pm o(k^2)},$$

where $\beta = \frac{3}{\pi}L\left(\frac{\pi}{3}\right) \simeq 0.32306594$ is the asymptotic entropy per unit tile of regions with nearly constant boundary height function, as given by Lemma 4.8.

PROOF. For the lower bound, apply part (2) of Lemma 4.8 to the region on the right of Figure 12.

For the upper bound, let $f(k)$ be any function such that, asymptotically, $1 << f(k) << \sqrt{k}$ (for example, $f(k) = k^{1/4}$). Let S_k be a tiling of T_k into a triangular grid of about $k^2/f(k)^2$ triangles of size about $f(k)$.

For each lozenge tiling of T_k, we cut T_k without breaking tiles but otherwise as close as possible to the tiling S_k. If a tile overlaps two cells of S_k, we choose, for instance, to give that tile to the bottom of the two. The total perimeter of the cells in S_k is clearly in $\Theta(k^2/f(k)) \subset o(k^2)$. Hence, the number of possible ways of cutting T_k produced in this way is in $2^{o(k^2)}$. and will not affect the final asymptotics. Our task is to bound the number of lozenge tilings compatible with a specific cutting.

For this, we consider independently the cells of S_k that only contain lozenges and those that contain at least a triangle in the lozenge tiling. Although this is not relevant for the asymptotics, observe that which cells contain triangles (and how many of them) is fixed by the cutting: for a specific cell, the number of triangles is the difference between upward and downward triangles of the triangular unit grid contained in it.

Since at most k cells contain triangles, we do not need to care much about their number of tilings. The easy argument that each downward triangle must be matched to one of at least three upward triangles shows that the number of tilings of each cell is at most $3^{f(k)^2/2}$. Hence, the cells that contain triangles produce a factor of at most $3^{kf(k)^2/2} \in e^{o(k^2)}$ in the final number and can be neglected.

For the cells that are tiled only with lozenges, we are in the situation of Lemma 4.8: the number of lozenge tilings of each is at most $e^{\beta\frac{f(k)^2}{2}+o(f(k)^2)}$ and the combined number is at most

$$\left(e^{\beta\frac{f(k)^2}{2}+o(f(k)^2)}\right)^{k^2/f(k)^2} = e^{\beta\frac{k^2}{2}+o(k^2)}.$$

□

5. Tropical polytopes

Develin and Sturmfels have recently started developing the theory of polytopes in tropical space [**5**]. We here give a brief account of their main results, specially in their relations to subdivisions of the product of two simplices.

The tropical projective space of dimension $l-1$, denoted $\mathbb{TP}^{l-1}$, is the quotient of $\mathbb{R}^l$ by the equivalence relation $v \sim v + (\lambda, \dots, \lambda)$, for every $v \in \mathbb{R}^l$ and every $\lambda \in \mathbb{R}$. By normalizing one of the coordinates (say the first one) to be equal to zero we can identify $\mathbb{TP}^{l-1}$ to $\mathbb{R}^{l-1}$.

The tropical hyperplane defined by a vector $(a_1, \dots, a_l) \in \mathbb{R}^l$ is the set of points $v \in \mathbb{TP}^{l-1}$ such that the minimum of the numbers $v_i + a_i$ is achieved twice. Clearly, the hyperplanes defined by $(a_1, \dots, a_l)$ and by $(a_1 + \lambda, \dots, a_l + \lambda)$ coincide, so we may say that a hyperplane is defined by a point $a \in \mathbb{TP}^{l-1}$. The hyperplane defined by $(0, \dots, 0)$ is the set of points $v \in \mathbb{TP}^{l-1} = \mathbb{R}^{l-1}$ such that v either lies in the boundary of the positive orthant or has minimum coordinate negative and repeated at least twice. Said in a more compact (and invariant) form, it equals the $l-2$-skeleton of the normal fan of the simplex with vertices $O, e_1, \dots, e_{l-1}$. The translation of this hyperplane by the vector $-a \in \mathbb{R}^l$ gives the hyperplane defined by a.

If H is the hyperplane defined by a point $v \in \mathbb{TP}^{l-1}$, here we call $-H$ the *anti-hyperplane* defined by v. For the purposes of this paper, the following consequence of the results in [**5**, Section 3] can be taken as a definition of tropical convex hull:

PROPOSITION 5.1. *Let $v_1, \ldots, v_k$ be a finite set of points in tropical $l-1$ space. Then, its tropical convex hull* $\mathrm{tconv}(v_1, \ldots, v_n)$ *equals the union of all bounded cells in the polyhedral arrangement of tropical anti-hyperplanes given by $v_1, \ldots, v_k$.*

The left part of Figure 13 shows an example of this. The tropical convex hull of the five dots equals the shaded region, including its boundary and the horizontal segment that reaches to point number 5. Develin and Sturmfels make no clear distinction between the tropical convex hull as a subset of $\mathbb{TP}^{l-1}$ and the polyhedral complex in the above statement, and use the term "tropical polytope" referring to both. Here we will use "tropical polytope" referring to the region and call "tropical order type" of the point set the polyhedral complex. Two point sets are *combinatorially equivalent* if they have the same bounded complex, in a labeled sense (with the label of each cell indicating its relative position in each of the anti-hyperplanes. This is essentially what Develin and Sturmfels call the "type" of a cell).

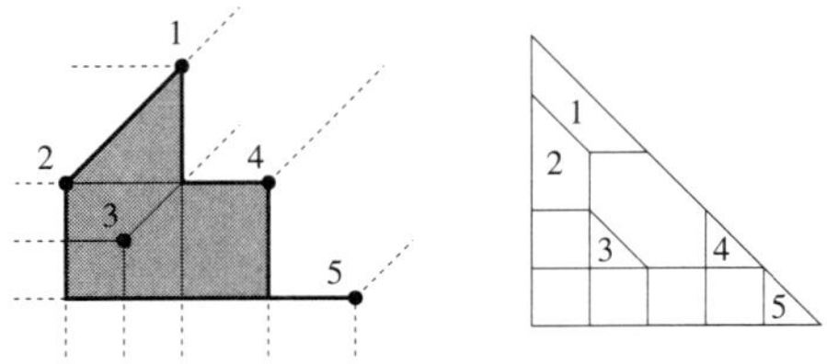

FIGURE 13. A tropical point configuration, and its associated regular mixed subdivision of $k\Delta^{l-1}$

The connection to mixed subdivisions is given in the following statement, paraphrased from Section 4 of [**5**]:

THEOREM 5.2. *Let M and M' be two $k \times l$ real matrices. Then, the columns of M and M' produce the same tropical order type if and only if they produce the same regular mixed subdivision of $k\Delta^{l-1}$ (where the ith column specifies the heights to lift the vertices of the ith copy of Δ^{l-1}).*

COROLLARY 5.3. *There is the same number of order types of k points in tropical $l-1$ space as coherent mixed subdivisions of $k\Delta^{l-1}$; that is, regular subdivisions of $\Delta^{k-1} \times \Delta^{l-1}$.*

Figure 13 illustrates the correspondence between a tropical point configuration with 5 points in 2-space and a mixed subdivision of $5\Delta^2$. As is easy to check, there is a 1-to-1 dimension (and order) reversing correspondence between the cells defined by the tropical point set and the cells in the mixed subdivision. Unbounded cells in the tropical point set correspond to boundary cells in the mixed subdivision. Corresponding cells are orthogonal to one another. Actually, the link of a cell in the tropical point set is the normal fan of the corresponding cell in the mixed subdivision.

As another example, Figure 14 shows all the fine mixed subdivisions of $3\Delta^2$, placed and numbered to exactly match the 35 types of "tropical quadrangles" as shown in Figure 6 of [**5**]. (This list was originally computed by J. Rambau [**15**], and its order is the one given as output by TOPCOM, that performs a "breadth-first search" on the graph of flips).

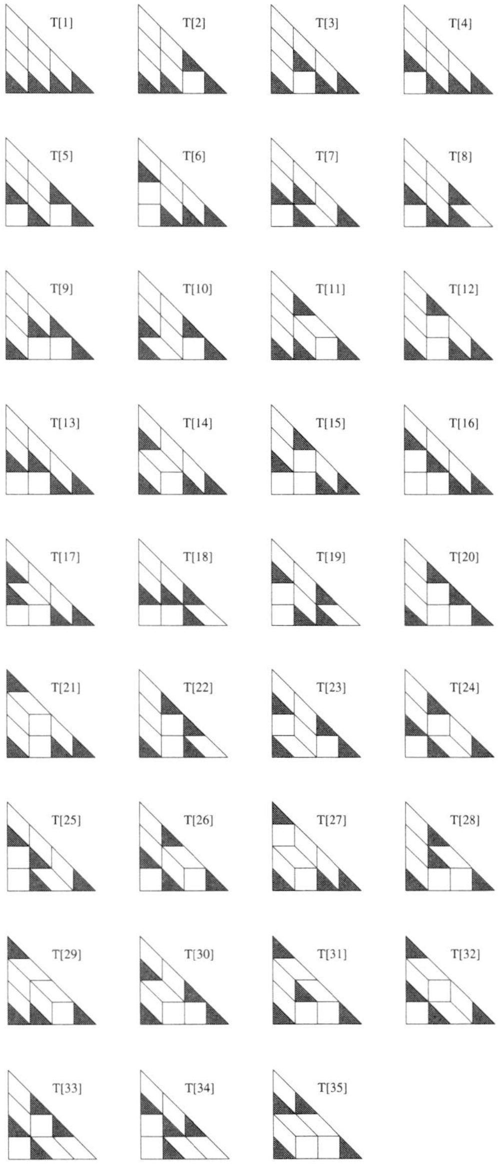

FIGURE 14. The 35 symmetry classes of triangulations of $\Delta^2 \times \Delta^3$

We now use Corollary 5.3 to give a bound on the number of regular subdivisions of $\Delta^{k-1} \times \Delta^{l-1}$. For this, we extend the tropical arrangement of k anti-hyperplanes that defines $\operatorname{tconv}(v_1, \dots, v_n)$ to a (usual) affine arrangement $\mathcal{H}$ of $k\binom{l}{2}$ hyperplanes. Indeed, the anti-hyperplane corresponding to a point $v_i = (v_{i,1}, \dots, v_{i,l-1}) \in \mathbb{R}^{l-1} \cong \mathbb{TP}^{l-1}$ is a $(l-2)$-dimensional polyhedral complex with $\binom{l}{2}$ maximal cells, lying respectively in the following hyperplanes:

$$
\begin{aligned}
&H(i;j,l) := \{x_j = v_{i,j}\}, && \text{for } j = 1, \dots, l-1, \text{ and} \\
&H(i;j,k) := \{x_j - x_k = v_{i,j} - v_{i,k}\}, && \text{for } 1 \le j < k \le l-1.
\end{aligned}
\tag{3}
$$

Clearly, point sets with different tropical order type produce different (labeled) hyperplane arrangements. Then:

THEOREM 5.4. *For any k and l the number of regular subdivisions of $\Delta^{l-1} \times \Delta^{k-1}$ is bounded above by:*

$$\left(\frac{\mathbf{e}}{2} kl\right)^{l(l-1)(k-1)}$$

PROOF. We can assume that $(k-1)(l-1) \geq 6$, because if this is not the case, the bound can be proved by direct inspection: if $l = 2$ then the number of triangulations is $k!$ and if $l = k = 3$ then the number of triangulations is 108 by Table 1.

We need to bound the number of different arrangements $\mathcal{H}$ that can be produced for varying $(v_1, \dots, v_k) \in \mathbb{R}^{kl}$. Two arrangements are "equal" if they have the same chirotope, that is to say, if every determinant of l of the $k\binom{l}{2}$ hyperplanes has the same sign in the two arrangements.

From the definition of the hyperplanes in equation (3) it is clear that each of the determinants that define the chirotope of $\mathcal{H}$ is a linear functional on the kl variables $(v_{i,j})$. Since the tropical order type is invariant under addition of a constant to a row or column of the matrix $(v_{i,j})$, we can assume $v_{0,j} = v_{i,0} = 0$ for every i and j, leaving only $(l-1)(k-1)$ variables.

Hence, the order type of $\mathcal{H}$ appears represented as a cell in a huge linear hyperplane arrangement of $\binom{k\binom{l}{2}}{l} \leq \frac{k^l l^{2l}}{2^l l!} \leq \left(\frac{\mathbf{e}}{2} kl\right)^l$ hyperplanes in $\mathbb{R}^{(k-1)(l-1)}$. This gives the statement, since the number of cells in an arrangement of N hyperplanes in $\mathbb{R}^D$ is maximal for simple arrangements, in which case it equals

$$\sum_{i=0}^{D} 2^i \binom{N}{i} \leq (D+1)2^D \binom{N}{D} \leq \frac{(D+1)2^D}{D!} N^D \leq N^D.$$

In the first inequality we assume that $D \leq N/2$, which always happens for $N = \left(\frac{kl\mathbf{e}}{2}\right)^l$ and $D = (k-1)(l-1)$; in the last inequality we have used our assumption that $D = (k-1)(l-1) \geq 6$. □

Observe that our bound is quite rough not only because different arrangements $\mathcal{H}$ may represent the same tropical order type, but also because the sign of many of the $\binom{k\binom{l}{2}}{l}$ determinants considered in the proof is constant (independent of the $v_{i,j}$'s). We believe the actual number of regular subdivisions to be in $(kl)^{\Theta(kl)}$. Anyway, for fixed l our bound gives the exact asymptotic behavior of the number of regular subdivisions:

COROLLARY 5.5. *For any fixed $l \geq 2$, the number of regular subdivisions of $\Delta^{l-1} \times \Delta^{k-1}$ is in $k^{\Theta(k)}$. For $l \geq 3$ the number of all subdivisions is in $2^{\Omega(k^2)}$.*

PROOF. For regular subdivisions, the upper bound follows from the previous theorem and the lower bound is trivial, since a single orbit of regular triangulations has already $k! \in k^{\Omega(k)}$ elements.

For all subdivisions, Theorem 4.9 gives the case $l = 3$ and the others follow immediately: any subdivision of a particular $\Delta^2 \times \Delta^{k-1}$ face of $\Delta^{l-1} \times \Delta^{k-1}$ can be extended to the whole polytope. □

Bibliographic remark: The writing of this paper has spanned an unusually long period of time, the first drafts dating back to 1998. Previous versions of it have been cited as "in preparation" under the title *Applications of the polyhedral Cayley Trick to triangulations of polytopes.*

References

1. M. Azaola, F. Santos, The graph of triangulations of a point configuration with $d+4$ vertices is 3-connected, *Discrete Comput. Geom.* **23** (2000), no. 4, 489–536.
2. M. M. Bayer, Equidecomposable and weakly neighborly polytopes, *Israel J. Math.* **81**:3 (1993), 301–320.
3. E. Babson, L. J. Billera, The geometry of products of minors, *Discrete Comput. Geom.* **20**:2 (1998), 231–249.
4. H. Cohn, R. Kenyon, and J. Propp, A variational principle for domino tilings, *J. of the Amer. Math. Soc.* **14** (2001), 297–346.
5. M. Develin and B. Sturmfels, Tropical convexity, *Documenta Math.* **9** (2004) 1–27.
6. J. A. de Loera, Nonregular triangulations of products of simplices, *Discrete Comput. Geom.* **15**:3 (1996), 253–264.
7. J. A. de Loera, J. Rambau and F. Santos, *Triangulations of polyhedra and point sets*, monograph in preparation.
8. I. M. Gel'fand, M. M. Kapranov and A. V. Zelevinsky, *Multidimensional Determinants, Discriminants and Resultants*, Birkhäuser, Boston,1994.
9. H. Haiman, A simple and relatively efficient triangulation of the n-cube, *Discrete Comput. Geom.* **6** (1991), 287–289.
10. B. Huber, J. Rambau and F. Santos, The Cayley trick, lifting subdivisions and the Bohne-Dress Theorem on zonotopal tilings, *J. Eur. Math. Soc. (JEMS)* **2**:2 (2000), 179–198.
11. D. Maclagan and R. R. Thomas, Combinatorics of the toric Hilbert scheme, *Discrete Comput. Geom.* **27**:2 (2002), 249–272.
12. D. Orden and F. Santos, Asymptotically efficient triangulations of the d-cube, *Discrete Comput. Geom,* **30** (2003), 509–528.
13. J. Propp, Enumeration of Matchings: Problems and Progress, in *New Perspectives in Algebraic Combinatorics (Berkeley, CA, 1996–1997),* L. J. Billera, A. Björner, C. Greene, R. E. Simion and R. P. Stanley, eds, Mathematical Science Research Institute Publications, 38, Cambridge University Press, Cambride, 1999, pp. 255–291.
14. J. Rambau, Triangulations of cyclic polytopes and higher Bruhat orders, *Mathematika* **44** (1997), no. 1, 162–194.
15. J. Rambau, personal communication based on a computation with his public software TOPCOM (www.zib.de/rambau/TOPCOM).
16. F. Santos, A point set whose space of triangulations is disconnected, *J. Amer. Math. Soc.* **13** (2000), no. 3, 611–637.
17. F. Santos, On the refinements of a polyhedral subdivision, *Collect. Math.* **52**:3 (2001), 231–256.
18. F. Santos, Non-connected toric Hilbert schemes, *Math. Annalen,* to appear; arXiv:math.CO/0204044.
19. B. Sturmfels, *On the Newton polytope of the resultant,* J. of Algebraic Combinatorics **3** (1994) 207–236.
20. B. Sturmfels, *Gröbner bases and convex polytopes*, American Mathematical Society, Providence 1996.
21. W. P. Thurston, Conway's tiling groups, *Amer. Math. Monthly* **97** (1990), 757–773.
22. G. Ziegler, *Lectures on Polytopes*, Springer-Verlag, New York, 1994.

Universidad de Cantabria, Departamento de Matemáticas, Estadística y Computación, Facultad de Ciencias, 39005 Santander, SPAIN.

E-mail address: `santosf@unican.es`

URL: `http://personales.unican.es/santosf`

Contemporary Mathematics
Volume **374**, 2005

Problems from the Cottonwood Room

Matthias Beck, Beifang Chen, Lenny Fukshansky, Christian Haase, Allen Knutson, Bruce Reznick, Sinai Robins, and Achill Schürmann

ABSTRACT. This collection was compiled by Christian Haase and Bruce Reznick from problems presented at the problem sessions, and submissions solicited from the participants of the AMS/IMS/SIAM summer Research Conference on Integer points in polyhedra.

CONTENTS

The fourth author was supported in part by NSF Grant #DMS-0200740.
The sixth author was supported in part by the USAF under DARPA/AFOSR MURI Award F49620-02-1-0325. Any opinions, findings, and conclusions or recommendations expressed in this publication are those of the authors and do not necessarily reflect the views of these agencies.

Questions about Ehrhart coefficients.

Presented by Matthias Beck (San Francisco State University).

Given a rational d-polytope P, Ehrhart's Theorem (see, e.g., [**3**]) states that the lattice-point enumerating function $i_P(t) := \#\left(tP \cap \mathbb{Z}^d\right)$ for $t \in \mathbb{N}$ is a *quasipolynomial*, that is, a function of the form

$$c_d(t)t^d + c_{d-1}(t)t^{d-1} + \cdots + c_0(t),$$

where $c_0, \ldots, c_d$ are periodic functions.

Conjecture 1.1 (Thomas Zaslavsky, unpublished). *For the Ehrhart quasipolynomial $i_P(t) = c_d(t)t^d + c_{d-1}(t)t^{d-1} + \cdots + c_0(t)$, denote the period of c_k by p_k. Then $1 = p_d \le p_{d-1} \le \cdots \le p_0$, or maybe even $1 = p_d \mid p_{d-1}| \cdots | \, p_0$.*

This conjecture was disproved during the conference independently by David Einstein (unpublished) and Tyrrell McAllister and Kevin Woods [**4**].

Problem 1.2. *Characterize those polytopes for which Zaslavsky's Conjecture is true.*

For the remaining problems, P will be a lattice polytope: the vertices have integral coordinates. In this case, i_P is an honest polynomial, i.e., $p_d = \cdots = p_0 = 1$.

Problem 1.3. *There are lattice (3-)polytopes whose Ehrhart polynomials have negative coefficients (see, e.g.,* [**5**, Example 3.5]*). Characterize those polytopes whose Ehrhart polynomials do not have negative coefficients.*

Problem 1.4. *The roots of the Ehrhart polynomial of the cross polytope*

$$\left\{(x_1, \ldots, x_d) \in \mathbb{R}^d : \ |x_1| + \cdots + |x_d| \le 1\right\}$$

all have real parts equal to $-1/2$ [**2, 6**]. *Find other polytopes whose Ehrhart polynomials exhibit such special behavior.*

Problem 1.5. *Improve the following bounds (proved in* [**1**]*):*

(i) *The roots of Ehrhart polynomials of lattice d-polytopes are bounded above in norm by $1 + (d+1)!$.*
(ii) *All real roots of Ehrhart polynomials of d-dimensional lattice polytopes lie in the half-open interval $[-d, \lfloor d/2 \rfloor)$.*

Remark 1.1.

(i) For $d = 4$, this upper bound can be improved to 1 (which is sharp).
(ii) We thought for a while that the upper bound in (b) could be conjectured to be 1, but then found a counterexample (a class of 0/1 order polytopes, described in [**1**]). Incidentally, the same counterexample got rid of another conjecture, namely that the Ehrhart polynomial of any 0/1 polytope has only positive coefficients.

Conjecture 1.6 (Beck-DeLoera-Develin-Pfeifle-Stanley). *All roots α of Ehrhart polynomials of lattice d-polytopes satisfy $-d \le \operatorname{Re} \alpha \le d - 1$.*

CONJECTURE 1.7 (Jesús DeLoera). *For the cyclic polytope $C(n,d)$ realized with integral vertices on the moment curve $\nu_d(t) := \left(t, t^2, \ldots, t^d\right)$,*

$$i_{C(n,d)}(m) = \operatorname{vol}(C(n,d))\, m^d + i_{C(n,d-1)}(m).$$

Equivalently,

$$i_{C(n,d)}(m) = \sum_{k=0}^{d} \operatorname{vol}_k(C(n,k))\, m^k.$$

References

[1] Matthias Beck, Jesús A. De Loera, Mike Develin, Julian Pfeifle, and Richard P. Stanley. Coefficients and roots of Ehrhart polynomials. In this volume.

[2] Daniel Bump, Kwok-Kwong Choi, Pär Kurlberg, and Jeffrey Vaaler. A local Riemann hypothesis I. *Math. Z.*, **233** (1):1–19, 2000.

[3] E. Ehrhart. Polynômes arithmétiques et méthode des polyèdres en combinatoire *International Series of Numerical Mathematics*, **35**, Birkhäuser Verlag, Basel, 1977.

[4] Tyrrell McAllister and Kevin Woods. The minimum period of the Ehrhart quasi-polynomial of a rational polytope. Preprint `math.CO/0310255`, 2003.

[5] M. Vergne. Residue formulae for Verlinde sums, and for number of integral points in convex rational polytopes. *European women in mathematics (Malta, 2001)*, 225–285, World Sci. Publishing, River Edge, NJ, 2003.

[6] Fernando Rodriguez Villegas. A Riemann hypothesis for Hilbert polynomials. Preprint, see `http://www.ma.utexas.edu/users/villegas/`.

A conjecture about the Euler characteristic of algebraic varieties over the fields $\mathbb{F}_q$, $\mathbb{R}$, $\mathbb{C}$ and the division ring $\mathbb{H}$.

Presented by BEIFANG CHEN (Hong Kong University).

Let X be an algebraic variety (either affine or projective) whose defining equations have integral coefficients. Then X is well-defined over the finite field $\mathbb{F}_q$ of q elements, the field $\mathbb{R}$ of real numbers, the field $\mathbb{C}$ of complex numbers, and the division ring $\mathbb{H}$ of quaternions. We denote the corresponding varieties by $X_{\mathbb{F}_q}$, $X_{\mathbb{R}}$, $X_{\mathbb{C}}$, and $X_{\mathbb{H}}$, respectively. Let $f(X,q)$ be the number of elements of the finite set $X_{\mathbb{F}_q}$. In many cases, the function $f(X,q)$ turns out to be a polynomial function of q. For instance, for the projective variety $\operatorname{Gr}(n,k)$, the Grassmanian of k-subspaces of the n-dimensional vector space $\mathbb{F}_q^n$, the function $f(\operatorname{Gr}(n,k),q)$ is the Gaussian polynomial

$$\binom{n}{k}_q = \frac{(q^n-1)(q^n-q)\cdots(q^n-q^{k-1})}{(q^k-1)(q^k-q)\cdots(q^k-q^{k-1})},$$

the q-analog of the binomial coefficient $\binom{n}{k}$; see [**4**].

CONJECTURE 2.1. *Let X be a smooth projective variety whose defining equations have integral coefficients. If $f(X,q)$ is a polynomial function of q, then*

$$\chi(X_{\mathbb{F}}) = f(X, \chi(\mathbb{F})),$$

where $\chi(X_{\mathbb{F}})$ is the Euler characteristic of the variety $X_{\mathbb{F}}$ over the field $\mathbb{F}$, and $\chi(\mathbb{F})$ is defined by

$$\chi(\mathbb{F}) = \begin{cases} q & \text{if } \mathbb{F} = \mathbb{F}_q \\ -1 & \text{if } \mathbb{F} = \mathbb{R} \\ 1 & \text{if } \mathbb{F} = \mathbb{C}, \mathbb{H}. \end{cases}$$

The conjecture was verified by the author for some special cases such as Grassmanians, partition varieties [**2**], quiver varieties, and toric varieties, etc. In particular, for the complex projective variety $X_{\mathbb{C}}$, the formula $\chi(X_{\mathbb{C}}) = f(X, 1)$ can be derived from the Weil conjecture [**3**].

Of course there is no need to restrict ourselves to projective varieties and to assume the smoothness condition in the conjecture. The same conjecture can be formulated for affine and singular varieties as long as $f(X, q)$ is a polynomial function. However, if X is an affine variety whose defining equations have integral coefficients, then the Euler characteristic $\chi(X_{\mathbb{F}})$ should be understood as the combinatorial Euler characteristic defined in [**1**], the alternating sum of the number of cells in a triangulation of $X_{\mathbb{F}}$.

One may only assume that $f(X, q)$ is a polynomial function for $q = p^k$, where p is a fixed prime. One may also consider solutions over $\mathbb{Z}$ and divide the integers into some classes. For each of these classes we obtain some equations to define a variety. The varieties obtained in this way correspond to some branches of the variety $X_{\mathbb{R}}$ over the real field $\mathbb{R}$. The conjecture can be similarly formulated for each of these classes.

References

[1] B. Chen, On the Euler characteristic of finite unions of convex sets, *Discrete Comp. Geom.* **10** (1993), 79-93.

[2] K. Ding, Rook placements and cellular decomposition of partition varieties, *Discrete Math.* **170** (1997), 107-151.

[3] R. Hartshorne, *Algebraic Geometry*, Springer-Verlag, New York, Heidelberg, Berlin, 1977.

[4] R. Stanley, *Enumerative Combinatorics*, Vol I, Cambridge University Press, 1998.

Lattice points in homogeneously expanding compact domains.

Presented by LENNY FUKSHANSKY (Texas A&M University).

A variety of interesting and important questions in geometric combinatorics and in geometry of numbers is connected to counting integer lattice points in compact subsets of a Euclidean space. In case such a subset is a rational polyhedron, the problem can be reformulated in terms of Ehrhart polynomial. In a general situation, however, one often has to rely on estimates of asymptotic nature. A good example of such an estimate is presented by S. Lang in Theorem 2 on p. 128 of [**3**]. We state it here. In the rest of this note we assume that $N > 1$ is an integer.

THEOREM 3.1 ([**3**]). *Let D be a compact subset of $\mathbb{R}^N$, and let L be a lattice of rank N in $\mathbb{R}^N$ with fundamental domain F. Assume that the boundary ∂D of D is Lipschitz-parametrizable. Then for each positive $t \in \mathbb{R}$ the number of points of L in tD is given by the following asymptotic formula:*

$$|L \cap tD| = \frac{\mathrm{Vol}(D)}{\mathrm{Vol}(F)} t^N + O(t^{N-1}), \tag{3.1}$$

where Vol *stands for volume in $\mathbb{R}^N$, and the constant in O depends on L, N, and Lipschitz constants.*

We recall that the condition that ∂D is *Lipschitz-parametrizable* means that there exists a finite collection of maps $\varphi_j : [0,1]^N \longrightarrow \partial D$, the union of images of which covers ∂D and there exists a constant K such that for all $\boldsymbol{x}, \boldsymbol{z} \in [0,1]^N$

$$|\varphi_j(\boldsymbol{x}) - \varphi_j(\boldsymbol{z})| \leq K|\boldsymbol{x} - \boldsymbol{z}|,$$

for each j, where $|\ |$ stands for the sup-norm on $\mathbb{R}^N$, i.e. $|\boldsymbol{x}| = \max_{1 \leq i \leq N} |x_i|$. The constant K is called the associated *Lipschitz constant.*

Notice that the main term in the upper bound in (3.1) is explicit and easily computable, but the error term is implicit. Loosely speaking, the main term of such an asymptotic estimate counts the number of "interior points" of L in D, i.e. points that are away from the boundary, and the error term accounts for the points near the boundary. For practical applications it is important to be able to explicitly estimate the error term. Such an estimate was carried out for instance by H. Davenport in [**1**] (see also [**5**] for a very nice account and generalizations of Davenport's theorem). However Davenport's bound on the error term depends on projection volumes of D onto certain subspaces of $\mathbb{R}^N$ as well as determinants of projections of L onto these subspaces. These are hard to compute. In some situations one would prefer perhaps cruder, but more tractable bounds on the error term. An alternative approach to this problem is to try to "quantify" the original argument in Lang's theorem. This has been partially done by P. G. Spain in [**4**]. We briefly outline this approach here and ask some further questions.

We start by sketching out the main idea of proof of Theorem 3.1. One proceeds by noticing that for each positive real number t

$$m(t) \leq |L \cap tD| \leq m(t) + b(t),$$

where

(i) $m(t)$ = number of $\boldsymbol{x} \in L$ such that $F + \boldsymbol{x}$ belongs to the interior of of tD,
(ii) $b(t)$ = number of $\boldsymbol{x} \in L$ such that $F + \boldsymbol{x}$ intersects $\partial(tD)$.

It is obvious that

$$m(t) \leq \frac{\text{Vol}(tD)}{\text{Vol}(F)} = \frac{\text{Vol}(D)}{\text{Vol}(F)} t^N,$$

which produces the main term. In order to produce the error term one needs to estimate $b(t)$. This unfortunately is not so easy. Lang only proves that $b(t) = O(t^{N-1})$ using the fact that the boundary $\partial(tD)$ of tD is Lipschitz-parametrizable, but does not exhibit any explicit upper bound. Although, as we discussed above, there are other methods for estimating the error term, the quantity $b(t)$ seems to be interesting in its own right. It can, for instance, be related to a covering problem, namely: how many translates of the closure of the fundamental domain F does it take to cover the the compact domain tD? Such a number can again be approximated by the expression $m(t) + b(t)$ as above. The following estimate for $b(t)$ in the special case when $L = \mathbb{Z}^N$ was produced by P G. Spain.

THEOREM 3.2 ([**4**]). *Let D be as in Theorem 3.1, so that the boundary ∂D is Lipschitz-parametrizable with Lipschitz constant K. Let $L = \mathbb{Z}^N$. Let F be the fundamental domain of L with respect to the standard basis, and let $b(t)$ as above be the number of translates of F that have nonempty intersection with $\partial(tD)$ where $t \geq 1/K$. Then*

$$b(t) \leq 2^N (Kt+1)^{N-1}, \tag{3.2}$$

and therefore

$$|\mathbb{Z}^N \cap tD| \leq \mathrm{Vol}(D)t^N + 2^N(Kt+1)^{N-1} = \mathrm{Vol}(D)t^N + 2^N \sum_{i=0}^{N-1} K^i t^i. \tag{3.3}$$

PROBLEM 3.1. *Provide an explicit bound on $b(t)$ as in Theorem 3.2 for a general lattice L.*

Perhaps one can modify Spain's argument to produce a solution to Problem 3.1. Notice that the upper bound on $|L \cap tD|$ as it comes out in (3.3) is a polynomial in t whose coefficients depend on volume of D and on the Lipschitz constant K. This suggests a certain analogy with Ehrhart polynomial: one may look for polynomial upper and lower bounds on $|L \cap tD|$ for some more or less general instances of L and D. Here is an example for a simple choice of D when L is any sublattice of $\mathbb{Z}^N$. Let

$$C_t^N = \{\boldsymbol{y} \in \mathbb{R}^N : \max\{|y_1|, ..., |y_N|\} \leq t\},$$

that is, C_t^N is a cube with side length $2t$ centered at the origin in $\mathbb{R}^N$.

THEOREM 3.3 ([**2**]). *Let $\Lambda \subseteq \mathbb{Z}^N$ be a lattice of full rank in $\mathbb{R}^N$ of determinant Δ. Then for each point $\boldsymbol{z}$ in $\mathbb{R}^N$ we have*

$$\begin{aligned} \left(\frac{2^N}{\Delta}\right) t^N &\leq |\Lambda \cap (C_t^N + \boldsymbol{z})| \\ &\leq \left(\frac{2t}{\Delta} + 1\right)(2t+1)^{N-1} = \left(\frac{2^N}{\Delta}\right) t^N + \sum_{i=1}^{N-1} 2^i \left(\frac{1}{\Delta} + 1\right) t^i + 1. \end{aligned} \tag{3.4}$$

In [**2**] an analogous bound for a rectangular box instead of a cube is also produced; the result of Theorem 3.3 is extended to lattices of not full rank and to certain modules over the ring of algebraic integers in a number field, viewed as $\mathbb{Z}$-modules.

PROBLEM 3.2. *Produce explicit polynomial bounds like (4) for more general choices of compact domain D.*

References

[1] H. Davenport. On a principle of Lipschitz, J. London Math. Soc. **26** 179-183 (1951).
[2] L. Fukshansky. Algebraic points of small height with additional arithmetic conditions, Ph.D. thesis, University of Texas at Austin (2004).
[3] S. Lang. Algebraic Number Theory Addison-Wesley Reading, MA 1970.
[4] P. G. Spain. Lipschitz: a new version of an old principle, Bull. London Math. Soc. **27** 565-566 (1995).
[5] J. Thunder. The number of solutions of bounded height to a system of linear equations, J. Number Theory **43** 228-250 (1993).

Reflexive polytopes in dimension 2 and 3, and the numbers 12 and 24.

Presented by CHRISTIAN HAASE (Duke University).

The objects of this problem are *reflexive polytopes*, which have a whole bunch of characterizations. In order to formulate these, we use the Ehrhart polynomial and its generating function $i_P(k) := \#(k\ P \cap \mathbb{Z}^d)$, $i_P^\circ(k) := \#(\mathrm{relint}(k\ P) \cap \mathbb{Z}^d)$, and $H_P(t) := \sum_{k \geq 0} i_P(k)\ t^k$. (See [**3**], and also Matthias Beck's section of this article). The Volume will always be normalized with respect to the lattice.

THEOREM 4.1. *Let $P \subset \mathbb{R}^d$ be a full-dimensional lattice polytope with unique interior lattice point* $\mathbf{0}$. *Then the following conditions are equivalent:*

(i) *The polar dual* $P^\vee = \{y \in (\mathbb{R}^d)^* \ : \ \langle y, P\rangle \geq -1\}$ *is integral.*
(ii) $\mathrm{vol}(P) = \sum \mathrm{vol}(F)$, *the sum ranging over all facets F of P.*
(iii) $i(P,k) = i^\circ(P, k+1)$ *for all k.*
(iv) $H(P, 1/t) = (-1)^{d+1} t H(P,t)$.
(v) *The projective toric variety X_P defined by P is Fano.*
(vi) *Every generic hypersurface of X_P is Calabi–Yau.*

The only non-elementary conditions (iv) and (v, vi) are due to Takeuchi Hibi [**4**] and Victor Batyrev [**1**] respectively. The polytope P is called *reflexive* if these conditions are satisfied. The last condition explaines the physicist's interest in reflexive polytopes.

In any given dimension, there is only a finite number of equivalence classes of reflexive polytopes. For example, these are the 16 reflexive polygons:

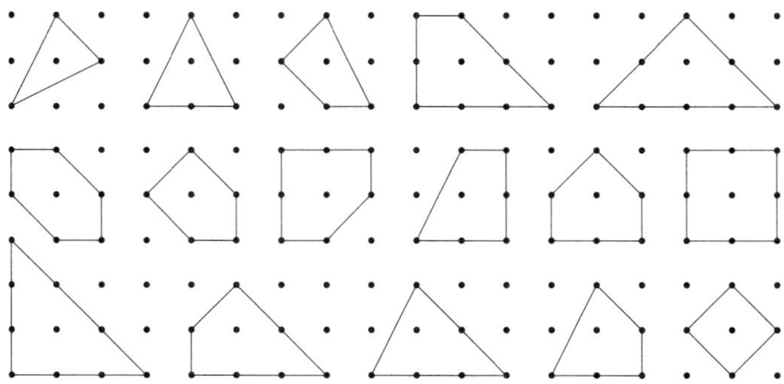

All reflexive polygons (up to $\mathrm{SL}_2\mathbb{Z}$).

Here is the first striking theorem.

THEOREM 4.2 (Cf. [**6**]). *The sum of the lengths of an admissible polygon and its dual is* 12.

In their beautiful paper [**6**], Bjorn Poonen and Rodriguez-Fernando Villegas give four proofs: exhaustion, walk in the space of reflexive polygons, toric surfaces, and modular forms. They also give an interpretation as a Gauß-Bonnet type theorem, where the curvature at a vertex is the length of the dual edge. Here comes the second striking result[1].

THEOREM 4.3. *If P is a* 3*-dimensional reflexive polytope, then*

$$\sum_{e \text{ edge of } P} \mathrm{length}(e) \cdot \ \mathrm{length}(e^\vee) = 24$$

"Proof": By the last characterization of reflexive polytopes, a generic hypersurface $Z \hookrightarrow X_P$ is a 2–dimensional Calabi–Yau, i.e., a $K3$ surface. Thus, $\chi(Z) = 24$. By [**2**], the above sum computes $\chi(Z)$. □

The only other proof known (to me) is the exhaustion proof. (There are 4, 319 reflexive 3–polytopes [**5**].) I am not satisfied with this proof. There should be an

[1]Communicated by Dimitrios Dais

elementary proof out there. Again, one could interpret the length of the dual edge as a curvature, and it looks like a Gauß-Bonnet formula.

References

[1] Victor V. Batyrev. Dual polyhedra and mirror symmetry for Calabi–Yau hypersurfaces in toric varieties. *J. Alg. Geom.*, 3:493–535, 1994.
[2] Vladimir I. Danilov and Askol'd G. Khovanskiĭ. Newton polyhedra and an algorithm for computing Hodge–Deligne numbers. *Math. USSR Izvestiya*, 29(2):279–298, 1987.
[3] E. Ehrhart. Polynômes arithmétiques et méthode des polyèdres en combinatoire *International Series of Numerical Mathematics*, **35**, Birkhäuser Verlag, Basel, 1977.
[4] Takayuki Hibi. Dual polytopes of rational convex polytopes. *Combinatorica*, 12(2):237–240, 1992.
[5] Maximilian Kreuzer and Harald Skarke. Classification of reflexive polyhedra in three dimensions. *Adv. Theor. Math. Phys.*, 2(4):853–871, 1998.
[6] Bjorn Poonen and Rodriguez-Fernando Villegas. Lattice polygons and the number 12. *Amer. Math. Monthly*, 107(3):238–250, 2000.

Are smooth toric ideals quadratically generated?

Presented by Allen Knutson (UC Berkeley).

A lattice polytope $P \subset \mathbb{R}^d$ defines an ample line bundle L_P on a projective toric variety X_P. (See, e.g., [**4**, §3.4].) If X_P is smooth (the normal fan of P is unimodular), then L_P is very ample, and provides an embedding $X_P \hookrightarrow \mathbb{P}^{r-1}$, where $r = \#(P \cap \mathbb{Z}^d)$. So we can think of X_P canonically sitting in projective space. The following question about the defining equations of $X_P \subset \mathbb{P}^{r-1}$ has been around for quite a while, but its origins are hard to track.

Question 5.1. *Let P be a lattice polytope whose corresponding projective toric variety is smooth. Is the defining ideal I_P generated by quadratics?*

There are two variations of this question (of strictly increasing strength).

- Is the homogeneous coordinate ring $\Bbbk[X_1, \ldots, X_r]/I_P$ Koszul?
- Does I_P have a quadratic Gröbner basis?

The last version has a combinatorial interpretation. It asks for the existence of very special, "quadratic" triangulations of P. These are regular unimodular triangulations whose minimal non-faces have two elements (they form a flag- or clique-complex) [**7**, Ch. 8].

Partial results were obtained by Robert Jan Koelman [**5**] who showed that Question 5.1 has an affirmative answer for polygons. Later, Winfried Bruns, Joseph Gubeladze and Ngô Viêt Trung [**2**] showed that smooth polygons even have quadratic triangulations. Lindsay Piechnik shows that smooth reflexive 4-polytopes have quadratic triangulations [**6**]. (See Christian Haase's section of this article for more about reflexive polytopes.)

Günter Ewald and Alexa Schmeinck [**3**] answer Question 5.1 affirmatively if the number of facets is at most $d+2$. They reformulate the problem as a pebble game, and provide a winning strategy. Given a set of n red and n green pebbles on the lattice points in P (possibly several pebbles at the same point) so that the barycenters of the red and the green set agree, is it possible to remove all the pebbles using the following two moves?

- If a red and a green pebble sit at the same lattice point, remove them from the game.
- Two red (or green) pebbles can move in opposite directions (so as to preserve the barycenter), provided they stay inside of P.

This is actually the first reference we could find where Question 5.1 was asked explicitly.

Rikard Bögvad announced (and later withdrew) a general proof of 5.1 [**1**].

References

[1] Rikard Bögvad: On the homogeneous ideal of a projective nonsingular toric variety. Withdrawn preprint `alg-geom/9501012`.

[2] Winfried Bruns, Joseph Gubeladze, and Ngô Viet Trung: Normal polytopes, triangulations and Koszul algebras. *J. Reine Angew. Math.*, 485:123–160, 1997.

[3] G. Ewald, A. Schmeinck: Representation of the Hirzebruch-Kleinschmidt varieties by quadrics; Beiträge zur Algebra und Geometrie 34 (1993), 151-156.

[4] W. Fulton. *Introduction to Toric Varieties*, volume 131 of *Annals of Math. Studies*. Princeton University Press, 1993.

[5] R. Koelman: A criterion for the ideal of a projectively embedded surface to be generated by quadrics; Beiträge zur Algebra und Geometrie 34 (1993), 57-62.

[6] Lindsay C. Piechnik: Smooth reflexive 4-polytopes have quadratic triangulations. Undergraduate thesis, Duke University, 2004.

[7] Bernd Sturmfels: *Gröbner Bases and Convex Polytopes*, volume 8 of *University Lecture Series*. A.M.S., 1996.

The classification of 1-point lattice tetrahedra.

Presented by BRUCE REZNICK (University of Illinois at Urbana-Champaign).

A k**-point lattice** n**-simplex** is a simplex $P = conv(v_0, \dots, v_n)$ with $v_j \in \mathbb{Z}^n$, and so that $P \cap \mathbb{Z}^n$ consists of the vertices and exactly k interior lattice points. For $n = 2$, Pick's Theorem says that a k-point lattice triangle has area $k + \frac{1}{2}$; conversely, any plane triangle with that area, lattice point vertices and no other lattice points on its edges is a k-point lattice triangle. Thus, the first interesting case is $n = 3$. We restrict our attention mostly to $k = 0$ ("empty lattice tetrahedra") and to $k = 1$.

For $(a, b, c) \in \mathbb{Z}^3$, let

$$T_{a,b,c} := conv((0,0,0), (1,0,0), (0,1,0), (a,b,c)).$$

In 1957, J. Reeve [**9**] noted that for $n \in \mathbb{N}$, $T_{1,1,n}$ must be empty, and since it has volume $n/6$, there is no upper bound on the volume of empty lattice tetrahedra. There are several equivalent, but somewhat different, characterizations of empty tetrahedra, due to G. White, R. Howe and H. Scarf (see the discussion in [**10**]), to independently to D. Handelman [**1**] and D. Morrison and G. Stevens [**7**]. One way of saying it is that the tetrahedron $T = conv(v_0, v_1, v_2, v_3)$ is empty if and only if there is a unimodular affine transformation taking T to $T_{0,0,1}$ or to $T_{1,b,c}$, with $1 \le b < c$ and $gcd(b, c) = 1$.

In 1983, D. Hensley [**2**] proved that for each fixed $k \ge 1$ and n, there is an upper bound on the volume of k-point lattice n-simplices. His bound was improved in 1990 by J. Lagarias and G. Ziegler [**4**]. Lagarias suggested [**3**] at Snowbird that this counterintuitive phenomenon regarding the volume of empty and non-empty simplices is worthy of new attention.

If P is an n-simplex, then the set $\{v_j - v_0 : 1 \le j \le n\}$ spans $\mathbb{R}^n$, hence for any $w \in \mathbb{R}^n$, there exist unique $\lambda_j \in \mathbb{R}$ so that $w - v_0 = \sum_{j=1}^n \lambda_j(v_j - v_0)$. Putting $\lambda_0 := 1 - \sum_{j=1}^n \lambda_j$, we see that each $w \in \mathbb{R}^n$ has a unique representation $w = \sum_{j=0}^n \lambda_j v_j$ with $\sum_{j=0}^n \lambda_j = 1$. The λ_j's are called the **barycentric coordinates** of w with respect to the vertices of P. Note that $w \in P$ if and only if $\lambda_j \ge 0$ for $j = 0, \dots, n$ and $w \in int(P)$ if and only if $\lambda_j > 0$.

Suppose P is a k-point lattice n-simplex. The barycentric coordinates of any interior lattice point are necessarily positive rational numbers. The author proved in [**10**] that there is an upper bound to the denominator of the barycentric coordinates, and so for fixed (k, n), only finitely many different sets of barycentric coordinates are possible. It was also shown in [**10**] that, up to permutation, there are exactly 7 possible barycentric coordinates for the interior point of a 1-point lattice tetrahedron, namely:

$$\left(\tfrac{1}{4}, \tfrac{1}{4}, \tfrac{1}{4}, \tfrac{1}{4}\right), \left(\tfrac{2}{5}, \tfrac{1}{5}, \tfrac{1}{5}, \tfrac{1}{5}\right), \left(\tfrac{3}{7}, \tfrac{2}{7}, \tfrac{1}{7}, \tfrac{1}{7}\right), \left(\tfrac{5}{11}, \tfrac{3}{11}, \tfrac{2}{11}, \tfrac{1}{11}\right),$$

$$\left(\tfrac{5}{13}, \tfrac{4}{13}, \tfrac{3}{13}, \tfrac{1}{13}\right), \left(\tfrac{7}{17}, \tfrac{5}{17}, \tfrac{3}{17}, \tfrac{2}{17}\right), \left(\tfrac{7}{19}, \tfrac{5}{19}, \tfrac{4}{19}, \tfrac{3}{19}\right).$$

A natural (and open!) question is to characterize, up to unimodular transformation, the 1-point lattice tetrahedra. This would also determine the maximal possible volume. The estimate from [**4**, p.1023] gives the maximum volume as $\le 14^{48} \approx 10^{55}$. This has been very recently improved by O. Pikhurko [**8**, p.23] to $< \frac{85}{6}$. It is noted in [**10**, p.235] that $T_{-13,-9,20}$ is a 1-point lattice tetrahedron, so the volume can be as large as $\frac{20}{6}$. This question was also discussed in a non-explicit way by J. Lawrence [**5**].

This question appears to be large, but tractable. Let T have vertices $\{v_j\}$ and interior point w. Then $T' = conv(v_0, v_1, v_2, w)$ is necessarily an empty tetrahedron, and so can be identified via White-Scarf-Howe-Morrison-Stevens. Since w has one of the barycentric coordinates listed above, we can then solve for v_3 and check whether T contains any unwanted lattice points.

Here is the simplest case. Suppose $T' = T_{1,1,n}$, and suppose that $w = (1, 1, n) = \frac{1}{4}(v_0 + v_1 + v_2 + v_3)$ is the centroid of T. Then $v_3 = 4w - (v_0 + v_1 + v_2)$ and $T = T_{3,3,4n}$. Since T lies between the planes $x_1 - x_2 = \pm 1$ with a single vertex on each, any non-vertices (a, b, c) in $T \cap \mathbb{Z}^3$ must have $a = b = 1$ or $a = b = 2$. An easy computation shows that

$$(1, 1, c) = \tfrac{5c-4n}{4n}(0,0,0) + \tfrac{4n-3c}{4n}(1,0,0) + \tfrac{4n-3c}{4n}(0,1,0) + \tfrac{c}{4n}(3,3,4n),$$

hence $(1, 1, c)$ is in $T \cap \mathbb{Z}^3$ provided $\frac{4}{3}n \ge c \ge \frac{4}{5}n$. When $n \le 2$, this gives only $c = n$, but for $n \ge 3$, $c = n + 1$ satisfies the inequality, hence $(1, 1, n + 1)$ is in T and T is not a 1-point lattice tetrahedron. Similarly,

$$(2, 2, c) = \tfrac{5c-12n}{4n}(0,0,0) + \tfrac{8n-3c}{4n}(1,0,0) + \tfrac{8n-3c}{4n}(0,1,0) + \tfrac{c}{4n}(3,3,4n)$$

is in $T \cap \mathbb{Z}^3$ provided $\frac{8}{3}n \ge c \ge \frac{12}{5}n$. For $n = 1$, there is no such point, but for $n = 2$, $c = 5$ satisfies the inequality. Thus, in this simplest case, the only 1-point lattice tetrahedron is $T_{3,3,4}$, which is unimodular, under $(x_1, x_2, x_3) \mapsto (x_1 - x_3, x_2 - x_3, x_3)$ to $T_{-1,-1,4}$. This tetrahedron also appears in [**10**, p.235].

More has already been done in this direction. As a 2002 undergraduate research project at the University of Michigan under the direction of S. Bullock, B. Mazur [**6**] analyzed the centroid case in complete detail. He showed that if T is a 1-point lattice tetrahedron whose interior point is the centroid, then it is unimodularly equivalent either to $T_{3,3,4}$ or to $T_{7,11,20}$.

Note added in proof. The characterization of 1-point lattice tetrahedra has been completed by Alexander M. Kasprzyk in the paper "Toric Fano 3-folds with terminal singularities", `arXiv.math.AG/0311284`, 17 Nov 2003. Each of the 6 non-centroid sets of barycentric coordinates has, up to unimodular equivalence, a unique representative 1-point lattice tetrahedron. Thus, there are 8 different examples in all. The author is grateful to Julian Pfeifle for this reference. The author had independently reached the same conclusion, in a forthcoming paper to be called "Clean lattice tetrahedra".

References

[1] D. Handelman, *Integral body-building in* $\mathbb{R}^3$, J. Geom., **27** (1986), 140–152.
[2] D. Hensley, *Lattice vertex polytopes with interior lattice points*, Pac. J. Math., **105** (1983), 183–191.
[3] J. Lagarias, personal conversation, July 13, 2003, mid-afternoon.
[4] J. Lagarias and G. Ziegler, *Bounds for lattice polytopes containing a fixed number of interior points in a sublattice*, Canad. J. Math. **43** (1991), 1022–1035.
[5] J. Lawrence, *Finite unions of closed subgroups of the n-dimensional torus*, DIMACS Ser. Discrete Math. Theoret. Comput. Sci., **4** (1991), 433–441.
[6] B. Mazur, *On lattice barycentric tetrahedra*, preprint available on the Mathematics ArXiv at http://front.math.ucdavis.edu/math.CO/0401037.
[7] D. Morrison and G. Stevens, *Terminal quotient singularities in dimensions three and four*, Proc. Amer. Math. Soc., **90** (1984), 15–20.
[8] O. Pikhurko, *Lattice points in lattice polytopes*, Mathematika, **48** (2001), 15–24.
[9] J. Reeve, *On the volume of lattice polyhedra*, Proc. London Math. Soc., (3)**7** (1957), 378–395.
[10] B. Reznick, *Lattice Point Simplices*, Discrete Math. **60** (1986), 219–242.
[11] J. Zaks, A. Perles and J. M. Wills, *On lattice polytopes having interior lattice points*, Elem. Math. **37** (1982), 44–46.

A conjecture on lattice tiles.

Presented by Sinai Robins (Temple University).

We first define a tiling of the n-dimensional integer lattice $\mathbb{Z}^n$ by "lattice tiles" and then formulate a conjecture about the structure of the tiles for $n = 2$.

We begin by writing the lattice $\mathbb{Z}^n$ as a finite, disjoint union of subsets:

$$\mathbb{Z}^n = \bigcup_{j=1}^{N} \{L_j + v_j\}$$

for some N, some collection of sublattices $L_j \subset \mathbb{Z}^n$, and some collection of integer translation vectors v_j. Each of these subsets $L_j + v_j$ is called a **lattice tile**.

For $n = 1$, there have been several papers [**1, 2, 3**] on tiling (or covering) $\mathbb{Z}$ by such lattice tiles, which are of course arithmetic progressions in the 1-dimensional case. Here we extend the notion of covers of the integers to tilings of the lattice $\mathbb{Z}^n$.

A recent paper of Zhi-Wei Sun [**4**] proves, using group-theoretic considerations, that if a group is given as the disjoint union of various translates of subnormal subgroups, then at least two of those subgroups must have the same index. Applying this theorem to the group $\mathbb{Z}^n$ given by a lattice tiling, we immediately see that at least two of the lattice tiles must have the same index. In other words, we know that at least two of the lattice tiles must have the same volume for their fundamental domain. We conjecture that for $n = 2$ more is true:

CONJECTURE 7.1. *Given any lattice tiling of $\mathbb{Z}^2$, at least two of the tiles must be translates of each other.*

That is, while the result of Sun [**4**] guarantees the existence of two tiles with the same area for their fundamental domain, experimentation suggests a refined statement about the shape of the corresponding fundamental domains.

References

[1] Erdős, P. Problems and results in Number Theory, in: H. Halberstam and C. Holley (eds), Recent progress in Analytic Number Theory, Vol. 1, Academic Press, New York, 1981, 1-13.
[2] Newman, M. Roots of unity and covering systems, Math. Ann. 191, 1971, 279-281.
[3] Novak, B. and Znam, S. Disjoint covering systems, Amer. Math. Monthly, 81, 1974, 42-45.
[4] Sun, Zhi-Wei, On the Herzog-Schönheim conjecture for uniform covers of groups, J. Algebra 273 (2004), no.1, 153–175.

Covering minima, free planes and deep holes

Presented by ACHILL SCHÜRMANN (Magdeburg).

In [**3**] Kannan and Lovász introduced the *covering minima*

$$\mu_i(K, \Lambda) = \min\{\mu > 0 \ : \ \Lambda + \mu K \text{ meets every } (n-i)\text{–dim. aff. subspace in } \mathbb{R}^n\},$$

$i = 1, \ldots, n$, of a (full rank) lattice Λ and a convex body K in $\mathbb{R}^n$, generalizing the previously known *covering radius* $\mu_n(K, \Lambda)$.

A problem, arising for example in the study of *free lattice planes* (cf. [**1**], [**2**], [**4**]), is to compute $\mu_i(K, \Lambda)$. A free lattice plane with respect to Λ and K is an affine subspace in $\mathbb{R}^n \setminus (\Lambda + K)$. Thus a free plane of dimension $(n - i)$ exists if and only if $\mu_i(K, \Lambda) > 1$. Of particular interest is the case where $K = B^n$ is the (solid) unit sphere.

It is known (cf. [**3**]) that

$$\mu_i(K, \Lambda) = \max\{\, \mu_i(K|L^\perp, \Lambda|L^\perp) \ : \ L \text{ is a } (n-i)\text{–dim. sublattice of } \Lambda \,\},$$

where $\cdot|L^\perp$ denotes the orthogonal projection onto the linear subspace $L^\perp$, orthogonal to L. Therefore, it is possible to show that $\mu_i(K, \Lambda)$ can theoretically be obtained by computing finitely many covering radii in dimension i. This approach is unpractical though.

Maybe another interpretation helps here. If $K = B^n$, then the covering radius $\mu_n(B^n, \Lambda)$ has two nice geometrical meanings: On the one hand, the covering radius is equal to the maximum (Euclidean) norm $\|v\|$ among the vertices v of the *Dirichlet–Voronoi–polytope*

$$\{\, x \in \mathbb{R}^n \ : \ \|x\| \le \|x - y\| \text{ for all } y \in \Lambda \,\}.$$

These vertices are called the *deep holes* of Λ. On the other hand, the covering radius is equal to the circumradius of some *Delone–polytope* of Λ — a Λ–polytope with its circumball containing its vertices on the boundary and beyond that no other lattice points (see Section "Lattice polytopes: some open problems" by Jean–Michel Kantor).

Both, the Dirichlet–Voronoi–polytope and the finitely many Delone–polytopes yield classical tilings of $\mathbb{R}^n$ — dual to each other — by lattice point translations. Thus an affine subspace attaining a covering minimum runs necessarily through a Dirichlet–Voronoi–polytope and through some Delone–polytopes. This gives reason to propose

PROBLEM 8.1. *Find a geometrical description of affine subspaces attaining* $\mu_i(B^n, \Lambda)$, *depending on* Λ*'s Dirichlet–Voronoi–polytope or its Delone–polytopes.*

In the planar case for example, lines attaining $\mu_1(B^2, \Lambda)$ run through midpoints of neighboring facets (bounding segments) of the Dirichlet–Voronoi–polytope or contain a facet. Depending on such a geometrical description or not, it would be nice to solve

PROBLEM 8.2. *Find a way to compute the covering minima of a lattice, e.g. of the root lattice* E_8 *or the Leech lattice.*

References

[1] Martin Henk, *Free planes in lattice sphere packings*, Adv. Geom., to appear.
[2] Martin Henk, Günter M. Ziegler and Chuanming Zong, *On free planes in lattice ball packings*, Bull. London Math. Soc. 34 (2002), no. 3, 284-290.
[3] Ravi Kannan, László Lovász, *Covering minima and lattice-point-free convex bodies*, Ann. of Math. (2), 128 (1988), no. 3, 577–602.
[4] Chuanming Zong, *From deep holes to free planes*, Bull. Amer. Math., 39 (2002), no. 4, 533–555.

DEPARTMENT OF MATHEMATICS, SAN FRANCISCO STATE UNIVERSITY, SAN FRANCISCO, CA94132
E-mail address: `beck@math.sfsu.edu`

DEPARTMENT OF MATHEMATICS, HONG KONG UNIVERSITY OF SCIENCE AND TECHNOLOGY, CLEAR WATER BAY, KOWLOON, HONG KONG
E-mail address: `mabfchen@ust.hk`

DEPARTMENT OF MATHEMATICS, TEXAS A&M UNIVERSITY COLLEGE STATION, TX 77843-3368
E-mail address: `lenny@math.tamu.edu`

DEPARTMENT OF MATHEMATICS, DUKE UNIVERSITY, DURHAM, NC 27708-0320
E-mail address: `haase@math.duke.edu`

UNIVERSITY OF CALIFORNIA AT BERKELEY, BERKELEY, CA 94720-3840
E-mail address: `allenk@math.berkeley.edu`

DEPARTMENT OF MATHEMATICS, UNIVERSITY OF ILLINOIS AT URBANA-CHAMPAIGN, URBANA, IL 61801-2975
E-mail address: `reznick@math.uiuc.edu`

DEPARTMENT OF MATHEMATICS, TEMPLE UNIVERSITY, PHILADELPHIA, PA 19122
E-mail address: `srobins@math.temple.edu`

DEPARTMENT OF MATHEMATICS, UNIVERSITY OF MAGDEBURG, 39106 MAGDEBURG, GERMANY
E-mail address: `achill@math.uni-magdeburg.de`

Titles in This Series

374 **Alexander Barvinok, Matthias Beck, Christian Haase, Bruce Reznick, and Volkmar Welker, Editors,** Integer points in polyhedra—geometry, number theory, algebra, optimization, 2005

373 **O. Costin, M. D. Kruskal, and A. Macintyre, Editors,** Analyzable functions and applications, 2005

372 **José Burillo, Sean Cleary, Murray Elder, Jennifer Taback, and Enric Ventura, Editors,** Geometric methods in group theory, 2005

371 **Gui-Qiang Chen, George Gasper, and Joseph Jerome, Editors,** Nonlinear partial differential equations and related analysis, 2005

370 **Pietro Poggi-Corradini, Editor,** The p-harmonic equation and recent advances in analysis, 2005

369 **Jaime Gutierrez, Vladimir Shpilrain, and Jie-Tai Yu, Editors,** Affine algebraic geometry, 2005

368 **Sagun Chanillo, Paulo D. Cordaro, Nicholas Hanges, Jorge Hounie, and Abdelhamid Meziani, Editors,** Geometric analysis of PDE and several complex variables, 2005

367 **Shu-Cheng Chang, Bennett Chow, Sun-Chin Chu, and Chang-Shou Lin, Editors,** Geometric evolution equations, 2005

366 **Bernhelm Booß-Bavnbek, Gerd Grubb, and Krzysztof P. Wojciechowski, Editors,** Spectral geometry of manifolds with boundary and decompositon of manifolds, 2005

365 **Robert S. Doran and Richard V. Kadison, Editors,** Operator algebras, quantization, and non-commutative geometry, 2004

364 **Mark Agranovsky, Lavi Karp, David Shoikhet, and Lawrence Zalcman, Editors,** Complex analysis and dynamical systems, 2004

363 **Anthony To-Ming Lau and Volker Runde, Editors,** Banach algebras and their applications, 2004

362 **Carlos Concha, Raul Manasevich, Gunther Uhlmann, and Michael S. Vogelius, Editors,** Partial differential equations and inverse problems, 2004

361 **Ali Enayat and Roman Kossak, Editors,** Nonstandard models of arithmetic and set theory, 2004

360 **Alexei G. Myasnikov and Vladimir Shpilrain, Editors,** Group theory, statistics, and cryptography, 2004

359 **S. Dostoglou and P. Ehrlich, Editors,** Advances in differential geometry and general relativity, 2004

358 **David Burns, Christian Popescu, Jonathan Sands, and David Solomon, Editors,** Stark's Conjectures: Recent work and new directions, 2004

357 **John Neuberger, Editor,** Variational methods: open problems, recent progress, and numerical algorithms, 2004

356 **Idris Assani, Editor,** Chapel Hill ergodic theory workshops, 2004

355 **William Abikoff and Andrew Haas, Editors,** In the tradition of Ahlfors and Bers, III, 2004

354 **Terence Gaffney and Maria Aparecida Soares Ruas, Editors,** Real and complex singularities, 2004

353 **M. C. Carvalho and J. F. Rodrigues, Editors,** Recent advances in the theory and applications of mass transport, 2004

352 **Marek Kubale, Editor,** Graph colorings, 2004

351 **George Yin and Qing Zhang, Editors,** Mathematics of finance, 2004

350 **Abbas Bahri, Sergiu Klainerman, and Michael Vogelius, Editors,** Noncompact problems at the intersection of geometry, analysis, and topology, 2004